Y0-AAA-525

PICTURE LANGUAGE MACHINES

PICTURE LANGUAGE MACHINES

Proceedings of a Conference held at the Australian National University, Canberra on 24–28 February, 1969

Edited by

S. KANEFF

Research School of Physical Sciences
The Australian National University, Canberra

1970

ACADEMIC PRESS · LONDON · NEW YORK

ACADEMIC PRESS INC. (LONDON) LTD
Berkeley Square House
Berkeley Square
London, W1X 6BA

U.S. Edition published by
ACADEMIC PRESS INC.
111 Fifth Avenue
New York, New York 10003

Library of Congress Catalog Card Number: 78–145679
ISBN: 0 12 396250 1

PRINTED IN GREAT BRITAIN BY
William Clowes and Sons Limited, London, Beccles and Colchester

PREFACE

This volume is the outcome of a Conference on Picture Language Machines jointly sponsored by the Division of Computing Research, Commonwealth Scientific and Industrial Research Organisation, and the Research School of Physical Sciences, Australian National University and held in Canberra in February 1969. The topic, and accordingly the selection of speakers, arose out of the related research interests of two research groups located in the sponsoring organisations. The rather specialised nature of the topics suggested that a general audience would be assisted by some presentations of a more or less tutorial nature, and the organisers are grateful to their overseas guests, Professor R. Narasimhan, Dr. D.G. Bobrow and Mr. E.L. Jacks, for agreeing to carry some of this tutorial load.

The organisers hoped, by bringing together workers from a number of disciplines, ranging widely in objectives and view-points, and actively engaged in research in Picture Languages, Graphical Communication, Natural Language Interaction Systems, Pattern Recognition, and relevant aspects from Linguistics and Psychology, that there would emerge useful transdisciplinary concepts, recognition of which would give better integration and understanding to the field. In addition, it was hoped that discussion would elucidate the significance, strengths and weaknesses of the research reported, and to point to possible future lines of development. Accordingly, the format of the Conference was intended to provide ample opportunity for discussion of papers: those discussions which took place within range of our recording equipment have been reported with minimal editing in order to retain the original flavour of the proceedings - unfortunately missing are the delightful and often brilliant elements of repartee which have reluctantly been expunged from the record.

I record the appreciation of the participants for the interest and support of Dr. G.N. Lance, Head, C.S.I.R.O. Division of Computing Research; Professor Sir Ernest Titterton, Director, Research School of Physical Sciences; and Professor Gordon Newstead, Head, Department of Engineering Physics.

Financial support is gratefully acknowledged from the C.S.I.R.O. Division of Computing Research and the A.N.U. Research School of Physical

Sciences (which also kindly provided facilities for holding the Conference). Finally, we acknowledge with sincere appreciation the assistance of Miss Jenny McCann and Mrs. Jean Hardham, of the Department of Engineering Physics, in preparing the typescript.

Department of Engineering Physics,
Research School of Physical Sciences,
The Australian National University,
Canberra, A.C.T.
Australia.

S. Kaneff

CONTENTS

FOREWORD

The topic of man-machine communication is a major preoccupation of research into Artificial Intelligence and more widely into the advanced uses of the digital computer. The 'communications interface' between a computer program and a user necessarily involves language-like forms. For example, a program to calculate statistical parameters of some corpus of numerical data will require that input data be formatted in a rather precise manner - as specified by one or more ⟨Format Statement⟩ s in the program. From a certain standpoint a language <u>is</u> just a set of format conventions. However the languages in which we express our ideas and concepts are not adequately captured by the sorts of ⟨Format Statement⟩ which algorithmic computing languages provide. Thus the challenge of man-machine communication is to identify the 'formats' of natural languages by seeking to describe them in a way which can be successfully embodied in a computer program. The problems encountered in trying to do this for English have been illuminated by attempts to write 'Question-Answering' (QA) programs and independently by linguists studying 'generative grammars' of English. That other natural forms of communication - specifically diagrams and pictures - might be conceived of in a similar fashion was the insight behind seminal papers of Narasimhan and Kirsch in the early 1960's. It was these papers which first applied the developing concepts of generative grammar and syntactic structure to the consideration of how we understand pictures and diagrams. Indeed it was in the paper by Kirsch that the phrase 'Picture Language Machine' was used to describe programs with a capacity to respond to pictures from a defined corpus (e.g. circuit diagrams) in the same general way that we might expect QA programs to respond to English statements i.e. by making a 'parse' of the input the basis of an understanding of the statement. A significant impetus to Kirsch's proposals was provided by the development of Computer Graphics where the bulk of the communication between program and user is mediated by graphical images. The problems encountered in providing representations of these images which could be manipulated by program amply confirmed - if confirmation were needed - that the structure or organisation of a picture is intimately bound up with its meaning.

While the quest for picture language machines (thinking of sentences now as a special kind of picture: a one-dimensional picture) need reflect nothing more than the requirements of more effective man-machine communication, it can and usually does have another significance. The need to embed a description of the language in a computing system demands that this description be, or carry with it, an effective procedure for its use. For without such a procedure the system will be unable to utilise the description to parse and understand the sentences and images offered to it. This requirement can be construed as a criterion for the adequacy of theories of language, so that picture language machines can be seen to express theories of language as well as being potentially useful fragments of a technology. This criterion can be applied to theories of other forms of intelligent behaviour and as such forms the basic principle of, and at the same time provides a methodology for, Artificial Intelligence.

The papers in this volume are all concerned in one way or another with this notion of a picture language machine, conceived of either as a piece of technology or as a putative theory of the language 'understood' by the machine. Their authors differ widely however in the extent to which they would accept a linguistic model as the best way to understand the work being reported. One of the objectives of the Conference was to provide a forum in which such issues could be discussed. While it would be misleading to suggest that such issues were resolved, a central theme of the discussions turned out to be the question of whether adequate descriptions of languages can be non-procedural in character (see for example the discussion of 'neutral descriptions' following Stanton's paper).

The first four papers were intended to be primarily tutorial in character to provide an introduction to the four topics singled out above as contributory to the notion of a picture language machine: Pictures (Narasimhan), QA programs (Bobrow), Graphics programs (Jacks), Generative Grammars for English (Clowes, Langridge and Zatorski). The remaining papers represent the current research interests of the authors and are usefully discussed under these four headings.

Pictures. In O'Callaghan's RECØG program which classifies characters drawn with a light pen, the list of connected points read in by the program is described relative to the models (AV's) of the classes of character which the program contains. The description which most nearly corresponds to the AV which determined it, is taken to be the 'correct' description, thus assigning the character to the same class as the AV. The basis of description is a sequentially ordered list of segments, 'straight lines', ' +ve corners' etc. In an extension to handle characters containing two or more strokes, spatial relationships between strokes (e.g. between their end points) are added to this descriptive scheme.

The same problem (letter recognition) is tackled by Clowes from a 'linguistic standpoint' treating strokes and the relations between them, e.g. coincidence (of stroke-ends say) as abstractions whose expression may be given different forms. Thus where strokes are depicted as a list of connected points created with a light pen, coincidence may be expressed either as a local maximum of curvature or as proximity of the terminal point of one sublist of connected points to another (possibly non-terminal) point of another sublist. Similarly a straight stroke may be expressed by a list of connected points having a wide range of geometries, a range which can be expressed functionally. In consequence where RECØG may use a number of distinct AV's to define a single class, PICNTERP will use a single 'strokes/coincidences' characterization of the class, handling variability in terms of a more or less complex mapping of picture structure onto strokes and coincidences between strokes.

The concern for pictorial relationships and for structure evident in PICNTERP is echoed in Macleod's extensive analysis of the requirements to be met by a system capable of inferring 'the structure of the situation underlying the picture'. The variety and complexity of structures described by Macleod, e.g. 'the contribution to density structure of interdependent aspect, pigmentation and illumination structures' lead him soberly to 'wonder whether it will ever be possible to assemble such a mechanism at all'.

While accepting the need for structured descriptions of pictures and more generally a knowledge of the objects represented pictorially, a major unsolved problem concerns the nature of systems capable of acquiring such knowledge. It is this problem to which Kaneff addressed himself in his paper on the role of learning in picture processing, in which he considers a number of requirements to be met by intelligent systems and especially the requirement that they be problem solving devices. He observes that the structure and knowledge that present day picture processing systems incorporate is, in the natural systems being emulated, the 'end result .... of the interaction of sets of (often complex) constraints and action capabilities'. It is these constraints (imposed by the real world) and the action capabilities (of the natural system) which we should seek to elucidate.

Graphics. To suggest that man-machine communication 'involves language-like forms' tells us nothing about why it does. In the study reported by Stanton, a case is made for thinking that it must involve language-like forms if the communication is to involve a common understanding by both parties (man and machine) of the data which expresses the communication. Where this data takes the form of a picture, Stanton remarks: "communication is possible only if the machine can see the same things in a picture that the man does". Seeing the same things will necessitate the machine recovering a structural description of the picture in a language which distinguishes between objects, relations and attributes. The language is in effect a generalisation of the notational apparatus, e.g. generative grammars, necessary to characterize the structure of sentences in, say, English. Stanton goes on to spell out this theory of communication in detail in the form of a graphics program which provides a simple map editing facility.

The fact that a great deal of engineering design involves calculations with respect to entities and relations naturally represented graphically, is the starting point for the computer graphics systems designer. In his paper, "Designer's Choice at the Console", Jacks discusses the logical design of a programming system in which the language and especially the programmer's view of the data structures of a program is strongly pictorial. Thus conventional programming languages constrain the programmer to write subroutines, etc., which operate on data whose only representation is numeric, while in a graphical programming language these data are given additionally a pictorial representation natural to the user. For a variety of reasons not least that the pictorial representations are generated by the machine (and not by the user for subsequent interpretation by the machine) the unsolved descriptive problems discussed in earlier papers can be bypassed. Jacks' paper exhibits the anatomy of a system which enables the user to create operators whose arguments and result are displayed graphically. These operators are then, by a technique called 'sequence recording', given the status of subroutines which can then be incorporated into a more or less conventional program. Given the enrichment of picture manipulation - both in the variety of classes of picture and in the capacity to interpret pictures as well as generate them - which will hopefully result from research into picture description languages, the prospect for natural language programming opened up by Jacks' work is very exciting.

Question Answering. In more conventional usage 'natural language programming' connotes the use of English, say, as the medium for man-machine communication. The four papers by Bobrow and Klatt, Narasimhan, Barter, and Zatorski all discuss different aspects of the problems to be overcome in achieving such communication.

Bobrow and Klatt address themselves to natural language utterances in a context where communication can be thought of as transmitting one of a finite (and in practice quite small) set of messages where the internal structure of any message has no communicative role. Despite this immense simplification, achieving reliable communication is still a complex task primarily because of the variation in the acoustic form of different instances of the same message. The author's catalogue of the principal types of variation is particularly valuable in that it is so often omitted from accounts of speech recognizers. Recognition is effected by describing each message in terms of the changes in state of a number of features, e.g. loud(ness), voice, etc., during the speech event. Recognition rates in excess of 90% are reported.

Recognising individual words be they uttered or typed, is of course only the beginning of the communication process. The central thesis of Barter's paper and also that of Zatorski is that such systems must have an adequate representation of the 'model of the world' or 'referential domain' to which communications address themselves. Barter argues that this 'model of the world' is reflected in the data basis of many Fact Retrieval systems where these data bases contain that information which input questions are intended to retrieve selectively. Every question contains an implicit specification of a world situation together with a query directed to that situation. Barter's basic point is that "the situation specification should direct the restructuring of the data base in order that the relationships implied by the question are explicitly represented in the data base". As presently constituted, data bases have a fixed structure which is reflected in the stereotyped form which questions are permitted to take.

Narasimhan's interest in natural language is primarily directed towards understanding language behaviour. He argues that Chomsky is fundamentally wrong in thinking that people acquire a language: rather they learn to put a particular variety of behaviour to use. Behaviour comes, he suggests in paradigmatic and syntagmatic forms. The former would be characterized typically by reasoning by analogy, the latter by formal deductive reasoning. In suggesting a formal system of rules to characterize what is acquired by a native speaker, Chomsky assumes, in error, that language behaviour belongs to the syntagmatic mode. To illustrate this concept of paradigmatic function, Narasimhan describes recent work by Ramani on a program designed to solve problems expressed in natural language by reasoning analogically. In a second part of the paper, Narasimhan provides an elaboration of some of the consequences of his position for the design of language systems.

What emerges from these papers is a general consensus of the relevance of descriptive methods in the characterization of the use of natural language for communication purposes. In a speculative essay which formed the last paper of the Conference, Clowes attempts to show how such descriptive methods might lead to a more plausible characterization of board games and especially of the notion of 'creative play'.

It is perhaps worth saying in concluding this Foreword that one of the unquestioned assumptions throughout the Conference was the essential role of the digital computer as a means of rigorously testing theories in this area. It is this involvement with machines which, more than anything else, is the new contribution that Artificial Intelligence can make to what are after all rather well known problems in Psychology, Linguistics and Philosophy.

Laboratory of Experimental Psychology,
University of Sussex,
Brighton, England.

M. Clowes

PICTURE LANGUAGES

R. NARASIMHAN

Computer Group
Tata Institute of Fundamental Research
Colaba, Bombay 5, India.

1. THE CLASSIFICATORY APPROACH

1.1. Pattern Recognition as a Categorizing Activity

Traditionally, picture processing with computers has been almost exclusively concerned with the problem of discrete symbol recognition. The compelling motivation for these studies has, of course, been the possibility of developing a device that could read alpha-numeric characters: machine-printed characters, hand-printed characters, cursive writing, etc. In this well-delimited context, it seems quite natural to view the recognition problem as one of categorization. The totality of characters to be recognized are tokens (i.e. images) belonging to a finite set of prototypes (i.e. the letter and/or number alphabet). The problem confronting a character reading device, then, is a straightforward one: given a finite set of picture prototypes, and a token of one of these prototypes, the task is to assign the token to the correct prototype.

The problem so stated is seen to be quite analogous to the problems that confront a statistician. Given a population of samples known to belong to a finite set of distinct classes (i.e. prototypes), how does one assign any specified sample to one of these classes? Statistical decision theory at once suggests itself as a powerful tool to deal with this problem. With each prototype associate a list of attributes, and with each attribute, a set of attribute values. Partition the space of attribute values into mutually disjoint regions, and assign each region to one of the prototypes. Given a token, now, compute its attribute values, and using these computed values, determine to which region in the property space it belongs; (we shall consistently use the term 'property' to refer to values of attributes). Accordingly, categorize the input token as belonging to this or that prototype.

In the last ten years, this decision-theoretic approach to pattern recognition has developed into a very flourishing area of research. The problem-formulation has been refined and considerably generalised so as to apply to a variety of contexts (many of them nonpictorial) where pattern recognition construed as classification is assumed to apply. For instance, Kovalevsky (1965) in a recent review, states the problem as follows:

"It is necessary to recognize a set of situations, k, by coming to various decisions, d. Decisions are selected depending upon observed results, v, of an experiment conditioned by a situation. This dependence is characterized by a conditional probability distribution, p(v/k). Quality of reached decision is evaluated by a magnitude of loss, L(k/d), which is specified for each situation, k, and for each decision, d. It is necessary to find such a rule, d(v), for reaching decisions based on observations, v, which results in minimal mathematical expectation of losses."

According to Kovalevsky: "the great variety of problems relating to reading devices, speech recognition, automatic control, medical and technical diagnosis, etc., are considered as pattern recognition ...". However, in this paper, we shall restrict our consideration to recognition of pictorial data. When applied to picture recognition, the set of situations, k, are the tokens. The decisions, d, are the prototypes. v are the set of property measurements made on the given tokens.

As we remarked earlier, this tendency to pose the recognition problem in this manner as a problem in minimal error decision-making, has grown out of the traditional preoccupation with the design of character recognition devices. Procedures that have been developed, based on this approach, to read multi-font printed characters, hand-printed characters, etc., have had a measure of success. Commercial character readers are now available, presumably functioning on these principles. A good summary of the considerable amount of work that has been done within this property-space framework may be found, for example, in Uhr (1963, 1966).

One of our primary aims in this paper is to consider whether this classificatory approach to picture recognition is a viable one for picture classes other than alpha-numeric characters and similar discrete symbol sets. A second aim is to discuss whether recognizing pictures is all there is to picture processing. In the next subsection, we shall argue that answers to both these questions must be in the negative. If one accepts this conclusion, one is left with the following problems:

What are the various aspects of picture processing using computers, and what is an appropriate framework within which all these aspects of picture processing can be adequately discussed? What is an appropriate formulation of the recognition problem within this framework? Are there any connections between this formulation of the recognition problem and the traditional one in terms of classification?

We shall discuss these issues in the rest of this paper, and try to arrive at some tentative answers.

1.2. Some Inadequacies of the Classificatory Approach

It is clear from our earlier description that the classificatory approach tends to treat a picture - a prototype or a token - as a single, atomic, entity. The methodology of attribute assignments and attribute value computations is guided primarily by the desire to optimize the partitioning of the property-space and to devise efficient inference techniques to make minimal error decisions on the basis of the computed statistics. The attributes do not play any other conceptual role: structural, semantic, or pragmatic.

While this simplified approach to a picture may be satisfactory in the case of identification of often-used alphabets and simple geometric shapes, it is seen to break down the moment one is confronted with a picture having complex features; for example, a picture of a neuron, a tree with heavy arborization, or even symbols of the Chinese alphabet. One can only identify such complex pictures by articulating their several aspects. That is, one can only identify a tree as a tree by recognizing that it has a part that looks like a trunk, from which issue parts which look like branches, which, in turn, are covered by parts that look like leaves, and so on. That is, to recognize a tree it is necessary to analyse it into its structured subparts and the relations between them.

The same is true of pictures which are of the nature of "scenes", i.e. pictures which contain a multiplicity of objects spatially distributed in some organized way. Photographs of terrains, faces, cloud covers, etc., biomedical slides, bubblechamber pictures, circuit diagrams, architectural drawings, cartoon pictures, textual illustrations, mathematical formulae, and a variety of other graphic data are instances of scenes. Such scenes can only be recognized (identified) by articulating their descriptions. An articulated description of a scene would typically seek to answer the following kinds of questions: Does a specified object occur in the scene? Does a particular configuration of objects occur in the scene? Does a specified object have a specified property? etc.

This need for analysing the structure of an input picture, and for articulating its several aspects on the basis of such a structured analysis is, again, seen to be essential in adaptive and/or problem-solving contexts. A powerful tool in problem-solving is the ability to compare two situations, say, two pictures, and perceive similarity, identity, and other transformational relationships between them. The ability to do this presupposes a methodology for articulating descriptions of these pictures.

One primary aspect of learning is the capacity to generalize available behavioural repertoire to new situations by perceiving structural analogies between these new situations and the already known situations. This applies equally well to pictures. As Clowes (1968) points out, without some such capacity to analyse input pictures into their aspects and see the correspondences between them, one cannot recognize that two novel patterns are alike.

We have advanced two kinds of arguments, so far, to emphasize that pictures cannot be dealt with intelligently except by analysing their structures and articulating their aspects. We showed this to be a necessary part of scene analysis and we argued that complex objects like trees and neurons can only be identified when analysed as if they were scenes. Notice that in certain kinds of problem-solving situations this argument would remain valid for even simple objects - in fact, for symbols like alpha-numeric characters for which the classificatory approach seemed so natural earlier.

Consider an artificial intelligence (a learning automaton) that functions in a problem-solving context. Assume that it has been exposed to a sequence of tokens of English letters. It is now called upon to answer questions of the following sort: In what sense are an A and an R similar? Are B and P similar in the same sense as E and F? Is this picture more like an A or an R? etc. Notice that unless the automaton has internalized aspects of the tokens on the basis of some kind of structure analysis performed on them, it would find it hard, if not impossible, to answer any of these questions. And it seems plausible that solving problems involving pattern discrimination demands the capability to cope with questions of at least this order of complexity. A good example of such a situation may be found in Evans' geometry-analogy problem-solving (Evans, 1963).

We are, thus, led to conclude that, as a general methodology, the unstructured classificatory approach is inadequate for picture recognition - especially recognition of pictures with complex features and pictures of the nature of scenes. We are also led to the view that recognition itself is a secondary operation which presupposes the capability to analyse and articulate the structural aspects of input pictures. It is only on the basis of such articulated descriptions that it is possible to acquire general problem-solving behaviour concerned with pattern discrimination.

Thus, computer processing of pictures must be concerned with picture descriptions and not merely with picture classification. Once adequate descriptions are available, pictures could be readily classified by processing further these descriptions. Notice that a description could be of two kinds: (1) either a generative description such that given this description an instance of the described picture could be generated; (2) or an interpretative description which is the outcome of the analysis of a given, specific, input picture. It is not essential, except in very special circumstances, for these two classes of descriptions to be identical. Classification is, in fact, one extreme type of interpretation.

The picture processing frameworks that we are looking for, then, are descriptive schemata for classes of pictures using which we could generate instances of picture tokens belonging to the class (generative schemata), or generate interpretations of given tokens belonging to this class (interpretative schemata). Such descriptive schemata concerned with picture classes are called picture languages. The principal problems to study are: What are plausible structures for picture languages? Given a class of pictures, how does one construct a picture language to describe this class? Given a picture language, how does one use it in specified contexts? In the next section, we shall consider these problems in some detail.

2. THE DESCRIPTIVE APPROACH

2.1. The Nature of Picture Descriptions

Our arguments in the last section have led us to the conclusion that a descriptive approach is essential in order to deal intelligently with classes of picture tokens. But this conclusion is only a partial solution to the picture processing problem. For, we still have to determine what kinds of descriptive schemata are feasible, and which ones are likely to be useful.

To get some insight into the nature of picture descriptions that are likely to prove useful, let us consider recognition of patterns once again. Although our immediate concern is with pictorial patterns, our remarks below apply to other kinds of patterns with suitable modifications.

As we have already emphasized several times, recognizing a pattern, if it is to be a non-trivial activity, clearly must be based on an analysed concept of what a pattern is. A pattern is an organization: it is an organization of subpatterns, or objects, or elements, or whatever. A subpattern is, again, an organization of further subpatterns, or objects. And so on. An organization is a complex of relationships that subsist between the elements which are organized. Thus, one method of recognizing a pattern is to see it as a particular organization, i.e. see it as particular objects satisfying particular relationships.

There is also a second way of recognizing a pattern. And this is done not by analysing it into its primitives and their mutual relationships, but by seeing this pattern as being related to some other pattern in some specific way. For example, one could recognize pattern A by seeing that it is, in fact, pattern B rotated by 90° clockwise; or that it is a mirror image of B, or that it is a projection of B on a horizontal plane, and so on.

We can call the first method compositional description, and the second method transformational description. It is clear that these two modes of description could be interrelated within a single schema by looking upon a transformation of a pattern as being realized, in practice, in terms of transformations defined over its components.

Thus, a complete schema for describing patterns should have incorporated in it the following aspects: (1) there must be a method of describing the elements of a pattern; (2) there must be a method of articulating relationships between pattern elements; and (3) there must be a system of transformations defined over the patterns which are realized in terms of transformations applied to components of a pattern.

A generative picture language is a computational language which consists of all these components, and which is defined in such a way that procedures (or programs) in that language define particular pictures belonging to the class to which the picture language relates. In the next section, we shall return to this topic of picture language specification and deal with it somewhat more formally. We shall also discuss some specific examples there. But in the rest of this section, let us consider a few important points about picture language usage.

2.2. Some Remarks About Picture Language Usage

(1) A picture language suitably embedded in a graphical data generation device could be used to generate instances of pictures belonging to the class described by the language. Such generative devices have been designed so far only for rather specialized classes of pictures such as alpha-numeric characters, mathematical symbols, simple geometric patterns, etc.

Table 1 gives some typical examples of picture classes that have been studied through computer simulation. Articulated generative descriptions of tokens of these classes may be found in the references cited. The specificational details as well as the notations differ from one implementation to another, but they will all be seen to conform to the metaformulation we shall discuss in the next section.

Table 1

Cursive English Handwriting	(Eden, 1962)
Handprinted Numerals	(Clowes, 1967)
Handprinted English Words	(Narasimhan and Reddy, 1967)
Handprinted Fortran Characters	(Narasimhan, 1968)
Nevada Cattlebrands	(Watt, 1966)

(2) A picture language suitably embedded in a graphical data processing device could be used to interpret, i.e. analyse and describe, a specific picture belonging to the class described by that language and, thus, presumably, also recognize it. When this interpretation procedure is strictly (and completely) under the control of the explicitly stated syntax rules of the embedded (picture) language, the process is sometimes called syntax-directed interpretation. When the interpretation procedure makes use of augmented contextual and other external informa-

tion in addition to what the syntax rules supply, the process may appropriately be called syntax-aided interpretation.

Our earlier discussions in Section 1 emphasized that nontrivial picture processing - in particular, recognition of complex pictures - necessarily requires syntax-aided interpretation techniques. It is our view, however, that syntax-directed techniques could only be meaningfully applied to artificial classes of pictures, i.e. a closed class of pictures explicitly defined through a formal specification schema. Such specifications could arise, for example, when a computer generated output is used as an input for further processing by the same or another computer. Naturally given pictures, i.e. pictures generated by people, photographs of natural objects, etc., constitute an open class, by definition, and can only be approximately delimited by a closed specification schema. Hence, interpretational procedures for open classes of pictures can only be aided by such approximate closed specificational schemata, and not be directed by them exclusively.

It is clear, however, that in special circumstances open classes may have a very simple structure and, hence, may be specifiable by formal descriptive schemata more or less completely in all their details. In such cases, syntax-directed interpretation schemes could effectively be made use of. Table 2 lists some computer simulation studies of picture analysis that are based on syntax-directed (in some cases, syntax-aided) techniques. Most of these studies are concerned with analyses of scenes composed of naturally given objects.

Table 2

Geometric Objects (3-dim)	(Guzman, 1967a, 1967b, Roberts, 1963)
Geometric Patterns (2-dim)	(Evans, 1963)
Mathematical Formulae (2-dim)	(Anderson, 1968)
Nuclear Particle Interaction (2-dim)	(Narasimhan, 1964, 1966; Nir, 1967; Shaw, 1968)
Biomedical Photographs (2-dim)	(Hilditch, 1968; Ledley et al, 1966; Lipkin et al, 1966)
Fingerprints (2-dim)	(Crasselli, 1968)
Photographs of Human Faces (2-dim)	(Roberts, 1963)

(3) Where syntax-directed schemes are applicable, it is attractive to investigate whether it is feasible to build a single interpreter that could accept tokens belonging to several different picture classes. It will be readily seen that this problem is quite analogous to the problem of constructing a single compiler that is able to interpret (translate) programs written in a variety of input languages. Such generalised compilers are called compiler-compilers. They require, aside from the particular token to be interpreted, an additional input about the specificational language to which the token belongs. Several quasi-formal procedures have been developed in the systems programming area for implementing such generalised compilers.

Recently, Evans (1968) has discussed a "grammar-controlled pattern analyser" which he has implemented in LISP. "The inputs required by the analysis program consist of: (1) a grammar, and (2) an input pattern in the form of a list of lowest level constituents, with any desired information attached for later use in the analysis process. The output will be a list of all the objects defined by the grammar which can be built out of the list of constituents forming the input pattern."

The following example taken from Evans' paper should give an idea of how such an analyser would function: "Supposing we have a straight-line drawing as in Fig. 1, and wish to find all the triangles (16, in the case shown). Suppose

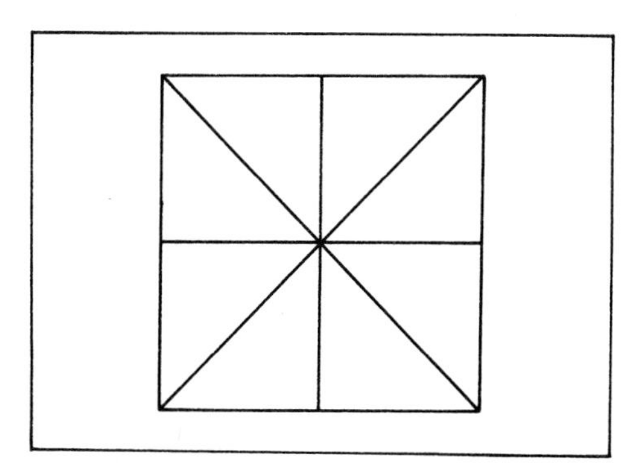

Figure 1 - Input to Evans' picture analyser.

the input is to be a list of vertices (9, in this case) with, attached to each, a list of the others to which it is connected by a line of the drawing. The grammar rule we need to define a triangle might look like:

(TRI(XYZ) ((PT X) (PT Y) (PT Z) (ELS XY) (ELS XZ)
(ELS YZ) (NONCOLL XYZ)) NIL)

which says a triangle consists of 3 points XYZ such that the predicate "exists a line segment between" (ELS) holds, pairwise, between them, and that the 3 points are noncollinear (NONCOLL). When this grammar was supplied along with the input, the program found the 16 triangles."

Feder has been trying to develop, independently, similar syntax-directed pattern analysers which would accept a variety of classes of patterns. He refers to them as parametrized pattern analysers. A preliminary description may be found in his report (Feder, 1966).

(4) Finally, it is clear that a descriptive schema associated with a class of pictures must be pragmatically adequate. That is, in the generative case, the schema must provide adequate descriptions using which acceptable instances of specific pictures can be generated. In the analysis stage, the schema should yield descriptions which are adequate for the purpose for which the analysis is being carried out. From this it follows that, depending upon the pragmatics of the situation, one might want to use different descriptive schemata with the same class of pictures.

3. PICTURE LANGUAGES*

3.1. Some Preliminary Considerations

Our main aim in this section is to discuss the structure of <u>specification languages</u>, i.e. generative picture languages. But, to begin with, it is essential to clarify some important points concerning descriptive schemata in general.

Firstly, consider the problem of generating a picture token given its specification in a specification language. Since a specification that is given must

* Much of this section is based on Section 3 of Narasimhan (1968).

actually be coupled to an action mechanism of some sort before a picture can be physically realized, it is clear that a particular method of specification must necessarily presuppose some particular action mechanism. It is not tenable to argue that specification schemata could be studied independently of action mechanisms and that, in fact, one could think in terms of setting up a universal specification schema whose output could be coupled to any particular action mechanism. Such an approach, instead of solving the picture generation problem in terms of actually realizable schemata, merely shifts the basic problem to a different level. For, now, having designed a presumed universal specification schema, we would be still left with the problem of having to *translate this specification* into a form that can be made use of by a given, particular action mechanism. There is no *a priori* guarantee that this could be done at all, or could be done uniformly for all action mechanisms. It is evident, analogous remarks apply to the interpretational part. Thus, we conclude that a picture language (that is concerned with both the generative and interpretational aspects), if it is to be completely specified, must take explicitly into account the action primitives available to the generating and analysing mechanisms. Notice also that, except in special circumstances, these two sets of primitives need not be identical.

Secondly, it is to be noted that although qualitatively the analysis aspect is the converse of the generative aspect in language usage, it need not necessarily be true that generation and analysis are inverse operations. And hence, in particular, it need not be the case that the analysed output is identical to the generative specification. The reasons for this are the following: The analysis aspect is actually concerned with analysis *and* description of the given picture. The analysed output is, thus, actually intended for generating descriptions. Hence, the nature of the output would depend on the nature of the descriptions to be generated. The nature of the descriptions, in turn, would depend on the pragmatics of the situation calling for these descriptions. Only in the special case where the description is required for generating a copy of the picture would there be a need for the analysed output to be identical to the generative specification. For reasons analogous to those discussed earlier, it would not be admissible to argue that the analyser could generate an output (universal in some sense) quite independently of the nature of the descriptions to be generated.

3.2. Picture Language: Generative Aspect

We shall be concerned, primarily, with classes of picture tokens of finite spatial extent in 2-dimensions. The cardinality of the class could, nevertheless, be infinite as we shall see later. A specification language, G, for a class of picture tokens is a finite system of rules in terms of which specific tokens of this class can be generated. G, as we remarked earlier, is a computation language. We shall discuss G informally in terms of its components. A formal description of G would consist of a formal definition of its components (in terms of some metaformalism), and a formal definition of a program in G. We shall not go into these technical details here.

We structure G in terms of 5 components: a set of primitives, P, a set of attributes, A, a set of relations, R, a set of composition rules, C, and a set of transformations, T, i.e. G = G(P, A, R, C, T). These components will be individually discussed now.

P consists of a set of primitive actions, the performance of which generates primitive picture fragments. We shall refer to these as picture atoms (or atoms for short), and identify them by the same names as their corresponding primitives. Each primitive (and hence, the respective atom) has associated with it a set of attributes (i.e. specificational parameters), A, which can assume values from well-defined ranges. Fixing the values of all (or some) of the attributes of a primitive results in an assignment (or a partial assignment) to the primitive. A primitive with an assignment (or a partial assignment) generates an atom with attribute values completely (or partially) determined.

P and A together, thus, allow us to generate atoms with assigned properties (i.e. with specified attribute values). We now require a machinery in G that would allow us to put together atoms with specific properties to form larger picture fragments. We shall call this procedure composition and the rules which allow us to compose picture fragments in this manner composition rules. Composition rules are specified making use of the set of relationships, R. Each relation is an m-ary predicate defined over the attribute values of the constituents of a picture fragment.

Let p_1, p_2 be atoms with assigned properties. Let r be a binary relation defined over some subset of the attribute values of p_1 and p_2. Then

$$r(p_1, p_2)$$

defines a picture fragment whose constituents are the atoms p_1 and p_2, and whose assigned properties satisfy the relationship, r. Let us give this fragment the name f. Then

$$f \leftarrow r(p_1, p_2)$$

is a composition rule which specifies that f is composed of p_1 and p_2 in such a manner that their attribute values satisfy the relationship, r. Since, in general, a picture would contain more than two atoms, it is necessary to be able to repeat this procedure of binary composition with at least one of the constituents as a picture fragment larger than an atom. To enable us to do so, a composition rule must actually be extended to include a specification of an assignment (of properties) to the fragment resulting from the application of that rule. That is, the original schema should actually be written as follows:

$$f(\ ) \leftarrow r(p_1(\ ), p_2(\ ))$$

where the parentheses ( ) exhibit the explicit property assignments on both sides of the arrow. This, then, immediately allows us to generalize a composition rule to have the form:

$$f_3(\ ) \leftarrow r(f_1(\ ), f_2(\ ))$$

where, in particular cases, f_1, f_2 could either or both be atoms.

So far, we have not restricted in any way the range of values any given attribute may assume. In particular, we have not excluded the possibility that some or all of the attributes assume infinitely many distinct values. In such a case, it is clear that the cardinality of the token classes can be infinite although the cardinality of C, the set of composition rules, is finite, and the tokens, themselves, are finite in spatial extent.

If, however, we also specify that the tokens have a finite <u>resolution</u> in all their attributes (i.e. the set of attributes is finite and also each attribute has a finite range), then the cardinality of the class becomes strictly finite. It must be emphasized that practically all real-life applications of picture generation and interpretation belong to this restricted variety. But this does not trivialize the problem. In fact, the finiteness of the token class has intrinsically very little relevance to the picture analysis and description problem. This is a fundamental assumption in our entire approach as presented in this paper.

One other point that requires some discussion is the possibility of a recursive occurrence of a constituent in a composition rule. A constituent, $f_1(\ )$, is said to occur recursively in a rule, if there is a composition rule of the following kind involving f_1 and some non-null $f_2(\ )$:

$$f_1(\ ) \leftarrow r(f_1(\ ),\ f_2(\ )).$$

Notice that there is nothing in what we have said about the structure of composition rules so far to preclude recursive occurrences of elements in such rules. The only condition that has to be met, for such rules to be feasible, is that the attribute sets of f_1 and f_2 must be compatible. But our original definition of picture classes restricts our considerations to picture tokens of finite extent. If we further restrict them also to be of finite resolution, then it is evident that recursive rules (i.e. rules in which an element occurs recursively) can occur in a specification language only if some kind of stop condition is available to terminate the application of such rules during composition. However, there may be pragmatically significant motivations for incorporating recursive composition rules in a specification language.

To see what is involved in being able to construct a generating mechanism which can generate picture fragments through the use of composition rules as defined above, let us consider the special case with composition rules having forms only of the following two kinds:

$$f_2(\ ) \leftarrow r_1(f_1(\ ),\ p(\ ))$$

$$f_3(\ ) \leftarrow r_2(p_1(\ ),\ p_2(\ )).$$

The mechanism must first be capable of performing the primitive actions, P. Given a composition rule of the first kind with $f_1(\ )$ available, or of the second kind with $p_1(\ )$ available, it must know how to "append" $p(\ )$ to the first, or $p_2(\ )$ to the second, so that the relationships, r_1 in the first case and r_2 in the second case, are satisfied. Analogous problems arise in the more general case where what is "appended" is not an atom but is itself a fragment. Since we are assuming that primitively what are generated are only the atoms, in this more general case, presumably, the two fragments must be independently generated as described earlier, and then composed together so that the given relationship, r, is satisfied between their attribute values.

Notice that this ability of the generating mechanism described above can be restated in a somewhat more constructive way. Let us assume that the generating process consists in performing one primitive action at a time, i.e. creating one atom at a time. To create an atom some or all its attributes must be assigned values. This value assignment is done so that the relationship, r, is satisfied between the assigned properties of this new atom and the already existing properties of the given fragment to which this atom is being appended. Thus, attribute values are assigned to the new atom in a certain context that is determined by the available attribute values of the given fragment. Let us refer to this as the generative context. The generative context at any given stage in the generating process, would depend on the properties of the fragment so far generated; it could, in general, depend also on auxiliary information of a global sort. In the next subsection, we shall generalize this notion of context so as to make it more widely applicable.

The last component of G that remains to be discussed is the set, T, of transformations. We shall distinguish two types of transformations:

(1) Transformations that alter properties of a given picture fragment without spatial change: colouring or changing the colour, texturing, shadowing, etc., are transformations of this type.

(2) Transformations that delete components of a fragment or introduce spatial changes in part or whole of the fragment: Deletion, translation, rotation, projection, mirror inversion, contraction, dilatation, distortion, etc., are transformations of this type.

We shall not treat transformations in any greater detail here.

In closing this subsection, it must be noted that in a formal definition of G, notational conventions would have to be introduced so that the primitives with attribute assignments, relationships and transformations can be explicitly represented and can be uniformly interpreted into action sequences that result in the creation and manipulation of picture fragments. Such a notational scheme would include a formal description of a program or procedure in the specification language, G.

3.3. An Example (Summary)

In Narasimhan (1968) a specification language for handprinted FORTRAN characters is described in considerable detail. We shall present a summary of that here in order to give some idea of the way the concepts discussed in the last subsection are actually applied in practice.

A character is handprinted in a given visual field - i.e. the ground - which is delimited by a frame (a rectangular area). A region within this frame is a connected set of points: a convex, say, circular, neighbourhood. The size of the region (i.e. the number of points in it) is determined by the size of the frame. As shown in Fig. 2, seven distinguished regions are identified in a frame by the names $P_1, P_2, \ldots, P_7$.

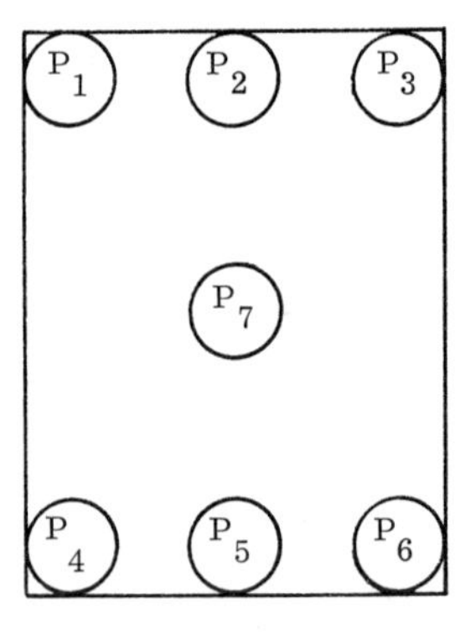

Figure 2 - The 7 primitive regions of a frame.

The basic set of action primitives (discounting region identification) consists of drawing road segments (roads) of specified types between, or through, named regions. Examples of such primitives are:

$$ST(P_i, P_j; P)$$

$$HZ(P_i).$$

ST (STraight) and HZ (HoriZontal) are names of road types. In the first case, regions P_i, P_j are joined by a straight road; its middle region is identified by the name P. In the second case, a horizontal road is made to pass through P_i. The road in this case, normally, extends from edge to edge of the basic frame.

Each primitive road segment has, in general, certain distinguished regions as its attributes: starting region, terminal region, middle region. Whenever required for further reference, these are identified by specific names (like P in the example ST above). In the case of roads specified by only one region, length is an attribute. Each road segment has a thickness or width parameter as an attribute.

The relations that are made use of in the composition rules are: left, right, above, below, slightly above, slightly below. Except for these, the only other relation used is identity of regions occurring as arguments in the primitives. This is indicated by referring to such regions by the same name.

No transformations are used to generate the set of FORTRAN characters. But a simple set of permutation transformations enables one to generate an augmented set including the left-right mirror images of each character.

The following examples of composition rules should give some idea of how the generative process is carried out. The primitives on the right of the arrow ($\leftarrow$) are executed one by one from left to right to generate the character named on the left.

$$'7' \leftarrow HZ(P_1) + ST(P_3, P_5)$$

$$'+' \leftarrow HZ(P_7) + VT(P_7)$$

$$'L' \leftarrow VT(P_1) + HZ(P_4)$$

This specification language has been implemented in a CDC 3600 computer to generate FORTRAN texts.

4. THE DESCRIPTIVE APPROACH: SOME FURTHER OBSERVATIONS

4.1. The Recognition Problem

We shall consider now, briefly, the connection between the traditional formulation of the recognition problem as a problem in categorization, and the analogous formulation within the descriptive framework that we have been discussing. As we saw earlier, the pattern recognition problem, as it is usually posed, consists in assigning a given picture token to one of a finite number of categories. Notice that, within the descriptive framework, this problem is well-posed only when applied to picture classes generated through a finite set of schemata. In this case, one can identify each schema with a given picture category, and the recognition problem can be stated as follows: given two tokens, decide whether they have been generated by two different schemata, or by the same schema with possibly different assignments to its constituent atoms (i.e by the same schema but possibly under different generative contexts)

A uniform approach to this recognition problem is through a comparison of the generative specifications of the two tokens. An algorithm that delivers a generative specification of a given token is called a parsing algorithm. Even if a specification language for a picture class is available, constructing efficient parsing algorithms, in general, need not be straightforward.

Clearly, a recognition procedure need not necessarily make use of a parsing algorithm. It may be possible to set up a battery of tests which determine whether a picture token has this or that attribute, and/or what the actual value of this or that attribute is. Based on the outcome of these tests, a decision procedure might be set up to assign the token to one of the predetermined categories. The properties in the property lists and the attribute sets of the atoms need not be the same. They may overlap completely, partially, or not at all. We can see the reason for this as follows:

We saw earlier that each step in the generative process (of a token) is determined by a context - the generative context - defined in a particular way. Let us now enlarge this notion of generative context into a wider concept which we shall call situational context, or just situation for short. We do this by appending to the picture language, G, a set, G*, of computable functions and predicates defined over the picture fragments of G. These added functions and predicates could, otherwise, be arbitrary.

Given a token or a picture fragment at some stage of the generative process, and a particular function or predicate, g*, from G* defined over the former, the value obtained by evaluating g* using the given token or fragment as the argument, is called an aspect of the situation at that stage of the generative process. Without loss in generality we can assume that G* includes functions that compute attribute values of atoms and predicates that evaluate relationships between them. This assumes, of course, that these are computable. This would, in general, be true if we restrict our considerations to picture classes finite in extent and resolution as discussed earlier. Thus, the generative context, as defined earlier, can be reduced to a subset of the aspects of the situation at any given stage in the generative process.

Let us refer to an aspect of a situation that is not part of a generative context, as an inferred property of the token or picture fragment at that stage of the generation process.

A recognition procedure for a class of pictures could conceivably be successfully set up, in special cases, through the use of property lists involving inferred properties alone. In such a scheme, the underlying picture language, G, clearly plays no direct role in recognition. A recognition device, thus, need not know how to generate a picture in order to be able to recognize it.

The crucial differences, then, between the decision-theoretic approaches to categorization and the approaches to categorization based on descriptive schemata are these. The property lists used in the decision-theoretic approaches are ad-hoc and arbitrary, and do not have any necessary structural, semantic, or pragmatic significance. The property-space which is partitioned is a point space set up on the basis of attribute values alone; the attributes are considered to be completely independent of one another. In the descriptive approach, on the other hand, the properties used are based on an attribute set which has a basic generative significance. The attributes form interrelated subsets on the basis of underlying relations that enter into the generative mechanism. Thus, the property space is no longer a simple point space but a highly structured relational space. It is our view that such spaces are akin to the semantic spaces or the space of concepts as these notions are used informally in problem-solving.

Clowes has independently been advocating (Clowes, 1968, 1969) a similar approach to picture interpretation based on relational picture grammars. He has been trying to exploit the syntactic machinery worked out by Chomsky in connection with grammars for natural languages. Clowes argues that it is essential to develop a methodology for the systematic study of "the varieties of pictorial relationships involved in our intuitive grasp of pictorial form and shape". We shall not discuss his ideas in further detail here since they would, presumably, be considered by him and others, from the Verbigraphics Project, in this Conference.

Shaw, at Stanford, has developed a Picture Description Language, PDL, (Shaw, 1968) which is capable of describing the structure of pictures formed of primitives each having two distinguished points, a head and a tail. Primitives may be joined to one another only at their tail and/or head points. A picture in this language is a connected, directed graph with labelled edges denoting primitives. PDL allows pictures to be manipulated through algebraic operations applied to their descriptions. Shaw has constructed descriptions in PDL and parsers for sparkchamber pictures, flow charts, etc.

4.2. Guzman's Program SEE

A good example of a recognition program based on the descriptive approach is SEE developed by Guzman (1967b), and tested on the PDP-6 machine of the Artificial Intelligence Group of Project MAC. We shall summarize below, very briefly, the important characteristics of this program. For a more detailed account, Guzman's report should be consulted.

SEE is designed to identify solid objects that form a scene (Fig. 3 represents a typical scene). It does this by looking at a 2-dimensional projection of a scene. SEE cannot name objects as, say, cube, parallelepiped, wedge, etc., but it identifies a collection of edges (lines), vertices, and plane surfaces as a plausible object. This it can do even if the objects are partially occluded.

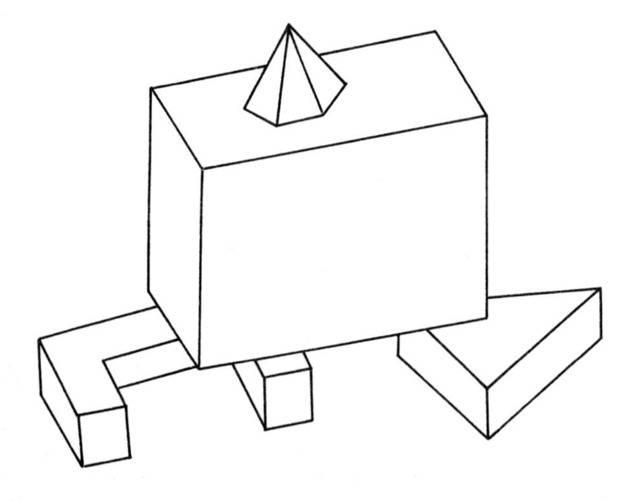

Figure 3 - A typical Guzman 'scene'.

The input to SEE is a scene specified in a special format. A scene is fundamentally a collection of regions, and vertices. A scene, together with its regions and vertices, must be specified with the appropriate attribute lists as follows:

INPUT

A scene has a name and the attributes
Region, Vertices, Background.

A region has a name and the attributes
Neighbours, Kvertices, Foop.

A vertex has a name and the attributes
Xoor, Yoor, Nvertices, Nregions, Kind.

The attributes have the following interpretations:

"Neighbours" is a counterclockwise ordered list of all regions that are neighbours to the atom "region".

"Kvertices" is a counterclockwise ordered list of all vertices that belong to the atom "region".

"Foop" is a counterclockwise ordered list of alternating neighbours and Kvertices of the atom "region".

"Xoor", "Yoor", are the X- and Y-coordinates of the atom "vertex".

"Nvertices" is a counterclockwise ordered list of vertices to which the atom "vertex" is immediately connected.

"Nregions" is a counterclockwise ordered list of regions to which the atom "vertex" belongs.

"Kind" is a counterclockwise ordered list of alternating Nregions and Nvertices of the atom "vertex".

First, a program, TYPEGENERATOR, classifies each vertex according to its slope, disposition, and the number of lines which form it. The vertex types into which the classification is done are:

L | FORK | ARROW | T | K | X | PEAK | MULTI.

Fig. 4 illustrates a typical member of each of these classes. For each type a set of properties called DATUM is computed. This set is type-dependent.

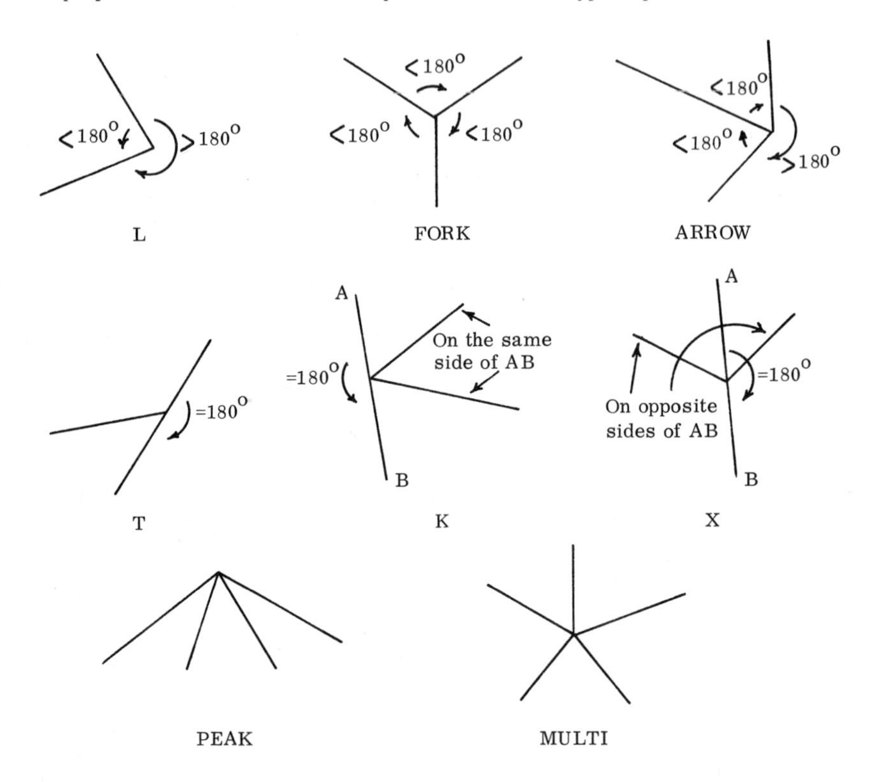

Figure 4 - Vertex types used by Guzman.

SEE, then, operates on the output of the TYPEGENERATOR, and the input information about regions, and outputs a list of objects. SEE consists of a sequence of subprograms which are sequentially executed. There is little recursion and no search.

An analysis is made of vertices, regions, and associated information, in search of clues that indicate that two regions form part of the same body. If such an evidence exists, the two regions are linked by a common tag. These tags could be either strong (global) or weak (local). Relations between properties of vertices and regions are used in computing these links. For further details of the several heuristics employed, Guzman's report earlier referred to, must be consulted.

4.3. Concluding Remarks

In this paper our primary concern has been to argue the need for picture languages to process pictures using computers. Recognition of complex pictures, problem-solving involving picture discrimination, generation and manipulation of pictures, all of these activities require, in an essential way, adequate descriptive schemata dealing with specific picture classes. We considered a structure for generative descriptive schemata, and discussed the differences between decision-theoretic classification techniques and classification based on generative descriptions.

Most of the work that has so far been done in the design and computer implementation of picture languages has been restricted to classes of pictures made up of line-like elements. Attributes, relations and composition rules, as well as algorithms for computing values of these for given pictures have been studied more or less as they apply to these rather narrow classes of pictures. There is a danger that intense preoccupation with a rather narrowly delimited class of pictures would result in the development of tools, notations, and procedures which are technique-oriented rather than concept-oriented.

There exists a rather large variety of picture classes of great practical importance which can only be described meaningfully in terms of picture atoms that are regions or domains with colours, textures, shadings of various sorts. Photographs of terrains, cloud covers, human faces, landscapes, etc., are examples of such classes of pictures. Picture languages set up to deal with line diagrams do not readily generalize to these latter classes of pictures. Much work remains to be done in devising appropriate primitives and attributes to apply to pictures of these classes, and in designing notational schemes for representing composition rules appropriate for these primitives. Butt et al. (1968) discuss some preliminary studies in texture manipulation and synthesis. They also consider a plausible attribute set. But it is clear that only the rudiments of picture language studies have so far been tackled.

ACKNOWLEDGEMENTS

It is a pleasure to thank Miss F.J. Kotwal and Mr. V.S. Patil for their assistance in the preparation of the typescript.

5. REFERENCES

ANDERSON, R.H. (1968), "Syntax-directed recognition of two-dimensional mathematics"; Ph.D. Thesis, Div. Engg. and App. Phy., Harvard Univ., (Jan. 1968).

BUTT, E.B. et al. (1968), "Studies in visual texture manipulation and synthesis"; Tech. Rept. 68-64, Comp. Sc. Center, Univ. Maryland, (May 1968).

CLOWES, M.B. (1967), "Perception, picture processing and computers"; in Machine Intelligence, I, Collins and D. Michie (Eds.), Oliver and Boyd, London.

CLOWES, M.B. (1968), "Transformational grammars and the organization of pictures"; presented at the NATO Summer School on Automatic Interpretation and Classification of Images, Pisa, Italy, (Aug. 26 - Sept. 7, 1968).

CLOWES, M.B. (1969), "Pictorial relationships - a syntactic approach"; in Machine Intelligence, IV, Meltzer& Michie (Eds.), Edin. Uni. Press.

EDEN, M. (1962), "Handwriting and pattern recognition"; Trans. IEEE, IT-8, 160-166.

EVANS, T.G. (1963), "A program for the solution of a class of geometry-analogy intelligence-test questions"; Ph.D. Thesis, Dept. Math., MIT, Cambridge, (May 1963).

EVANS, T.G. (1968), "Descriptive pattern analysis techniques"; presented at the NATO Summer School on Automatic Interpretation and Classification of Images, Pisa, Italy, (Aug. 26 - Sept. 7, 1968).

FEDER, J. (1966), "Linguistic specification and analysis of classes of patterns"; Tech. Rept. 400-147, School Engg. and Sci., New York Univ., (Oct. 1966).

GRASSELLI, A. (1968), "On the automatic classification of fingerprints"; Proc. Int'l Conf. on Methodologies of Pattern Recognition, Hawaii, (Jan. 1968).

GUZMAN, A. (1967a), "Some aspects of pattern recognition by computers"; M.S. Thesis, Project MAC Rept., MAC-TR-37, MIT, Cambridge, (Feb. 1967).

GUZMAN, A. (1967b), "Decomposition of a visual scene into bodies"; Project MAC, MAC-M-357, MIT, Cambridge, (Sept. 1967).

HILDITCH, J. (1968), "An application of graph theory in pattern recognition"; in Machine Intelligence, III, D. Michie (Ed.), Edinburgh Univ. Press.

KOVALEVSKY, V.A. (1965), "Present and future of pattern recognition theory"; Proc. IFIP Congress 65, Vol. I, Sparton Books, Inc., 37-43.

LEDLEY, R.S. et al. (1966), "Pattern recognition studies in the biomedical sciences"; Spring JCC, 28, 411-430.

LIPKIN, L.E. et al. (1966), "The analysis, synthesis, and description of biological images"; Ann. NY Acad. Sci., 128, 984-1012.

NARASIMHAN, R. (1964), "BUBBLESCAN I Program"; Rept. 167, Digital Comp. Lab., Univ. Illinois, (Aug. 1964).

NARASIMHAN, R. (1966), "Syntax-directed interpretation of classes of pictures"; Comm. ACM, 9, 166-173.

NARASIMHAN, R. and REDDY, V.S.N. (1967), "A generative model for handprinted English letters and its computer implementation"; ICC Bulletin, 6, 275-287.

NARASIMHAN, R. (1968), "On the description, generation and recognition of classes of pictures"; presented at the NATO Summer School on Automatic Interpretation and Classification of Images, Pisa, Italy, (Aug. 26 - Sept. 7, 1968).

NIR, M. (1967), "Recognition of general line patterns with application to Bubble-chamber photographs and handprinted characters"; Ph.D. Thesis, Moore Sch. Elec. Engg., Univ. Penn., (Dec. 1967).

ROBERTS, L.G. (1963), "Machine perception of 3-dimensional solids"; Tech. Rept. 315, Lincoln Lab., MIT, Cambridge, (May 1963).

SAKAI, T. et al. (1968), "Line extraction and pattern detection in a photograph; Dept. Elec. Engg., Kyoto Univ., Kyoto, (June 1968).

SHAW, A.C. (1968), "The formal description and parsing of pictures"; Ph.D. Thesis, Comp. Sc. Dept., Stanford Univ., (Dec. 1967); also Tech. Rept CS94, (Apr., 1968).

UHR, L. (1963), "Pattern recognition: Computers as models of form perception"; Psych. Bull., 60, 40-73.

UHR, L. (1966), "Pattern recognition"; John Wiley, New York.

WATT, W.C. (1966), "Morphology of the Nevada cattlebrands and their blazons, Pt. I"; Rept. 9050, App. Math Div., NBS, Washington, D.C., (Feb. 1966).

6. DISCUSSION

Clowes: You have said "Classification is in fact an extreme kind of interpretation". I think of picture interpretation as frequently, if not always, a problem of translation. What I am not very clear about is if I am right in thinking that interpretation is the same as translation. How do you relate translation to classification?

Narasimhan: Translation is from object to object - for simple objects. I have objects 1, 2 and 3, and so on and I have a bundle of objects. I take one of the bundle and translate it to there. Then I take another and translate to there - that is classification: it is also translation, so that it is interpretation.

Clowes: I was hoping you would say more. It is well known that that kind of trick was played in machine translation ten years ago and produced the most awful mess.

Narasimhan: That's why I argue that this approach will not work if you go to complicated objects. In machine translation, people have been trying to cheat, by considering complicated objects as simple objects, so that they can in fact classify by dictionary look-up - you have a sentence segmented into words, and you have a dictionary which classifies these words in terms of other words: You just plug them in, and try to get a translation - it turns out it doesn't work.

Clowes: It seems to me that translation is so named because it attempts to map a structure in one language into a structure in the other, and the presence of objects is, in a sense, incidental. The real core of the problem is to grasp whatever structure there is in the source language, and to provide an adequate representation of that structure in the target language. If one sticks with the idea that we are playing with objects all the time and not with relationships and so forth, namely structure between objects, then I don't think that classification comes into it at all. It is misleading to say that it is an extreme case.

Narasimhan: This it seems to me is a definitional statement because I consider that an object has nothing but a complex of relationships.

Clowes: Yes, but when you do translation, you never recognize that object. You merely translate the relationships that you have recovered. You do not need to recognize the object in order to provide a valid translation.

Narasimhan: A complex of relationships looked at over one level, looks like an object. If you go inside the object and open it up, then you just have the relationship; there is no object.

Vazey: I would like to disagree with your conclusion that a descriptive approach is essential in order to deal intelligently with classes of picture tokens. It appears that what you are saying is, providing we can describe the picture, all is well. I am questioning the fact that you can always describe - this seems an open question.

Narasimhan: You are saying something slightly different from what I was trying to say. You are asking, "how do I know that everything can be described?" That is not arguing against the stand that descriptions are essential. The question to answer is whether certain kinds of problems can be tackled without trying to generate descriptions. I have suggested several problems and I argue that these problems can be tackled only if one knows how to generate descriptions. If you feel that it is not a fact that descriptions can be generated, then you must concede that these problems can never be solved, or you must find other methods for solving these problems without asking for descriptions. So the burden of proof is on you. You can ask me "Can you give uniform procedures for generating descriptions?" That is another question, because on the basis of that you can decide whether this uniform procedure will in fact deliver descriptions for all kinds of objects that you may be interested in. On the basis of that you can criticize a particular approach that I might suggest for generating certain descriptions, but in this way one can at least progress towards producing theories of descriptions.

Bobrow: You talk about generating descriptions, but you don't talk about the purpose for doing so. You always have a particular purpose in mind, and perhaps for that purpose you can do things in another way. For example, you want to know why "A" and "R" are confused. From measurement systems (multi-dimensional) used for classification you can say "A" and "R" are confused because a line at a certain angle is involved, which gives a certain weight to a particular classification, so you can say that the classification approach actually gives a description of what is wrong. Another matter you don't talk about is the necessary use of classification to identify basic objects, even if only a "T" joint in a picture, or a line segment - so it is a question of, what are the levels of description, when do you go back to classification and description, and how do you do classification once you get a description? There are a whole host of problems where classification and description interact.

Narasimhan: In answer to your first question, I have said that descriptions must be pragmatic. They must be determined by the goal - it is conceivable that for a particular goal, a very simple classificatory scheme will be adequate, and it would be foolish to try to describe objects in detail which is completely irrelevant to the end in view. To answer your specific point about the confusion between "A" and "R", you probably can construct a property space and find characteristics in terms of which you are able to articulate this confusion aspect. Now I would ask you the question, if you have a pair of objects, "Is this pair of objects similar in the same way as that pair?" I construct a variety of questions like this. Then I would consider a way of talking about these objects as adequate, if it enables me to answer a whole variety of questions without going outside this system. This approach is severely limited and messy. I would expect any description that I construct to work at least as well as any classificatory approach, otherwise I would consider it to be inadequate. You may argue that you can construct a classificatory scheme which does less work and still delivers the result - it is conceivable that I may be doing too much work, but this point can be argued. I would be on your side when it comes to the current work in Linguistic Analysis in the context of Conversational Systems. There, I would agree that the current linguistic approaches are messy and irrelevant to what one wants to do - there are much simpler methods of doing it.

Bobrow: My question was to imply that there are kinds of articulation. I think articulation certainly is necessary. The point I am trying to make is that one has to recognise the purpose of the articulation. For example in the characterisation of an alphabet, one way to do that is to have a very complex articulated structure which has subparts and things of this nature, but I think that all of the questions that have been posed can be answered by a simple one-level property description, where the properties are complex, and that is a better way to do it for the kind of question that you want to answer. I think one must match the scheme to the purpose.

Macleod: How do you deal with the problem of degrees of grammaticality? That is, you have a generation scheme, and you generate letters; the aspects of the situation will measure with the aspects of a truly grammatical situation. You want to recognise things that are almost the same.

Narasimhan: There are two ways in which almost grammatical entities could arise. They could arise because the man who drew them knew only an "almost" grammar, or they could arise because there is output plus noise. In both cases, some question making is involved - I would classify this as a problem-solving act. In the case of noise, it is slightly easier to handle because if you know something about the structure of the particular noise, you know what kinds of distortions are produced and you may be able to do some pre-processing. On the other hand, if the trouble arises because of partial knowledge of the grammar, then you may have to do a little more work. In fact you may have to construct some kind of theory which says what is the partial grammar which the man knows - what is plausible and so on. Assuming that he makes sufficient mistakes, you can find out what kind of mistakes he makes and then argue about the possibilities. This is what presumably handwriting experts do.

Bobrow: In some cases you are talking about a metric of closeness of parsing and so on - you are saying "I can parse this picture by a grammar which says what some pictures are", and then we can look at the recognition problem that turns out, the transformations which distort the parsing, and distort the description, as opposed to learning the grammar.

Narasimhan: Yes you can do it at this level also. For example, you can see that from a partial knowledge of the grammar supposing you are confronted with, say, sloppy handwriting. A good part of sloppiness might just consist in not preserving the relational constraints all the way through; for example, if you write at high speed, certain kinds of relations and constraints will be dropped.

Lance: It seems to me that you have a certain amount of flexibility in that you can for example draw all the block letters of the alphabet by using only two primitives, if you are prepared to allow quite complicated transformations and relations - I am thinking of a straight line and a semi circle. When it comes to parsing, one would imagine that this is a simplification, in other words it is easy to identify primitives if you have got very few of them. On the other hand, you could obviously have many primitives, and then you are not going to get so involved in the relational transformations. Would you like to comment on the choice of these two techniques - lots of primitives and therefore simplicity in the relationships, or few primitives and complicated relations?

Narasimhan: I can't say very much about this - it is a very crucial problem in artificial intelligence, because I tend to think that intellectual sophistication in some sense involves being able to operate with reasonably complex primitives at various levels. This I would tend to think is what enables you to process information at a reasonably fast rate. Part of this could also be due to the fact that you have various contextual clues.

Speight: Do you think that in considering whatever the problem is, one should try to pick primitives that are at the right level of complexity and perhaps at a consistent level of complexity, and not have some very simple derivatives and some more complicated ones?

Narasimhan: Why would you want them to be uniform?

Speight: To simplify the relationships that you are going to have to deal with.

Narasimhan: But why don't you put it the other way round? Why don't we say that you would want to choose the primitives so that the relationships are simplified?

It may turn out that if you choose some primitives as complicated and some as simple, the relationships are simpler.

Bobrow: In some complex problem solving systems, what one likes to do is first classify the problem, and then use a different set of primitives even if problems are related, depending on what the kind of problem is or what the sub-part of the problem is: for example in the alphabetic classification you may want to note that there are no curved lines, and then throw away the set of primitives and use only the corners instead of the lines themselves.

Capon: When you were talking about your generative system, it seems to me that although you have a relation in it, you don't have a conditional. Does this mean for instance if you were trying to draw a picture which has hidden parts to it, you couldn't specify this because you can't say, "don't draw this line"? Is this related to the converse problem of trying to recognize a picture of which parts are obscured?

Narasimhan: Yes.

Capon: Has this been used in practice?

Narasimhan: I think in Guzman's problem, he does deal with occluded objects and I would tend to think that he would use conditional expressions at various times. I don't know that in a generative case it is necessary to have conditionals.

Clowes: I think your transformations, including deletions, must direct themselves to this very problem.

Capon: What I was wondering about is just where this sort of system could be applied. It seems that this is very good for systems where you essentially know what you are trying to recognize, which is presumably the whole object, but you could not apply it to a system where you had an implicit description of the picture which you wish to draw for the benefit of the user. You could get a totality of this without a conditional.

Narasimhan: What would you say is an implicit definition?

Capon: Where as a result of some calculation you have a forest of numbers and you have no idea of what this actually looks like. As an example, suppose you had a selection of cubes and you threw them up in the air and used a random process, you knew where they fell, you want to draw a picture. You would have to have a conditional to do that, but these systems we are discussing will not handle that situation.

Clowes: It would seem to me that when you did this throwing up exercise, what you have is a scene, and essentially you must have within the capacity of your descriptive system, some node, some object, which you call a scene object. In order to be able to generate that in the sense of draw it, you must have catered in your grammar for just the problems you are talking about. For example, supposing I have two cubes in my scene, what are the varieties of relations between them? How do I define the relation "partially obscured"? If you can define that relation, then your problem conditionals are satisfied, as essentially the relation will tell you as it were what should enter into the conditional. So essentially I think the difficulty is simply this, that if we are going to have this kind of situation where we can generate scenes, generated by throwing boxes in the air, then we must have within our description apparatus the capacity literally to articulate a scene and to recover all those relationships which are relevant to the drawing.

Narasimhan: It seems to me that it is securable in the scheme I have given. The conditional is implicit in the way the relations are applied.

Capon: Perhaps I am misunderstanding your system, but it seems that you have the facility to say that a given fragment is composed of two things which satisfy a given relation, but you do not have a facility to say that if this relation is not satisfied, then do nothing.

Narasimhan: The way I have defined it, I don't think that possibility would arise, because I am constructing step by step, and at every step I satisfy my relational constraints, so I would never encounter a situation where a particular relational constraint will not be satisfied. I think it will never arise.

Clowes: I think this is a very good question, surely one that we ought to come back to, to see whether in fact descriptive systems of this kind can apply to what is really a well-known problem: the hidden line that we were talking about is one manifestation of this problem.

Further points discussed included the questions of uniqueness in the selection of primitives and in construction in the purely grammatical approach. Moore contended that there might be several equivalent construction techniques to get the same symbol, for example. Narasimhan commented that no assertion was made that there is a unique way of constructing a grammar for the English alphabet - there are indeed a number of ways, this is no problem at all.

NATURAL LANGUAGE INTERACTION SYSTEMS

DANIEL G. BOBROW

Bolt Beranek and Newman Inc.
Cambridge, Massachusetts

1. INTRODUCTION

In this paper I describe six computer programs illustrative of many which have been designed for interaction with people in natural language (Simmons 1965, 1970). All of these systems interact in a restricted subset of English, and I use the terms, "English" or "natural language" to refer to this medium of communication. However, the reader should understand that this identification of "natural" language with a language such as English, French or German, belies the fact that they are not always the most natural medium for communication. A physicist describing the motion of objects in a gravitational field uses a differential equation because it is more natural, and, more important, more precise. An architect feels a picture is worth a thousand words; both use their own languages (mathematics and graphics) to interact with computers. Even in cases where English would do, when wordy message must be transmitted again and again or very rapidly, people use codes or push buttons to "converse with computers".

When are these natural languages "natural"? In cases where no code or jargon has been generated, with messages seldom repeated or where the ideas to be transmitted are not really precisely defined. An important aspect of the communication between people is that a listener, by asking the right question of the speaker, forces him to define more carefully the relationships he is describing. In our efforts to build computer systems to understand natural language, we cannot ignore this important type of interaction. A good natural language system must be active, not passive, and must work with the user to achieve understanding.

We have used the word "understand" in connection with a computer system. We cannot talk about a computer "understanding" in vacuo. The computer understands only in the context of a body of information and procedures which it contains. All the systems I shall describe contain a fairly complex store of information and set of procedures. For each system there is an assumed purpose and context underlying the interaction, and it is only on the basis of succeeding in its purpose that one can talk about a system understanding English. One of these systems is designed to maintain a conversation, and the important point is that the conversation be natural and "human". Other systems answer questions within a limited semantic domain and it is their question-answering capability which provides a measure of the understanding of these systems. For each system we describe, we will describe the data structure utilized and how the transformation of the input reflects the purpose of the desired communication.

2. THE ELIZA SYSTEM

The ELIZA system (Weizenbaum, 1966) is designed to interact conversationally. This conversation is designed to simulate one that might be held in the office of a Rogerian psychiatrist, one who never gives direct answers, but answers a question with a question. The purpose for this limitation was to define a context in which a limited form of conversation would seem natural. A basic assumption underlying a conversation with a psychiatrist is that he is trying to help you to think through your own problems. When he asks you a question, even if you think he knows the answer, you feel that he may be trying to prompt an association. If you say, "I like boats a great deal", and the psychiatrist responds "what's a boat?" you do not assume that he does not know about boats; rather you assume that he is trying to probe your feeling about boats.

Before describing the ELIZA system I would like to tell a story which shows one conversation in which ELIZA took part. It also illustrates how in a limited interaction the basic assumption need not be strictly true but can be replaced by a weaker one; namely that the person assumes he is communicating with another person rather than with a machine.

One Saturday morning about 9 a.m., a BBN Vice President in charge of selling our Telcomp commercial service arrived at our PDP-1 Computer Room hoping to be able to use the computer for a demonstration. He found a note on the PDP-1 teletype that read: "TYPE ALL MESSAGES INTO THIS TELETYPE, AND END THEM WITH A PERIOD." This note was left there by one of the programming bums who had just succeeded in making run a LISP version of the DOCTOR program (based on ELIZA as developed by Weizenbaum). However, the VP thought that this note indicated that this teletype was linked with the one in my home, since I often used the computer all night. Therefore, he started typing to "me", but in fact was typing to this version of the DOCTOR program. The following conversation ensued:

VP:	DO YOU THINK THAT I CAN USE THE COMPUTER THIS MORNING?
Teletype Response:	WHY DO YOU ASK?
VP:	I MIGHT BE ABLE TO MAKE SOME ADDITIONAL TELCOMP SALES.
Teletype Response:	WHY AREN'T YOU SURE?
VP:	MY PROSPECTIVE CUSTOMERS HAVE NEVER SEEN A DEMONSTRATION OF THE SYSTEM.
Teletype Response:	DOES THIS MEAN VERY MUCH TO YOU?
VP:	OF COURSE IT DOES.
Teletype Response:	YOU SEEM VERY POSITIVE ABOUT THAT.
VP:	THIS IS RIDICULOUS.
Teletype Response:	WHAT DO YOU MEAN THIS IS RIDICULOUS?
VP:	IS THERE ANY REASON WHY I CAN'T USE THE COMPUTER NOW?
Teletype Response:	TELL ME MORE ABOUT YOUR PROSPECTIVE CUSTOMERS.
VP:	PLEASE DIAL ME UP ON 491-1850

Note that after the last remark the VP did not obey instructions, and left out the period. Therefore, of course, the computer didn't answer him. This so infuriated the VP, who thought I was playing games with him, that he called me up, woke me from a deep sleep, and said:

VP:	Why are you being so snotty to me?
Bobrow:	What do you mean why am I being snotty to you?

The VP angrily read me the dialogue that "we" had been having, and couldn't get any response but laughter from me. It took a while to convince him it really was the computer.

Having shown what type of conversational response ELIZA makes, let us now indicate what mechanisms it uses to achieve these very natural responses. The first basic assumption made in the construction of ELIZA is that responses need not be sensitive to all of the input string. Certain words that the computer knows about are considered key words. Each input string is scanned for the occurrence of these key words and a stack of those occurring in any input is constructed. The position on the stack is determined first by a value associated with that word and secondly by the order from left to right in the input sentence. In the process of scanning, substitutions are made for a number of words and phrases to make the later transformation process easier. These include converting the sense of pronouns and verbs, that is changing "I" to "You" and "am" to "are" and vice versa; the scan also replaces by a standard form (e.g. mother) possible variants of a word, (e.g. ma or mommy).

Having made the standard transformations and isolated the key words, ELIZA selects a possible transformation of the input string on the basis of the most important key word. This selection is made from a set of possible transformations called a script, and it is this script along with the key words, which defines the conversational capability of the program. In the Figure 1 we show an input whose principal key word is "can" and the element of the script which was retrieved for this

INPUT: Can I get on the computer?

SCRIPT: ($ CAN YOU $) = (DOES IT MEAN A LOT TO YOU TO 4)

OUTPUT: Does it mean a lot to you to get on the computer?

Fig. 1 - Input with principal key word 'can', and the element of script retrieved for this key word.

key word. This transformation contains a left half which is a pattern which must be matched by the input, and a right half which describes the output to be generated if the left half matches. The left half defines the minimum context for which this response is reasonable, and gives a segmentation of the input string into significant pieces in case of a match. In the left half, the symbol "$" will match any string provided that the rest of the pattern is matched. The pattern shown will match any input which contains the two words "CAN YOU" in sequence; the left most $ will match the words up through these two constants, and the right-most $ will match all of the input past those two constants. Thus this left half pattern matches the input with the first $ matching a null string and the second one matching the string "GET ON THE COMPUTER". Note that the input string had the pronoun "I" which was replaced by the input substitution process by the pronoun "YOU". The output construction rule indicates that the output is to contain the constant string "DOES

IT MEAN A LOT TO YOU TO", and then that portion of the input string which was the fourth element of the parsing or segmentation found by the matching of the left half. Thus the output generated is "DOES IT MEAN A LOT TO YOU TO GET ON THE COMPUTER". The pattern action format shown as the transformation in this script is one that is used a great deal in a number of the systems that we shall describe.

An important element to maintain the naturalness of conversation is based upon the fact that people rarely make exactly the same response to a repeated input string, and people have a set of responses they make when they don't understand an input at all. In the ELIZA system, for any given pattern there are a set of actions for output formats which can be utilized for any one input pattern. This allows the system to make a random choice of these outputs to provide some variety within this limited conversational context. When the input is not understandable, ELIZA chooses from a set of standard default remarks such as "I don't understand you fully" or "hmmmm"; since one would soon tire of a conversationalist who never remembered what you said previously, ELIZA uses previously stored matched contexts in some cases where the current input cannot be matched to a pattern. Things which the "patient" talked about as being his (that is adjacent to the word "my") are probably important to him, and therefore reference to them will probably elicit another remark which is more meaningful to the computer. An example in the story above is shown when the VP asks "IS THERE ANY REASON I CAN'T USE THE COMPUTER NOW?" and the machine, not having any appropriate patterns which can match this input, responded with "TELL ME MORE ABOUT YOUR PROSPECTIVE CUSTOMERS" referring back to the earlier remark made by the VP.

The ELIZA system illustrates the minimal knowledge required to conduct the interaction between man and a computer which is natural in an assumed context. The computer has almost no knowledge of the world and the success of the program is based upon an (unfounded) assumption made by the user that if appropriate portions of his input stimulus is identified and an appropriate response selected, then there must be an underlying understanding. This may have some moral relevant to interactions between people which I refrain from drawing.

3. THE STUDENT SYSTEM

The STUDENT system (Bobrow, 1964; Minsky, 1969) operates in the context defined in a more limited semantic domain than that chosen for ELIZA. All of the input forms to STUDENT are assumed to be related to the process of solving algebra story problems such as those found in high school algebra books. One of these is illustrated in Fig. 2. In this limited domain STUDENT makes meaningful responses,

> IF THE NUMBER OF CUSTOMERS TOM GETS IS
> TWICE THE SQUARE OF 30 PER CENT OF THE
> NUMBER OF ADVERTISEMENTS HE RUNS, AND
> THE NUMBER OF ADVERTISEMENTS HE RUNS IS 50,
> WHAT IS THE NUMBER OF CUSTOMERS TOM GETS?

Fig. 2 - Typical algebraic STUDENT problem.

which implies that it translates the English input into algebraic equations, solves the set of equations, and responds giving the answer to the problem in English. By using a well known semantic domain (i.e. those facts representable by algebraic equations) the program emphasizes techniques involved in the translation of English into this representation and the utilization of appropriate problem-solving techniques to find the answer to the question asked. Two different mappings are used for English: one (called STUDENT) maps problems given to the computer into equations; the other (called REMEMBER) stores global information not specific to any one example which may be necessary for solving these story problems.

STUDENT starts with a set of transformations of the English language input into a set of standard forms or simple sentences which it can map directly into equations. These transformations, based on pattern action rules, are described in more detail below. During the transformation process STUDENT keeps a list of the answers required, a list of the units involved in the problem (e.g. dollars, pounds) and a list of all the variables (simple names) in the equations. Then STUDENT invokes a SOLVE program to solve the set of equations for the desired unknowns. If the solution is found, STUDENT prints the values of the unknowns requested in a fixed format, substituting in "variable is value" the appropriate phrases for variable and value. If the solution cannot be found, various heuristics are used to identify two variables (i.e. find 2 slightly different phrases that refer to the same object). If two variables, A and B, are identified, the equation A = B is added to the set of equations. In addition, the store of global information is searched to find any equations that may be useful in finding the solution to this problem. STUDENT prints out any assumptions made about the identity of variables, and also any equations that are retrieved because they might be relevant. If the use of global information, or equations from identifications leads to a solution, the answers are printed out.

If the solution is not found, and certain idioms are present in the problem, a substitution is made for each of these idioms in turn and the transformation and solution process is repeated. If the substitutions for these idioms do not enable the problem to be solved, then STUDENT requests additional information from the questioner showing him the variables being used in the problem. Any additional information is used to make another attempt to solve the problem.

Transformations used in STUDENT are keyed to words and phrases in the input string. These key words and phrases are divided into two classes, surface forms and operators. For each surface form it substitutes another string of words which may contain key words or dummies (i.e. words not understood by the system). Substitutions may be mandatory, such as the substitution of "2 times" for the word "twice" shown in the problem in Fig. 2. The phrases "square of" and "per cent" also require substitutions in that problem. Optional substitutions of phrases may be made if required to solve the problem.

The second class of key words are the operators which indicate the presence of a relationship which may be able to be represented in the underlying semantic model, that of algebraic equations. They may also be used in the parsing of the input. One such key item is the comma after the first clause in the long sentence in Fig. 2. Using a simple set of pattern-action rules, STUDENT divides this long sentence up into three simple sentences each of which can be directly mapped into an equation. In Fig. 3 we see the first of these simple sentences with each of the

THE NUMBER (OF/OP) CUSTOMERS
TOM (GETS/VERB) IS 2 (TIMES/OP 1)
THE (SQUARE/OP 1) 30 (PERCENT/OP 2)
ADVERTISEMENTS (HE/PRO) RUNS (PERIOD/DLM)

Fig. 3

operators labelled, and mandatory substitutions made. The principal key word in this sentence is "is" which is not marked because it is used uniquely in this problem context to define an equality between two arithmetic expressions in the variables of the problem. Each of the operators is a key to a set of pattern action rules which check for appropriate contexts to determine a transformation to be made to a form in the semantic domain. In Fig. 4 we see the transformation rule which

K/NUMBERP OF $ → (TIMES K 3)

Fig. 4

checks the context of a particular operator, the word "of", to determine whether it implies a transformation, i.e. whether the "of" indicates the function of multiplication. The test NUMBERP (associated with K/) determines if the item immediately preceding the word "of" is a number. Only in the contexts such as "Two of the girls who danced" or "3 of the number of advertisements" should the "of" be interpreted as a multiplication operator.

This illustrates another problem faced by STUDENT, that is the sensitivity to order among the transformations used. If this pattern were tried before the transformation for PERCENT which changes the number immediately preceding it to .3, the match would fail, "of" would not be considered an operator, and this appropriate transformation would not have occurred. The verb "gets" is marked because, in the context of STUDENT, sentences such as "Mary gets 2 fish" or "Tom gets 3 customers" must be transformed into the sentences "The number of fish Mary gets is 2" or "The number of customers Tom gets is 3". These transformations must take place only in an appropriate context where no other verb is present in the sentence.

Fig. 5 shows the results of the transformations which have converted this

```
(EQUAL (NUMBER OF CUSTOMERS TOM GETS)
   (TIMES 2
      (EXPT (TIMES .3 (NUMBER OF ADVERTISEMENTS HE RUNS))
         2)))
```

Fig. 5

sentence into a prefix representation of an algebraic equation. Variables in this equation are "number of customers Tom gets" and "number of advertisements he runs". These strings of words are not further interpreted by the STUDENT system. An assumption is made that a questioner will almost always use the same string of words in a problem to refer to the same concept. Therefore, in the domain with which STUDENT deals, a large class of words are simply considered dummy markers in a string to name variables. These strings are then used directly as variables solving the equations. The solution printed out by the STUDENT in this case is

THE NUMBER OF CUSTOMERS TOM GETS IS 450.

The assumption that the same string of words is used to represent the same concept each time is not strictly true, and a number of heuristic techniques are used in STUDENT to get around this problem. They are invoked only if the problem is not solvable with this assumption. In trying to identify two variables, STUDENT explores whether substitutions for pronouns will help - e.g. the substitution of "Tom" for "he" in "the number of customers he gets" or "the girl in the store" for "she" in "the number of books she has". In addition STUDENT looks for truncations of phrases, and assumes the identity of "the number of trips" and "the number of trips Dan made to Australia". Other identifications may require a "knowledge of the world" which must be added as an item of global information. For example, in one problem there was the phrase "the number of students who passed the examination" and later, the problem refers to "the number of successful candidates". Rather than build in a complete knowledge of the value set of our society, STUDENT utilized an ad hoc piece of global information in its store. If it had been unable to solve the problem using information previously stored, it would ask the user for relationships among the variables, and hopefully, the user would give the STUDENT program a clue.

Other global information allows use of schema such as "Distance equals speed times time", entered directly into STUDENT in just that form. When key words in the problem coincide with key words in one of these global schema, the schema is retrieved and added to the equation set. Note that the truncated phrase identification process must be used to identify the phrases "distance" and "distance between Sydney and Canberra". In addition to global information, STUDENT makes use of special heuristics to solve special classes of problems. Included in one such class are "age problems", such as "Mary is twice as old as Ann was when Mary was as old as Ann is now. If Mary is 24 how old is Ann?". This old nut usually takes people between 10 to 15 minutes; it took the IBM 7094 only about a half minute. However, if I tell people how the program works, it turns out that the STUDENT heuristics work well for people too, and this cuts solution time for humans down considerably.

4. THE CARPS SYSTEM

The CARPS system (Charniak, 1969) is designed to solve calculus rate problems such as that shown in Fig. 6. CARPS uses a generalization and expansion

(A SHIP IS 30.0 MILES SOUTH OF POINT 0 AND
TRAVELLING WEST AT 25.0 MILES PER HOUR.
HOW FAST IS THE DISTANCE FROM THE SHIP
TO 0 INCREASING?)

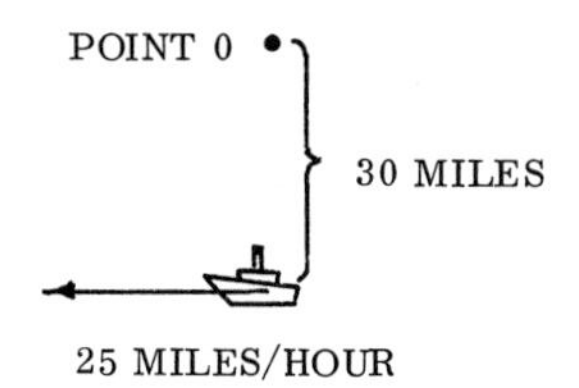

Fig. 6

of the techniques used in STUDENT; as in STUDENT, the problem statement is to be transformed to a set of mathematical equations. However, the problem which is encountered in trying to do this exactly as done in STUDENT is that the sentences do not directly reflect equations. Therefore, CARPS utilizes an intermediate semantic model, which Charniak calls a situation, where each situation reflects the structure of a different type of problem. CARPS analyzes a problem to determine which type of situation it is facing, and parses each of the simple sentences it derives to determine assignments of values to attributes in the model situation. As in STUDENT, CARPS starts by tagging key words as shown in Fig. 7. The simplified sentences shown in

```
(THE PROBLEM WITH TAGS ON IS)
(((A SHIP (IS VERB) 30.0 (MILE UNIT) (SOUTH PNOUN)
POINT 0 AND (TRAVELLING VERB) (WEST PNOUN)
(AT PREP) 25.0 (MILE UNIT) PER (HOUR UNIT)) (1.))
(((HOW QWORD) (FAST RWORD) (IS VERB) THE DISTANCE
(FROM PREP) THE SHIP TO 0 (INCREASING VERB)) (2.))
```

Fig. 7

Fig. 8 express relationships which must be mapped into the situation.

```
(THE SIMPLIFIED SENTENCES ARE)
(((A SHIP (IS VERB) 30.0 (MILE UNIT) (SOUTH PNOUN)
POINT 0) (1.))
((A SHIP (TRAVELLING VERB) (WEST PNOUN)) (1.))
((A SHIP (TRAVELLING VERB) (AT PREP) 25.0 (MILE UNIT)
PER (HOUR UNIT)) (1.))
(((HOW QWORD) (FAST RWORD) (IS VERB) THE DISTANCE
(FROM PREP) THE SHIP TO 0 (INCREASING VERB)) (2.)))
```

Fig. 8

In this problem, CARPS has determined that the problem is of type <u>distance</u> by using key words such as "travelling" and "distance". Four specific patterns are used for <u>distance</u> problems and in the table below we show these patterns, examples of matching kernels, and the meaning of these kernels in the problem domain. From the meanings one can deduce the operations which CARPS performs to insert values for attributes in its paradigm for the distance problem. For each pattern an appropriate set of action routines coded in LISP are invoked.

TABLE OF DISTANCE TRANSFORMATIONS

NVP - VERB - PATTERN - POSIT	A ship - travelling - north	Indicates direction of velocity.
NVP - VERB - AT - TIME INDICATOR	A ship - starting - at - 12 AM	Indicates zero time. The time indicator may also be "noon" or "midnight".
NVP - VERB - NUMBER - UNIT - POSIT - NVP	A ship - is - 30 - mile - north - point 0	Indicates value and direction of position with respect to a second object.
NVP - VERB - POSIT NVP	A ship - is - north - point 0	Indicates direction of position with respect to a second object. Only used in distance problems.

In Fig. 9 we see the mapping of simple sentences into the distance paradigm. There is an object which is a ship which has a position and a velocity. The labels

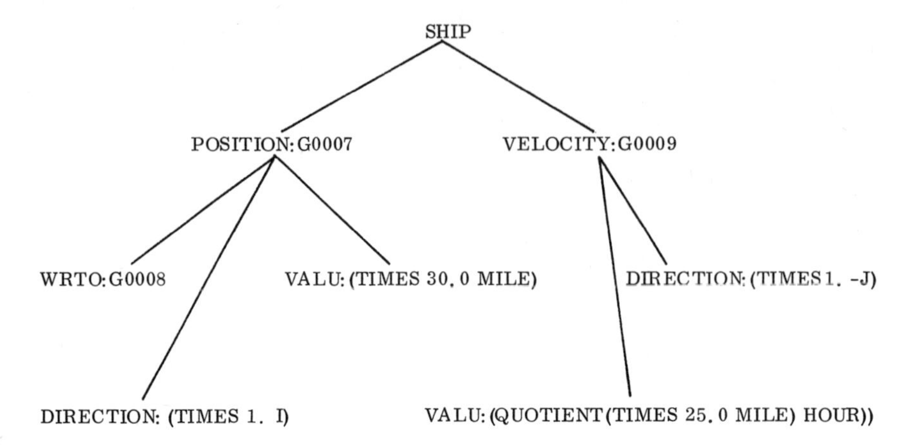

Fig. 9 - Mapping of simple sentences into the distance paradigm.

such as G0007 following the colon are to allow the ship to have more than one position and still have each position uniquely identified. Coordinates are given on a set of (I, J) axes; distance and velocity are shown as LISP prefix arithmetic expressions in units for the system.

The mapping from the intermediate semantic data structure of a situation to algebraic equations is again done by rules in a pattern action format. CARPS problem solver uses standard procedures for solving each problem with more sophisticated mathematical knowledge: taking derivatives, substituting boundary conditions, etc. CARPS utilizes a package of programs developed for SIN (Moses, 1967) for integration. The answer to the problem shown is Ø, which CARPS finds by going through its formal procedures, but which a person might see immediately from a diagram if he realized the direction of motion at the point given (West) is perpendicular to the direction to the origin (which is North).

Significant in the approach of both CARPS and STUDENT is the fact that the unit of analysis is more than a single sentence. Problems of generalized pronominal reference are faced. Attributes and values may be referenced across the length of the (specialized) paragraphs which are handled, and a scope of several sentences used to determine the specific context for interpretation of each individual sentence.

5. AN AIRLINE GUIDE QUESTION-ANSWERER

In building this airline guide question answering system, Woods attacked the problem of developing a uniform framework for performing the semantic interpretation of English sentences (Woods, 1967). In developing his system, Woods assumed two significant boundary conditions. First, he assumed that an input question could be analyzed into its deep structure parsing in terms of a transformational grammar (Chomsky, 1965). Although no such grammar exists for all of English, this is a problem that is being attacked by a number of linguists, and approximations to a deep structure parsing can be obtained.

The second assumption Woods makes is that his data base can be defined in terms of a set of primitive objects, primitive functions on these objects, and primitive relationships between these objects which can be tested by a set of predicates. He also assumes there exists a set of commands which can operate on objects and structures in the data base to provide output from a question-answering process. These items provide a procedural language for managing the data base.

As a sample data base, Woods chose the airline guide which describes the schedule of airplane flights and provides other information about such flights. The objects in this system are things like airplane flights, airports, cities, meals, dates, etc. Woods assumes a primitive predicate such as FLIGHT(X), TIME-ZONE(X), etc. which can check the type of any object, and also predicates which can compare two objects, e.g. GREATER (X1, X2) where X1 and X2 are times of day or numbers. The primitive functions in the system include such items as DTIME(X1, X2) which represents the departure time of flight X1 from place X2, EQUIP(X1) which is the type of plane on flight X1, and others. The primitive commands he uses are TEST(D1) which tests the truth of statement D1, and LIST(X1) which prints the name of the object X1. It is assumed that the primitives will be programmed to work with the data base, and the semantic interpretation is independent of the particular representation that is used in the computer. This separates the general process of determining the meaning of the question from the particular processes which are required to retrieve answers to questions from any implementation of the data base.

Let us now consider an example shown in Fig. 10 which is a deep structure parsing of the question "does American Airlines 57 fly from Boston to Chicago?". Interpretation of this structure is done using a set of pattern action rules. The patterns which are used to match the structure are partial tree structures such as those shown in Fig. 11. The pattern G2 matches the structure only if there is an S which

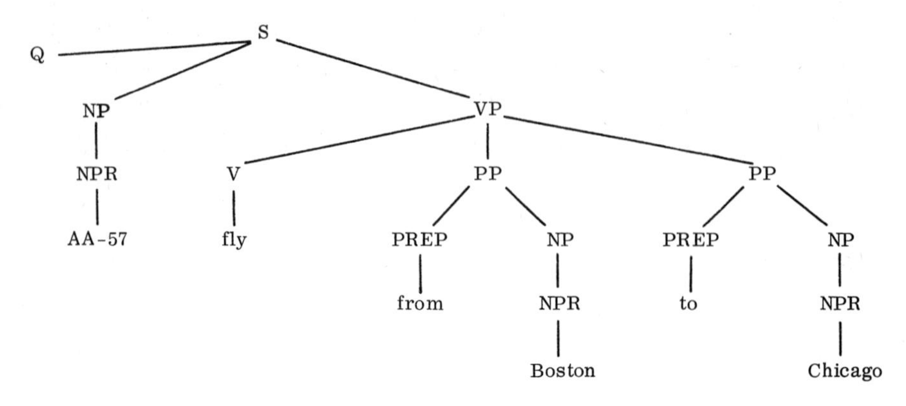

Fig. 10

immediately dominates a VP which has in it a V and a NP in that order (not necessarily contiguous). This pattern is a paradigm for the verb-object relationship.

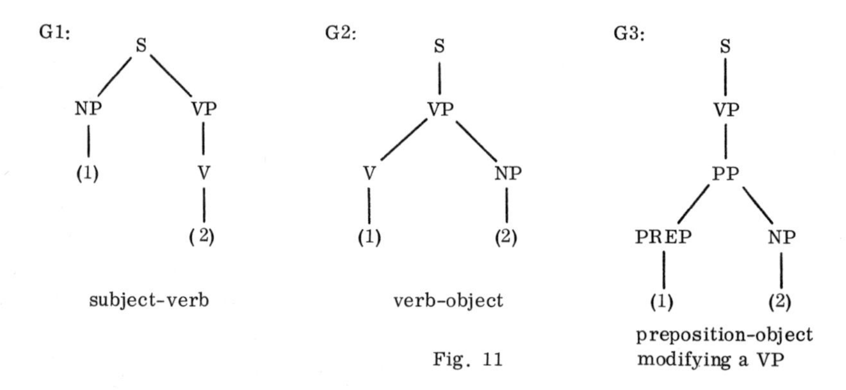

Fig. 11

Fig. 12 indicates the type of pattern action rule used to translate the tree structure into a logical representation in terms of a predicate, which in this case is "CONNECT."

1 - (G1: FLIGHT((1)) and (2) = fly) and
2 - (G3: (1) = from and PLACE((2))) and
3 - (G3: (1) = to and PLACE((2)))
$\Longrightarrow$ CONNECT (1 - 1, 2 - 2, 3 - 2)

Fig. 12 - Pattern Action Rule.

The pattern to be matched has three clauses; the first clause states that the tree must match pattern G1 shown in Fig. 11, and that the node labelled 1 in G1 must satisfy the predicate FLIGHT; that is, this noun phrase must be the name of a flight. Node 2 must be the word "fly"; the other two clauses are interpreted similarly. If the patterns have been matched, the rule states that this can be partially interpreted in terms of the predicate CONNECT with three arguments: the first argument is the first element in clause 1; the second argument is the second element in clause 2, and the third is the second element in clause 3.

Note that in matching the patterns G1, G2, and G3 no ordering is implied in the tree structure to be matched. Therefore either "AA57 flies from Boston to Chicago" or "AA57 flies to Chicago from Boston" will match this set of patterns. Woods thus finds the interpretation independent of some variations in the sentence structure.

All possible pattern matches are tried and the final interpretation of the sentence given earlier is shown in Fig. 13. To determine whether this is true, this

TEST (CONNECT (AA-57, BOSTON, CHICAGO)
AND DEPART (AA-57, BOSTON)
AND ARRIVE (AA-57, CHICAGO)).

Fig. 13 - Interpretation of Sentence.

predicate would be embedded in a test operator which would then check the data base to see whether all three clauses are satisfied.

An important contribution by Woods is his careful discussion of the role quantifiers play in English questions. His general form for a quantifier is:

(FOR QUANT X/CLASS: R();P(X))

where quant is a quantifier, X is the variable being quantified, class is the name of a set over which the quantification is to range, R(X) is a (possibly vacuous) further restriction on the range of quantification and P(X) is the proposition or command being quantified. For example:

(FOR EVERY X1/FLIGHT: DEPART(X1, BOSTON): LIST (X1))

is a quantified command which directs the retrieval component to print the name of every flight which leaves Boston. It is assumed that there are a relatively small number of sets that can take the place of CLASS in this quantificational form, and that for each such set, the retrieval component contains a successor function which enumerates the members of that set. More restricted ranges of quantification can be obtained by imposing additional restrictions R(X). Every question which requires a fill-in answer, that is which contains a wh word such as what, requires the use of a quantifier. In addition to the usual quantifiers of EVERY and SOME, Woods includes others such as MANY, NONE, and the one usually found in most questions THE which requires the unique answer.

Let us consider an analysis of the question, "What is the departure time of AA-57from Boston?". The deep structure of this sentence is shown in Fig. 14. Note

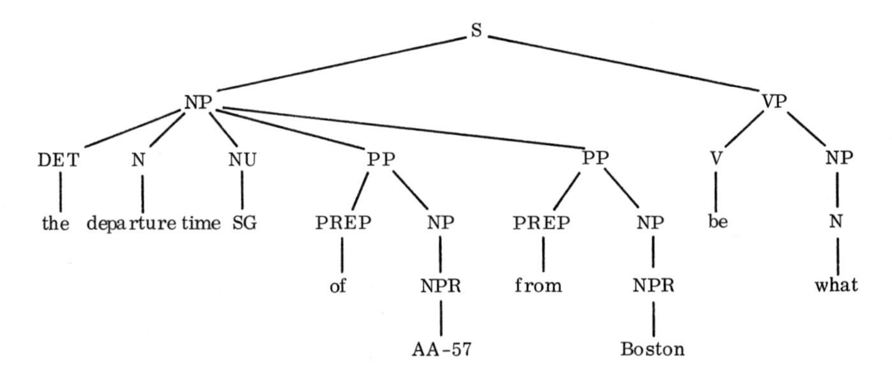

Fig. 14 - Deep structure of "What is the departure time of AA-57 from Boston?"

the conversion of the verb "is" to a standard form "be", and the reversal so that the questioned word is the object of the verb. Figure 15 shows one rule used in the trans-

S23 1 - (G1: (2) = BE) AND
2 - (G2: (1) = BE AND (2) = WHAT)
⟹ LIST (1 - 1)

Fig. 15

formation of this deep structure to the quantified form. The sentence requires an object to be listed, that is to be typed out to the user. The 1-1 refers to the interpretation of the noun phrase in the pattern, in this case "the departure time of AA-57 from Boston". In the interpretation of the noun phrase the quantifier is THE, which is internal to the structure of the noun phrase; however its scope is the entire sentence, since noun phrases satisfying the predicate indicated by the sentence must be generated and the result given to the command. The interpretation for the sentence above is

(FOR THE X1/DTIME(AA-57, BOSTON) :-; LIST(X1))

The problem of the scope of a quantifier may be perhaps better understood from the example "Tell me all flights leaving from Boston for Chicago". The interpretation is

(FOR EVERY X1/FLIGHT: CONNECT(X1, BOSTON, CHICAGO); LIST(DTIME(X1, BOSTON)))

This indicates how the "every" which was attached to one noun phrase within the sentence is brought out to the sentence level. The generating mechanism for such noun phrases must call each candidate to be tested by the predicate expressed by the sentence.

Utilizing the same semantic interpretation scheme but a different set of primitive functions, one can store information into the data base from an English language input. The interpretation of "AA-57 flies from Boston to Chicago" which is CONNECT(AA-57, BOSTON, CHICAGO) can be used to store a piece of connection information (rather than check it). It will put this information into the data base, if the routine for CONNECT stores rather than tests data. Thus, Woods' interpretation scheme is symmetric with respect to storage and retrieval, except for the fact that different underlying primitives must be used for the two operations. Woods' approach has shown how meaning can be operationally defined in terms of a procedural language with a well defined language base.

6. QUASI-NET

QUASI-NET (Lemmon, 1969) is designed to accept input declarative statements in English, and to store the information in a semantic network. It can then accept questions stated in English and retrieve answers to these questions from the network based on the information it has received. It is designed to be a general language comprehender and works well in this regard within the limitations imposed by its parsing strategy and the deduction rules embedded in the system.

The semantic deductive portion of the program was built first with the assumption that it would be able to utilize the output of a syntactic analysis system provided by Kuno (Kuno, 1966) the Multipath Predictive Analyzer. The predictive analyzer develops a parsing which provides labels for sentence components: Subject(S); verb(V); object(O); complement(C) which includes indirect objects; and other complement modifiers (E) such as adverbs. However, because of the difficulty involved in interfacing two large programs, Lemmon was forced to build his own syntactic analysis system which gave a similar type of output. It is designed to work interactively in the sense that if information is missing about any of the words and their parts of speech, the system requests the user to supply this information. It is limited to finding the first good parsing of a sentence rather than finding all possible parsings, but for the class of sentences Lemmon tried, this was sufficient. He used a relatively unusual algorithm which obtains the parsing from right to left in a sentence. Although this is unnatural as a psychological model for a parser for English, it has certain efficiencies. English has many more structures which branch to the right than branch to the left. Therefore, starting at the right and going to the left, one can often tell unambiguously where a constituent ends, more often than in the reverse direction. For example, scanning from right to the left one can find a noun which is the object of a prepositional phrase, and then find a preposition which delimits the entire prepositional phrase. Moving from left to the right, the preposition is found first, and then ambiguities between nouns which may be the object of the preposition and nouns which may modify a following noun must be decided.

The semantic analysis of a sentence is based directly upon the syntactic analysis found. Associated with each syntactic component is a semantic analogue. For each word there is a node in the network which uniquely represents that word. This mapping is done automatically for Lemmon within the LISP system, since a hash lookup is made for each word as it comes in, and there is a unique representation for each word within the system. On this atom or node Lemmon puts pointers to all the items associated with this word.

Each noun phrase in the input is mapped into a set which is defined by the properties associated with this noun phrase. An adjective modifying the noun head becomes a restriction on the types of elements of the set, all of which belong to the class defined by the noun head. A prepositional clause embeds this noun phrase in a restriction related to the type of preposition used. Each clause found in a sentence (including the sentence itself) is mapped into a proposition which is a relationship with arguments, with quantifiers inserted as special restrictions on the propositions.

We shall illustrate each of these forms with examples.

Each noun in the store has a special link called "CATEG" which points to another node which is a superset of the first node, e.g. candy has a link CATEG to food, that is, the set of candies is a subset of the set of foods. This important link allows Lemmon to perform certain types of deductions in the question-answering process. In Fig. 16 we illustrate the structure associated with the simple sentence

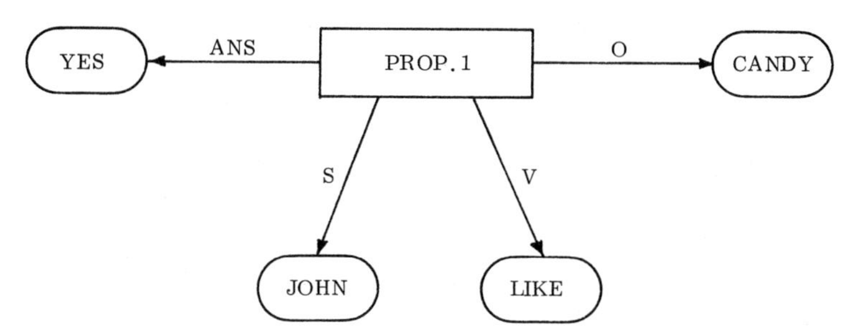

Fig. 16 - Structure associated with "JOHN LIKES CANDY".

"John likes candy". Each of the words shown are nodes in the network and as indicated earlier a word identifies a node or concept uniquely. For example, the word "candy" is a unique identifier for a node in the system which is called "candy" on the outside, and to which are attached all things which are candies. Lemmon assumes all primitive objects are represented by single words externally, and by nodes internally.

In the mapping of "John likes candy", "likes" is reduced to the internal form "like" where the inflection of "like" has been suppressed and the base form used. The proposition has three arguments labelled, S, V, and O which correspond to the subject, verb and object of the sentence. The argument ANS is used to indicate the truth value associated with this proposition; the "YES" indicates that this is a true proposition. If the sentence were "John doesn't like candy", the only thing that would change would be the "ANS" node which would contain "NO". If the sentence were "John likes candy under certain conditions", then ANS would contain the semantic analogue of "under those conditions". Any clause can be inserted for "ANS", and thus the truth value of the proposition can be computed by computing the truth value of the conditions which are other proposition(s).

Let us now consider the representation of a compound noun phrase, for example the object of the sentence "John likes foods which are sweet". This object is represented in the QUASI-NET as a dummy object which is a member of set 3 as shown in Fig. 17. The item "HASPROP" indicated by the argument "R" on propo-

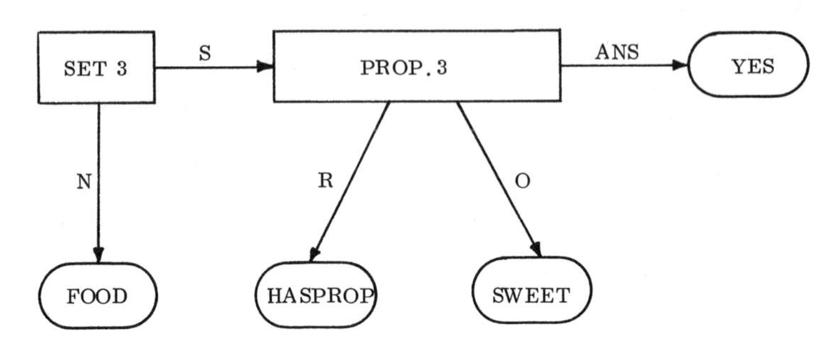

"SWEET FOODS"

Fig. 17

sitional 3 is the translation of the verb to be where the object is known to be an adjective. The mapping rules for creating this type of object maps both "sweet foods" and "foods which are sweet" into the same object. Fig. 18 illustrates the representation of "Foods which John likes".

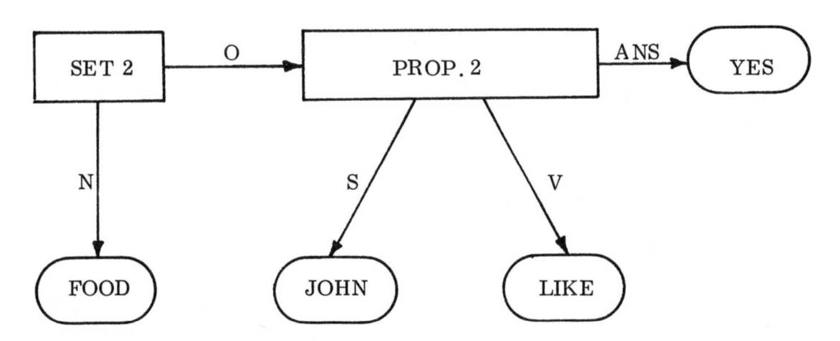

"FOODS WHICH JOHN LIKES"

Fig. 18

We have indicated the techniques for representing declarative statements within the network. Let us now consider the process of answering a question based on information stored in this network, for the question "does John like candy?". It is transformed into a form similar to that shown in Fig. 16 above, except that attached to the ANS node is a question mark. A search is made in the network to find an item which immediately matches this question (except for the node ANS). If a direct match is found (the retrieval is very rapid since propositions are indexed by every item), the answer is immediately decided by the constant that matches the question mark. However, consider what must happen if the direct answer is not stored in the network. Fig. 19 illustrates a network in which are stored three facts: "candy is a food", indicated by the CATEG link; "candy is sweet" stored by a link between candy and set 2 which is defined by proposition 4 to have a property sweetness; and the statement "John likes foods which are sweet" in which "foods which are sweet" is represented by a referent which is an element of set 1, whose noun head is food, and which has a property indicating that its elements are sweet. When the search for a direct match to the question "does John like candy" fails, a search is made for all items which are generalizations of either the subject "John" or the object of the sentence "candy". This generalization retrieves the statement "John likes foods which are sweet". This generates the two sub-questions "is candy a food?" and "is candy sweet?". These two sub-questions are answered by the same question-answering process, and in this case can be answered directly. That is, candy is a food can be determined straightforwardly from the CATEG link from candy to food and "is candy sweet?" is answered by noting the answer YES on the proposition which has as its elements "CANDY", "HASPROP" and "SWEET". Having retrieved the answers to these two sub-questions, the answer to the main question can be returned. This generalized matching process utilizes the assumption that things true of supersets are true of their subsets. This is the major deductive rule used in the system.

Lemmon talks about a number of techniques for handling quantifiers in questions. For many questions using quantifiers, QUASI-NET generates all examples of the items in a set, or enough of them until the answer can be determined, e.g. for "Does John like some sweet food" QUASI-NET will generate each of the sweet foods in turn until one is found that "John likes". For the question "Does John like most sweet foods?" all the "sweet foods" known by the system are considered and a count is made of those John likes and those he doesn't like; then the appropriate answer is generated. However, for the quantifier "all" in the question "Does John like all sweet foods?" or the simple form of this question which is "Does John like sweet foods?", the QUASI-NET procedure utilizes the intensional definition of the set of sweet foods rather than the extensional definition which is comprised of

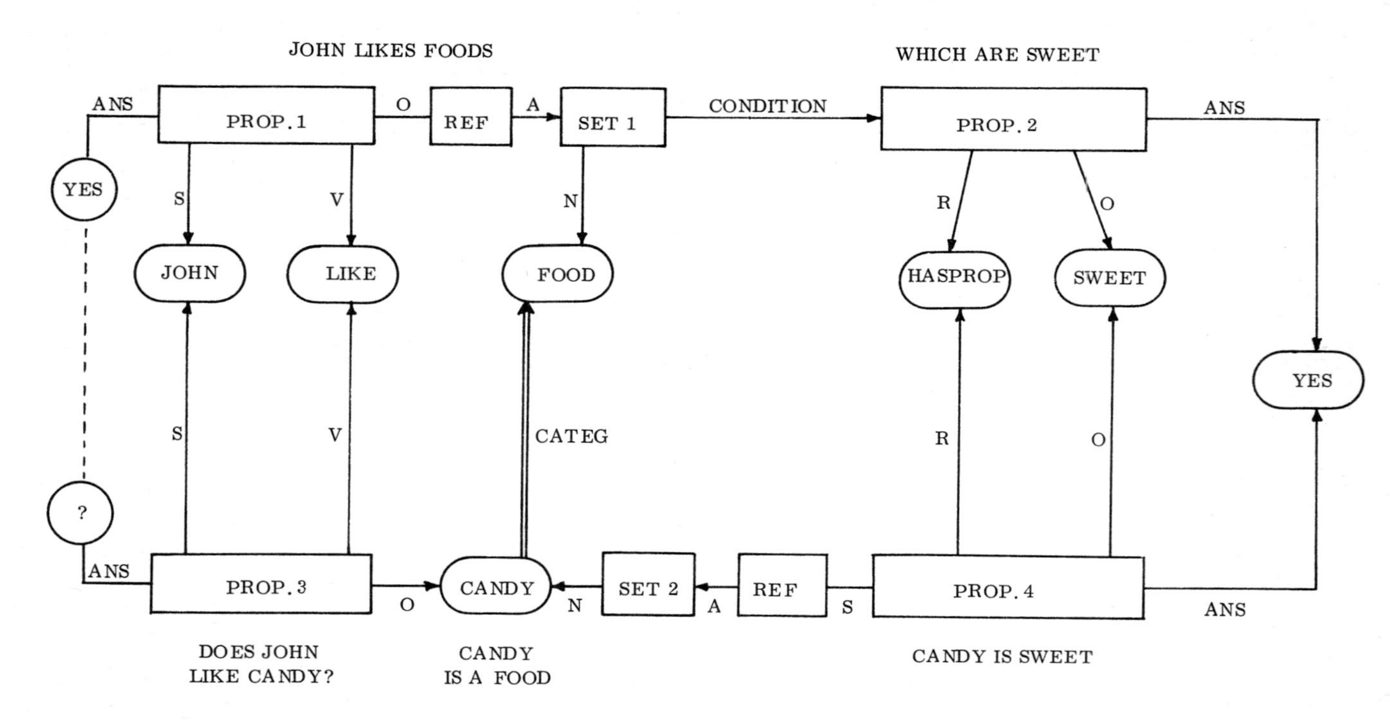

Figure 19 - Network storing "candy is a food"; "candy is sweet"; and "John likes foods which are sweet".

the elements which the system knows about (because more elements may be added later to the set and the system does not want to make hasty generalizations). This question is answered by finding out if a generic object which has only the properties of being sweet and being a food is liked by John. If so, John likes all sweet foods. This object, which Lemmon calls an "ideal", provides for stronger logical operations than might otherwise be possible.

The system is not limited to relationships on nouns. It can use a relationship between relationships. For example, in order to answer the question "does John hate candy?" given the information in the data base that "John likes candy", QUASI-NET requires a relationship between the relations "like" and "hate". In addition to these relationships which can be stored in network terms, QUASI-NET also allows a knowledgeable user of LISP to store in functions which may be invoked directly by the English language input. Thus QUASI-NET can answer questions such as "what is the product of 3 and 7?" without having to store individually all the relationships between numbers, but can compute the answer because it knows the meaning of the product.

The principal contribution of QUASI-NET is the specification of a widely applicable general data base. However its deductive procedures are limited, and assumptions such as unique parts of speech for words, and dependency on a peculiar parse structure make it unlikely to be immediately extendable.

7. A TEACHABLE LANGUAGE COMPREHENDER (TLC)

TLC (Quillian, 1966, 1969) is a program designed to be able to read encyclopedias or newspapers or any general text and to comprehend what is in that text. Comprehension, in this sense, means relating what is seen in the text to the things that the system already knows. The basic philosophy in the design of this system is that people do not understand very much that is new in any sentence; they only invoke previously known facts and tie them together in a slightly new way. In reading a new sentence, you draw analogies to what you already know without adding much that is new.

Fig. 20 shows how the TLC system operates. An important aspect of the design is the use of an on-line human monitor who watches the text comprehension process on a number of different levels, some output in English, and some in the

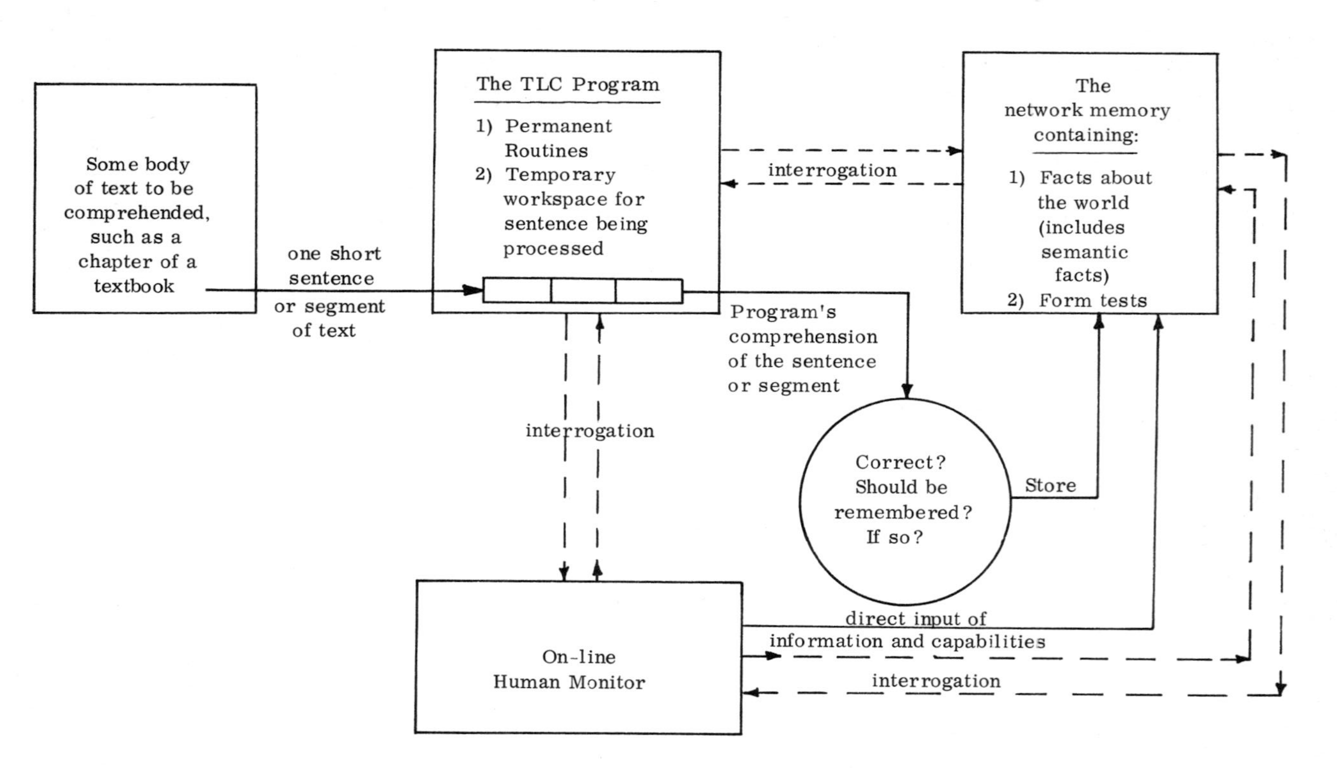

Figure 20 - Principal components and interactions of the TLC system.

data structure as it is being produced. When the monitor sees that the program is not understanding what he would like it to understand (i.e., it is not retrieving or understanding certain words or facts) he can help it. This is why the program is called Teachable. It may be that TLC does not know a fact that the monitor thought it knew; he can add the fact. This operation is similar to that of a teacher who asks a question of a student and notices that the student is not understanding the problem simply because he is missing a fact. It also may be that TLC does not know something about a word in a sentence, or it may not know something about the local syntactic structure of the sentence. Therefore, the monitor can add to the system the capabilities for parsing and for looking at new structures. Given that TLC understands a piece of text, the operator looks at it to see if it is correct and if TLC should store this information in its memory.

The memory consists of two types of items: facts about the world and facts about language called "form tests". Both of these need to be learned; TLC cannot learn them by itself. (Most of us are taught language; if we are left on our own we do not learn very much or very well.)

Basic to the TLC program is the memory structure for semantic data. This memory structure (a very highly interrelated network of information) is accessed through a dictionary that is outside the memory. The dictionary contains words, and associated with each word in the language are a number of definitions. Each definition is a structure in the memory and the dictionary contains pointers to the alternative definitions in the memory structure. The dictionary provides an index for going from language behaviour tokens to semantic tokens.

Fig. 21 illustrates a piece of the memory structure that talks about "client". The representation of noun is a unit which is shown in square brackets. The first thing in a unit is always a superset to which this noun belongs. A word or concept is always defined in terms of some superset with modifications which makes the definition specific to this word or concept. In the example shown in Fig. 21, the superset is "person" and the subclassification comes from the property "employs a professional". In the definition, the word "person" is not stored, but rather a pointer to an appropriate definition in the memory structure. In going into the concept of client, you have available all the things that make a person a person, all the relationships and facts which the program knows about persons.

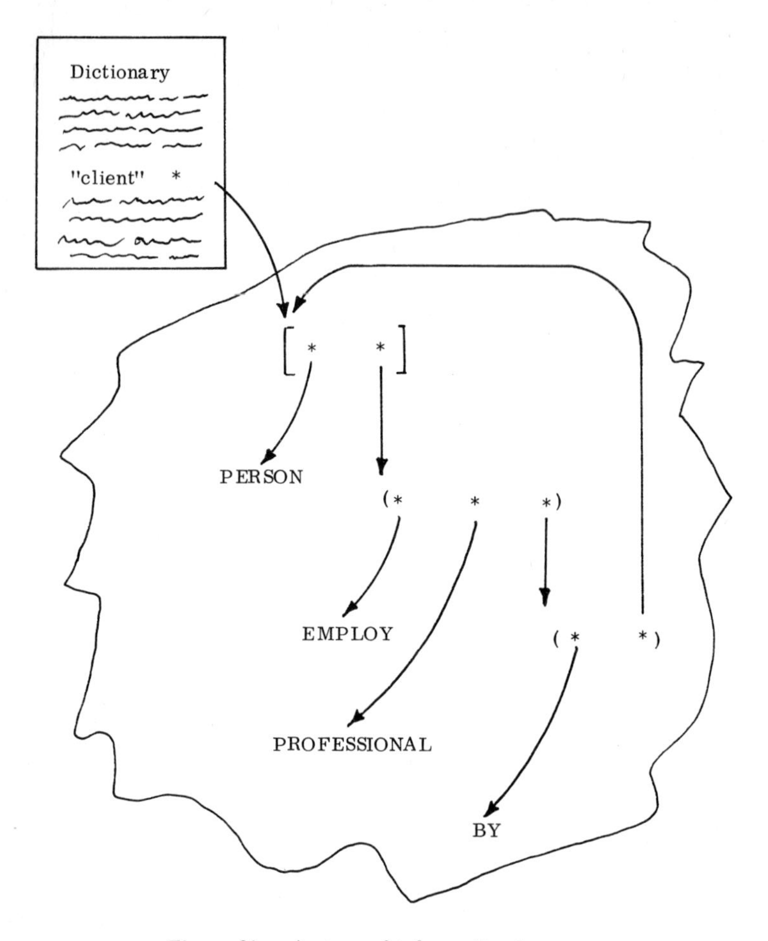

Figure 21 - A piece of information in memory.

A property always has two parts, an attribute and a value and defines a relationship between two units, the one it is attached to, and the value of the property. In this memory structure, the attribute-value pair has been generalized to include not only things like colour (red), but also to embody most of the relationships usually found in the English language; the range for the value of the attribute is defined in terms of the specific relationship - so that, for example, a verb as an attribute has as value its object; similarly the value of attribute preposition is its object. Each of those attributes fit in the format as a property that helps to refine the concept of a superset; in the example this is "person" as it is needed for this definition of clients.

There are also ways of refining the meaning of the attribute-value pair. For example, a client is a person who employs a professional; however, not only does he employ a professional, but also the employing of the professional is done by this client. Notice that in Fig. 21 there is a pointer back to this unit for "client". The TLC memory is a very tightly interrelated network of relationships.

Fig. 22 shows how TLC understands the phrase "client's lawyer". This phrase refers to a lawyer who is employed by a client; this lawyer represents or advises this client in a legal matter. In creating this new memory structure from the input, TLC has added information that was not explicitly in the phrase; that is, the phrase "client's lawyer" has evoked the information known about clients and about lawyers, and TLC has attached this information to its interpretation of the phrase.

Let us now consider the procedures underlying TLC's comprehension process. TLC starts with two words in a sentence and tags items in the "extended concepts" of these words. The concept of a word is not just what is in immediately the definition, but what can be found by going into the superset of the word, and looking at the properties that are on its definition, and by going into the superset of the superset of the word, etc. Following the extended concepts of two words one can eventually find a common part in the concepts of the two words. For example, if TLC finds that a "client is someone who employs a professional", "a lawyer is a professional who ... ". The paths to this intersection may indicate a relationship which holds between a client and lawyer, the intersection is on the word professional.

Having found a semantic connection between two words, TLC must then see if, in fact, this connection is accidental in this sentence, or whether the syntactic form of the sentence is such that TLC can use this semantic relationship which has been found. TLC uses a set of form tests to determine if the structure is appropriate; that is, having identified two words that are possibly related and knowing what their relationship is, TLC uses information associated with that relationship which tells different ways that that relationship can be represented in a sentence. So, for example, the test might be that the first word is separated from the second word by "'s". In this case, of course, the test would succeed and TLC would then say that the phrase means that the client is employing the lawyer. TLC then looks for some other properties that might also be applicable; for example, can it utilize the "by" property in the

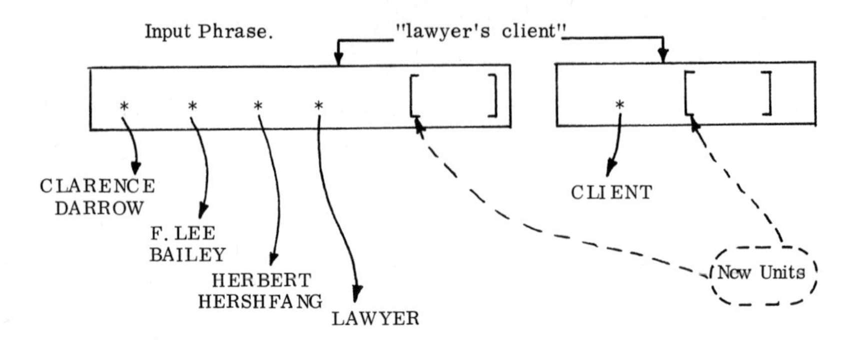

A. Initial steps.

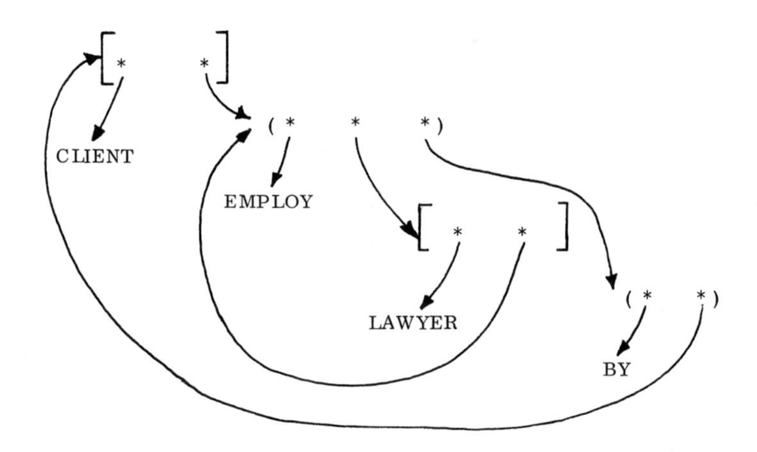

B. Output of comprehension.

Figure 22 - Stages in the comprehension of "... lawyer's client ...".

RT(E 3)

CLIENT'S LAWYER

```
CURRENT-TEXT:
   (LAWYER (EMPLOY (*THIS* . LAWYER)
          (BY (CLIENT)))
       ((AOR REPRESENT ADVISE)
          (*THIS* . CLIENT)
          (BY (*THIS* . LAWYER))
          (IN (MATTER (TYPE LEGAL)))))
```

(NOW WE ARE TALKING ABOUT A LAWYER WHO IS EMPLOYED BY A CLIENT ; HE IS A LAWYER WHO REPRESENTS OR ADVISES THIS CLIENT IN A LEGAL MATTER) data indicating the employing was done by the client? When another form test has been passed, this property can be used.

Note that the memory had "a client employs a professional". TLC modifies the property in its adaptation of the memory structure replacing "professional" by the particular professional being talked about, producing an analogy with the structure which was in the memory. In the memory, a specific new unit for client is created whose superset is the general node for client, so that TLC can use all the things about clients in general for this particular client who employs this particular lawyer.

In handling complex syntactic structures care must be taken to amalgamate inner portions of the sentence before relationships to distant words are tested. Fig. 23 shows an example where this is done. The sentence is "The client healed by the doctor employs the lawyer".

The program finds first that "client" and "doctor" are related; it knows that doctors heal (or cure) patients. TLC uses the property "curing patients" where for the object of "cure" it identifies "patient" with "client" since they are both persons, and from the definitions of "heal" and "cure" TLC decides that those two words can mean the same thing. In general TLC looks at two words and decides whether they are compatible to see if it can identify them. "Client" and "patient", for example, are compatible because there is no common property that has a different value for these items and they have the same superset "person"; therefore a "client" can be "patient". The figure shows that from "doctor" the system chooses form test T29. TLC amalgamates all of this information under a new structure, where this new structure can be treated as if it were a client; at the head of the new structure, when it is done, is the word "client". TLC eliminates the words that have been used between doctor and client. TLC then tests to see if it can believe that healing is done by the doctor. Another form test says that can be done. Then it considers "the client employs the lawyer", since it has amalgamated "client healed by the doctor" into "a client". This is understood and TLC prints its interpretation, in English, saying "we are discussing the employing of the lawyer by a client, who is a patient, and who is cured by a doctor". Inside, of course, the information is represented as one node, as illustrated.

The important points about the TLC system are, first, that it is driven by the semantics; that is, there is a technique for looking at a piece of text and deciding

```
RT(E36)

THE CLIENT HEAL ED BY THE DOCTOR EMPLOY S THE LAWYER


CLIENT: (NIL THE N AJ)
HEAL: (ED NIL V)
BY: (PREP)
DOCTOR: (NIL THE NAJ)
EMPLOY: (S NIL V)
LAWYER: (NIL THE N AJ)


USING:  CURE PATIENT. ATR*:  HEAL. ARG*:  CLIENT
FROM:  DOCTOR. PER:  ATRIB NESTED T29. HEAD:  CLIENT


USING:  BY DOCTOR. ARG*: DOCTOR
FROM:  DOCTOR. PER:  NESTED T21. HEAD: CLIENT


USING:  EMPLOY PROFESSIONAL. ATR*:  EMPLOY. ARG*:  LAWYER
FROM:  CLIENT. PER:  ATRIB CKBACK T17. HEAD:  EMPLOY


USING:  BY CLIENT. ARG*:  CLIENT
FROM:  CLIENT. PER:  NESTED CKBACK T18. HEAD: EMPLOY


CURRENT-TEXT:
   (TO ((*THIS* . EMPLOY)
        (LAWYER)
        (BY ((AND CLIENT PATIENT)
             ((CURE)
               (*THIS* . CLIENT)
               (BY (DOCTOR)))))))


WE ARE DISCUSSING THE EMPLOYING OF A LAWYER BY A CLIENT, WHO IS A
PATIENT, WHO IS CURED BY DOCTOR!
```

Figure 23 - Illustrating the amalgamation of inner portions of a sentence before relationships to distant words are tested.

it already knows some relationships about some of the objects in the text. In this sense, TLC does not do a syntactic analysis first, rather it tries first to find out what the text might have meant. This has some implications in cases where sentences are non-grammatical or where good guesses are sufficient. Secondly, TLC shows how syntax can be used only locally. Third, TLC has a generalization procedure. Once it learns that "client's lawyer" somehow can mean "the client employs the lawyer", TLC can then know that "an actor's agent" means "an actor can employ an agent". Conversely, if it learns that "the agent of the actor" means "the agent is employed by the actor", TLC can generalize backwards with the form test to know that "the lawyer of the client" is "the lawyer employed by the client". TLC can also generalize from properties on supersets, and it can make generalizations on these in terms of the classes of objects that it can handle. Finally, TLC is teachable. The user does not have to decide all things in advance; that is, the monitor can begin to type to the system and build up the knowledge in the system. When the system misbehaves, (since he is watching it behave), he can make appropriate changes and not before then.

8. CONCLUSION

In this paper, I have described a number of computer programs designed to interact with humans in a limited subset of English. In each case, limitations on the input and an assumed purpose and context have simplified the task of making these interactions natural. To extend the domain of such systems will imply a very large increase in the size of the data bases used, and expansions of all of the ideas and techniques illustrated.

9. BIBLIOGRAPHY

BOBROW, D.G. (1964), "A question-answering system for high school algebra word problems"; Proc. AFIPS Fall Joint Comput. Conf., Vol. 26, Pt. 1, Spartan Books, New York, pp. 591-614.

CHARNIAK, E. (1969), "Computer solution of calculus word problems"; Proc. Int. Jt. Art. Intel. Conf., Washington, D.C., pp. 303-316.

CHOMSKY, N. (1965), "Aspects of the theory of syntax"; M.I.T. Press, Cambridge, Mass.

KUNO, Susumu (1966), "The augmented predictive analyzer for context free languages"; Comm. ACM., 9 (Nov.).

LEMMON, Alan (1969), "QUASI-NET, a question-answering system with input of natural English text"; Ph.D. Thesis, Harvard University, Div. of Engineering and Applied Physics, Cambridge, Mass.

MINSKY, M. (1969), "Semantic Information Processing"; M.I.T. Press, Cambridge, Mass.

MOSES, Joel (1967), "Symbolic integration"; MAC TR-47. Project MAC, M.I.T., Cambridge, Mass., (Dec.).

QUILLIAN, M. Ross (1966), "Semantic memory"; Ph.D. Thesis, Carnegie-Mellon U., Pittsburgh, Pa., (Feb.).

QUILLIAN, M. Ross (1969), "The teachable language comprehender"; Comm. ACM., 12, 8 (Aug.), pp. 459-476.

SIMMONS, Robert F. (1965), "Answering English questions by computer: a survey"; Comm. ACM., 8, 1 (Jan.), pp. 53-70.

SIMMONS, Robert F. (1970), "Natural language question-answering systems: 1969"; Comm. ACM., 13, 1 (Jan. 1970).

WEIZENBAUM, J. (1966), "ELIZA - a computer program for the study of natural language communications between man and machine"; Comm. ACM., 9, 1 (Jan.), pp. 36-45.

WOODS, William A. (1966), "Semantic interpretation of English questions on a structured data base"; Mathematical linguistics and automatic translation, Rept. NSF-17, 1967, prepared by Computing Lab., Harvard U., Cambridge, Mass., (Aug.).

10. DISCUSSION

Clowes: You stated in your talk that there were several standard paradigms in the CARPS system: you have mentioned the distance paradigm - what are the others?

Bobrow: Distances, as such, is not necessarily defining one of the paradigms - where there is a three-dimensional situation as with a kite flying in the air, or a ladder slipping down the surface of a wall, these are situations involving distance but are very different from the boat problem. In the ladder problem, for example, you are interested in a fixed distance sliding between 2 perpendicular edges. There is the shadow paradigm, with a light in a certain position and a man walking away: what you are interested in is how fast his shadow is advancing. When you are flying a kite, you are interested in the rate of vertical ascent, for example, as the line is lengthening at a certain rate. There are several other paradigms, in fact a set of 14 different kinds of problems.

Barter: Could you say what the system uses to indicate what kind of a situation problem to select?

Bobrow: It is again this key word concept. This is very important in the preliminary extraction of information. It turns out that boats and trains and words like that, travelling and ship, are the key words for this particular problem. Given a set of these problems, it is easy to make up a minimum set of distinctions to be able to realize what is happening, and then you have a feedback loop - try a particular paradigm; if it does not work, see if there is another one there to try.

Barter: You can fool the system by having a boat leaning up against a wall?

Bobrow: That is correct! You can fool almost any of these systems - like people in fact. It turns out, by the way, that these systems do better than people in general.

Zatorski: In the Airline Guide Question-Answerer, what about a sentence such as "Does AA-57 fly from Boston to Chicago and Detroit?"

Bobrow: It would parse that by making it essentially, "Does AA-57 fly to Chicago?" and "Does AA-57 fly to Detroit?" It would now have that as two separate sentences, conjoined by the 'and'; there are two tests. Now the question is, "How does it make it into a question", and that comes into the problem of what is a quantifier, because quantifiers are the things which generate the questions. The question on Flight AA-57 is going to have a quantifier, something which looks at a flight which is equal to AA-57 - is it true that "connect", or is it true that "there is"? That is how this conjunction is done and then it is assumed that the deep structure is good enough to separate those things out into separate questions.

Zatorski: There are two connects now.

Bobrow: That is correct.

Zatorski: What of the language for the formalized system? Is it possible to ask, "Does Flight AA-57 stop at Detroit after landing at Chicago?"

Bobrow: Yes, it is possible to ask that kind of question; but if you want to ask that kind of question, you must formalize what "after" means, and you must formalize each of the concepts that is inside. So here is a case where you must know all about all the words - you have to, for deep structure parsing, but in addition, you have to specify in advance all the questions you are going to answer. For example, "Can you make a connection from Boston to Los Angeles through Chicago?"; you have to have this concept of making a connection which has to say that you had better take a plane that comes in 3 hours early to Chicago because it will probably be stacked. Well, depending on whether you want to put that in or not, whatever that program is you will have to put it in a predicate form.

Zatorski: In the TLC system, how do you prevent the system from learning that 'doctor's dilemma' is a dilemma employed by the doctor?

Bobrow: What you do is the following. If you believe that you see the system coming in and saying that it's a dilemma employed by the doctor, then you have to make the distinction, you want to be able to put into the form tests some semantic tests which distinguish between dilemmas and patients.

Zatorski: In other words, its another taxonomic problem?

Bobrow: That is correct, but what makes it much easier for the user, is that he does not have to decide these things in advance. A user of the system can start to talk to the system and build up the knowledge of the system: when the system misbehaves (you are watching it behave), and it says something very odd, then you can make the distinction, and not before. You do not have to right away do the taxonomy. The taxonomy does not have to be uniform throughout, because you can make the tests very specific for the words. We assume that we have a lot of information specific to words and that the generalizations are up through the superset, and that is where we get the taxonomy. We can get things which are specific to words which are not true of their superset - for example, we might want to know that 'birds fly' but an ostrich is a bird, and an ostrich does not fly; so you would have a basic semantic contradiction, and it can accommodate that.

Narasimhan: Can you say something on the dictionary in TLC?

Bobrow: The dictionary is essentially a word dictionary, but associated with each word in the language is a number of definitions; essentially the dictionary is a set of pointers into alternative definitions in the memory structure and is fairly simple in that sense. A lot of the information that you want would be stored in the data structure itself - the form tests are stored with the words, etc. The dictionary is essentially a look-up procedure which goes from language behaviour tokens to these semantic tokens.

Narasimhan: How much of the memory structure is put in - what is the take-off level?

Bobrow: The take-off level is the level of a string, delimited by spacers. We are just developing some techniques for multiword units, but at the moment let us consider the ones limited by spacers. Starting with a string, we get a pointer essentially into the alternative structures which that string can represent, and then through a semantic parsing process, we decide which ones of these are in fact reasonable and use those in building up a new structure.

Overheu: How would you handle a circular definition which can occur?

Bobrow: All of these word definitions are circular. You are talking about a client who is somebody who employs a professional, so there is a circularity. In fact the dictionary is very circular because lawyers are people who are employed by clients and that is in the definition of lawyer, and when I say 'lawyer', I really do not have just a word there, I have an entire structure which represents lawyer, the specific one that is relevant in this case. So the programs which handle these things know about when the circularity is being introduced and they follow this structure. That is why the program is able to put up 'this lawyer', because it knows that this was a pointer back up to where it had been before. What is unfortunately true is you can take practically any word in the dictionary - you sort of hang the dictionary off that word; you will probably find it goes into all the words and back into many, many loops. The real question is whether or not the loops themselves give you the information you need to understand what are particular properties you are interested in looking at in the input sets.

Smith: Do you envisage any procedure for selectively expunging past memory?

Bobrow: We have really not done very much about "What kind of things do you store and where?" We have been very non-selective, but are just starting to think about the large data base problem, so I cannot really say very much on that. Right now we are just storing on every word that is in the phrase. That obviously is wrong, but you have things like 'Battleships in World War I' - there were battleships in World War I that had 21 inch guns. Now, do you want to store that under 'guns' or under 'battleships', or under 'navies' or 'World War' or 'history' or what, or not at all?

That is a very hard problem - how one knows what to remember and what to forget.

Smith: There is also the problem that having got bad information in at some stage along with good information, how do you weed out the bad?

Bobrow: Well, part of that is done by the teacher, who can say, "Look that test is no good. Its not good enough. Let me give you a better test which encompasses that", or "That definition is not good enough; let me modify the definition". This is not a system which works in a vacuum; it is a system which works with a human teacher. There is a capability for expunging, and what should be expunged, the teacher will do.

I have talked about natural language interaction systems of great variety. The key thing to remember as I have said previously, is that one must look at the purpose for which we are trying to understand the natural language, what kind of processing you want to go through, and what kind of response is an adequate response to end with.

DESIGN PRINCIPLES FOR GRAPHICS SOFTWARE SYSTEMS *

EDWIN L. JACKS

Computer Technology Department
General Motors Corporation
Warren, Michigan

1. INTRODUCTION

This paper is aimed at defining and illustrating the environment of graphics systems used for industrial design, by referring to products off our own graphics system (without doing a survey of all the graphics systems in the United States of America). The environment is important because the system is a response mechanism; a system of people and machines.

Existing systems for producing graphic design date back to the ancient Egyptians. You may doubt that, but if you have seen pictures of the ancient Egyptians building the pyramids, you will have found triangles in those pictures, and you will have found pencils; the concepts of geometry to build the pyramids, and the current drafting systems, the current techniques for converting engineering ideas to mechanical hardware, date back so far, and are not that much different in the drafting rooms. On talking about the environment for graphic consoles and the environment for computers in industrial design, it must be realised that that environment has to be matched to the computer; the first part of this paper is addressing that problem, and illustrating what can be done. The second part can be summarised very quickly - it says that you _must_ understand list processing, you _must_ understand formal language structures for graphics, and you _must_ understand computer time sharing, to be in the graphics business for industrial design purposes.

2. THE ENVIRONMENT

I have a movie which shows the work we have been doing. This movie was made by three engineers in General Motors from our Chevrolet division. For the period of about two years they have been using our facilities, that is, a graphic console attached to a 7094 computer in a multi-programming mode - allowing one

* Edited transcript of lecture - Manuscript was not available at time of printing.

console plus background jobs to run. The system dates back to 1962; the console was installed in 1963 and much of the technology involved is really that of 1963-1966. I comment on this because although very little use of list processing is made, for instance, they are still getting a lot of work done. The approach of the movie is to explain to other engineers how they are using the console for one particular job, the design of windshield-wiper systems on automobiles - a job which normally takes about 15 days on a drafting board.

Their problem is to ensure that a certain minimum range of vision (subject to government regulations) is provided for the driver, and that when the wiper mechanism works, they have 100% of a certain part of the windshield wiped clean all the time, 95% of an additional part of the windshield wiped clean, and 80% of a further part, so their job is to do the drafting and the positioning of the mechanism, such that these conditions are satisfied. The film shows the steps and procedures involved using the graphics console.

I am going to outline very briefly the steps in our design cycle, from concept through to making the parts. This process varies considerably from industry to industry, as it is a function of production rates, part interchangeability, manufacturing methods, production volume, and so on. The cycle starts out with sketches and full scale renderings - these are called blackboard drawings because they are done on black sheets of paper that are 5' high and 20' long, like a blackboard. They give the Stylist the concept, so that he can see the side view on full scale. Clay models, then wood mock-ups, and best appearance drawings - these are drawings which say essentially, "this is what we want the manufacturing department to make."

We make a break after Styling - below that are the firms that have to do the job, like Fisher Body, they start out with essentially a preliminary part drawing - this is a drawing which is used to prove that the data they are getting is correct; it is a first guarantee that the part can be manufactured. If the part is a piece of sheet metal, they must make sure it can be stamped; a first step must be made at being sure they can assemble it together, weld it, rivet it or whatever else is necessary.

Next, they take all the individual part drawings and make an assembly drawing, from which they go to test models - these are wood, just to make sure that everything is right to date. This whole process requires co-ordinating 3000 draftsmen. Then they go to die models to start getting information ready for the manufacturing process in which there are dies, fixtures, and then you are in production. The time cycle on all of this is 3 years.

A feature of this apparently sequential string of processes is that there is in fact a very tight iteration between any two operations; the further apart these operations are on the list, the less chance there is for iteration. This means that the graphic system has to support continual change, and that is what the movie says, essentially. It must be possible to take a different approach at any time or the process is not as flexible as the manual technique on the board. The example I gave with the windshield was an iteration between the clay models and the release appearance drawing, where they had the clay model and then made up the drawings for release; the designer would then change his mind on the clay model. There must be that flexibility, or you just are not doing the job for design; this impinges on the whole concept of how to organize the graphics.

An important factor to be noted is an inherent addition of information by people. There are engineers and draftsmen inserting information into the system, and that is how the task proceeds to the end; this is exactly what graphics is - how you get information into the design from the man.

A further point is the time span over which the information is inserted - the real problem here is we have one man and one computer, and we must work them together. We are trying to link the two together - to the Electrical Engineers there is an impedance match problem - we have information coming into the man at some rate and with some characteristic, and we have to get a product out. The man is really a receiver - he has a processor and a memory, he has transmitters; the computer turns out to be the same thing: the graphic console can be used for most of the receiving and transmitting.

Let us take a very simple part; it is drawn 10 times full size and is, say, a moulding for a car. This is handled at the graphic console: there is a complete finished drawing out of the existing system to start with. What is needed is, say, not the finished drawing, but a 10 times size enlargement of the outline of the moulding, with certain dashed lines which are boundaries for inspection purposes. This drawing is used in a shadowgraph to show that the part has been properly made. The problem at the console is to get in the shape and construct the lines. What happens in fact at the console is that the centre line of the shape is constructed using the known dimensions; a series of circular arcs is constructed on the screen, the known radii are specified, and as soon as the centre line is positioned, a program is called which measures off from the centre line a constant distance and constructs all the rest of the lines. It is a very quick job once you have gone to that point. A drawing machine is used to produce the finished drawings.

After this application was worked out, the System Designers said, "Well why did we start with the finished drawing? Why can't we get this out of the computer too?" So they backed up and said, "let us try to make the finished drawing", and somebody said "well, we get the finished drawing from the group that is designing the parts to make this thing". Then they said, "what do you do when you design the parts?" It turns out that you start out with essentially a sketch of the moulding. This is what Styling says it should be; it is a freehand sketch made with french curves - they figure out how to manufacture it; this is done by designing rollers. You start with a flat sheet of metal, go through a series of rollers, and each roller gradually bends it around until it comes out the shape required. So what has happened is we no longer make this drawing as an object at the console; we send an engineer down with a die designer for these roll dies; he starts with a sketch of the shape that is desired, and about 4 hours later he has 10 rollers designed; he has the inspection drawings completed and he has the finished drawing for the whole operation. This is again the problem - if you are in graphics, you have to match with the existing system to get started; once you get started, the technique will start spreading.

After the assembly drawing is made, that is the drawing which shows all the parts which fit together, each part is taken off the drawing and a separate drawing is made for it: it is only when this has been done that it is dimensioned. It is only when this is done that it can actually be labelled so that it can be passed on to other people to be used in the process. Apart from that, the purpose of the assembly drawing is to make sure everything fits together. At the console, what happens is they have the lines that represent the drawing part, they pick them all up off the drawings into the computer, then they label and put arrowheads on little lines. There may be a line, a label, two points and a distance between it, dimensions, and so on. The designers sit at the console and specify this to occur. The reason this can be done at the console cheaply, is that the dimension between one point and another is required, the data has already been digitized - the designer merely specifies essentially whether he wants the true distance, the horizontal or the vertical distance, and indicates the one point and the other point, and the computer puts in all the other marks. There is a mark ratio of about 5:1, that is, for all the marks that end up on the drawing, you use about one mark for every 5 required in the output; this is a gain in efficiency essentially. This process goes on and the final drawings are obtained through the drawing equipment. We are doing practically none of this at present, however, for one reason - the payoff is lower than the payoff from other operations, such as the surfacing, so this job is never run on the console now, even though the software is there.

To return to the windshield design task, a grid of lines spaced five inches apart is established on a design 40" x 60". The process starts with three lines going into the computer. These three lines are from front view to across the windshield, down the side, across the front; all assumed symmetrical so there are really three lines to define the piece of glass to start with - that is what is loaded into the computer. The next thing that happens at the console, is to produce the one surface, and about twenty lines. That is, the man performs his construction operations on the data - this results in a surface being defined which is a developable surface; we have a technique for evaluating a function in any place on the surface, on the glass, and we can use it for such things as numerical control work.

The above operation gives the glass. Given the glass, it then must be fitted into the sheet metal that goes around the glass, and that results in about eight surfaces and a total of 60 lines. The man who usually does this work at the console is an instructor in drafting techniques, a skilled draftsman, so he knows a lot about what he is trying to do. That gives essentially enough information to say, "I know what the sheet metal looks like around the glass". The glass has thickness which has to be considered, and the moulding that goes around it has to be considered - that is carried out at the console; the definition of all these details is effected at the console. To finish the job, now that the sheet metal has been developed, you end up with about a hundred lines plus the surfaces, and one wood model - that is what I refer to as 'released appearance drawing to Fisher Body'. Fisher Body then have to concern themselves with all the sheet metal that surrounds this glass; they start from a master layout for one windshield. Based on <u>one</u> design coming out of Styling, they end up with 3800 lines on nine master layouts. So there is a tremendous explosion as the process proceeds through the computer; at all stages information is being added in, but certainly the basic drawing is not being put in, there is no drawing board operation to get to this point.

3. MAN-MACHINE RELATIONSHIPS

I would like to consider some very simple arguments regarding the relation between man and machine, applying in graphics. Some of the arguments are for those who wonder why time sharing, for instance, is used; some are for those who say "why don't you use teletypes and be done with it?" - there is a series of such arguments. Although such arguments may not be needed here, I hope that the following will at least give an understanding of why various things are being done and why resistance has been met all of the time - at least in the United States where, from the period 1964 onwards, there has been tremendous resistance to work in graphics. While you find it being pursued, it has been carried on in very few places, and essentially it is University work. It has been a hard struggle, and these arguments are aimed at producing the realisation that useful work comes out of this type of approach: not only is the work useful, but computer graphics is really a good match to man.

We have come to a very simple viewpoint on what we are trying to do with graphics, and how problems should be organised for it - how files and major records, should be organized.

Historically, when people first started using graphic consoles they said, "well, a console is nothing but an I/O device", and the first thing they did was write a program which would go directly to the display, and they claimed they were using graphics - you will find people still doing that. Then they woke up to the fact that they had a keyboard out there, and they brought that back, and then they realised they had a light pen, so they connected that in, and they then found they really had data but it was part of the program. Practically all of the early applications of graphics (and many are still carried out in this way) view the problem in this manner - there is a program with data buried in it, and the display is driven directly.

In the next development, somebody said you have some sort of intermediate display, and that intermediate display was a buffer for the things that you want on the screen, but the scale was wrong. As soon as the intermediate display was put in, however, it was found that they had lost control of the display; they also found that they wanted to add another set of programs; so they put in a set of programs which controlled the transformation from the intermediate display to the final display - the object was to do rotation, translation, scaling, and so on, independently of the main problem. When they did that, they really found they were in trouble because their structure in going from program to display was not too good to start with, and they had no back links to work with. To solve that problem, list processing really took control and thereby there were provided back links to the data.

From this point, development proceeded through other phases; just what they were depends on whose work is being studied. In all of our work we have dropped to an approach which says that in general, there is no main program, in general there is a large number of programs which are required to operate on data, and so we now talk about a model record, and we talk about essentially a display record, essentially a library of operations, and a library in which these operations can be written in say, PL/1 : they are the primitive operations of the system and have the property that if used by, say, the design people, and applied to the model, then the results are consistent with the model. The library is made up of these operations.

This makes a big step forward. Most graphic operations are very highly specialized and are not organized for flexibility, but by making the program set independent of the display operation, in a sense making it so that any one of these programs can be applied at any time, the main control code does not appear in the operations set, you then can work back and forth between the programs and there is great flexibility.

When this is done, the model record must be organized in a manner which is independent of the sequence of the operations, and the information can be analyzed. When we speak of a model, we consider normally a list processing structure: one of the key pieces of information in the model is the topological data, and for a design, for any piece of sheet metal, we have to keep track of all the different surfaces, all the boundaries on the surfaces, all the holes in the sheet metal, in that topological model, in order that constructions can be carried out on that model - for instance "construct a line normal to the surface". The surface may contain many other surfaces and you have to find which one of the sub-surfaces is being talked about. Normally the task includes problems such as, 'given a point in space, find the nearest surface of a group of surfaces, and then construct the normal to the nearest'; the data structure is normally organized for this type of operation.

On the question of what a model is, I am not going to try to define it further, but it involves inherently enough information so that the independent operations can come along and modify it: apply operations to surfaces and you still have surfaces; apply operations to lines, and you still have lines; apply operations to displays and you still have displays, so that the variables at that level are consistent.

The role of the display record is essentially the role of selection of variables: quite often the light pen is on the screen; the man is selecting a line, selecting a point; that is really the purpose of the display record, it is to bring the variables out to the man without the man having to know a name; all he has to do is to have a handle which is the image of the line, so that is the function we serve here. It is an analogue representation of what has been done internally, and you can grab hold of it.

The function of the library is the same as that of any library: you can write out a program and you can use it later. Quite often at the console we find the designer wants to study an approach to a problem, he does not know how to do the problem; he does not know what is the right geometric construction to give him a satisfactory design. So he sets up a temporary file; this is a temporary record in which operations from the library are applied to the model with the results always going in a temporary record. Linkage is being maintained between the two of them, but the linkages are always a reference so that the model is not destroyed, it is kept for historical purposes essentially, in this technique.

After the designer applies a sequence of operations, we set up another record - a sequence record. This record is a sequence of temporary operations, which gives the construction required and this is stored as a separate sequence. Another thing which can happen is to take that sequence of operations and connect it to the model record and make it one of those. The designer can do the same with the temporary sequences, because it turns out that it is the last version of these that he wants.

We have performed some correlation on how much it helps to be able to save a sequence of operations at the console. Is it worth anything, and how much do the list processing techniques of these models help the overall process? The reason we have been able to do the correlation is that we started out with very early technology with no list processing, and then we gradually added it in. Another way to put this is to say that the system did not work at the start, so we had to add list processing to make it work. Doing it this way, we were able to keep track roughly of different levels of performance of the system, and we correlated with two things. Let us call one quantity α_D, for which we do not have a numeric value, but if it was 100% , we would have as much structure as we can possibly file and we would have a complete definition of everything that had been contained therein. All we can say is, as we increase the amount of list processing, we can show what has happened. The other quantity is α_T which can tell us what happens as we increase the amount of sequencing, keeping track of sequences. What this amounts to is, as we have indicated changes by keeping the sequences, we can change the data and re-do the work without going through all the steps at the console. So you tend to make this system become automatic by making it less automatic by going to graphic consoles. I cover more of this in my further paper, "Designer's Choice at the Console", and discuss more of the organization we have achieved to bring in the whole concept of these records, these operation sequences and library operations and displays, as an integral part of the system. Consequently I am not going to pursue it further here, but it is important to realize that these factors must be separated. You can still see an awful lot of work in the United States where the system is one program, with one display, with no ability to carry information over a period of time, no structure over a period of time, and you just change the whole name of the game by effecting separation.

I hope now to address more about the man and the rate of transmission back to the man from the computer. There is an initial burst of data - the objectives data, the initial conditions for the design problem. At some time later, the designer is told that there is a change coming, and later again he is told, "here is the change". This sequence can run on quite a while. Relative to the man, one key thing we have found is that normally a good designer hears a statement that there is a change coming, and he will watch the rate at which the changes arrive. He will stop working at the console if the rate gets too high, and says "why should I bother fooling around trying to keep up with it?". Consequently he does a nice job of damping out a situation which could be essentially unstable. A situation where they did 9 designs on a windscreen was essentially unstable also; management in fact stopped it. *

A good designer recognises the need for damping and will stop doing the work

* See discussion

if the changes are coming too fast. On the other hand, he needs the ability to selectively modify, and that is one of the reasons why you have to separate the processes so that you can apply the operations independently of the sequence in which the work is coming through; that is, be able to start at any point on the design job.

The α_D and the α_T can be interpreted in a different way: the associative data is essentially a long term memory in the system of how things are structured, and it is this because it is there you have the ability to make changes later: α_T says if the same change comes through twice, you have the ability to handle that quickly, so these become parameters. If you want to give the computer a little humanoid characteristics you can say it is 'how well the computer remembers the subject'.

I remind you that the bit rate of a human being is around 10^7 bits/sec., that is why you use graphics.

One other thing which we do in the cycle, I can illustrate by referring to a simple operation on a circle tangent to two circles. The operation is specified first by the man: when he hits the screen, he has essentially said, "I want a variable of the type 'line', (it does not have to be a circle), I want a variable of the type 'line' to be what goes in the program", so at the same time as we intensify this line, a check is made to see if it is a variable that represents a line, so that he cannot give the program the wrong class of data in the operations - there is a syntax check being made on the operations. One of the big headaches arising in a written descriptive geometry language occurs where the type of variable going into a program is wrong, or in calling sequences in normal programming languages where there is no check on whether the variables are really right until execution time. This is checked as soon as the man selects a variable on the screen, in this approach.

One of the things we have been trying to discover is 'what is the difference in performance between a batch monitor operation for a job, and a time sharing system on the same job?' We have this information experimentally, of course. We have also tried to develop an argument for what happens in batch processing. We have said, "let's compare monitor operation with time sharing operations", and there are some very stringent assumptions. First we assume that we have a job that takes exactly so much computing time, (that is, if you do everything right, make no errors, submit it to the computer, it requires a certain amount of time). Now let us assume that we can break that job up into 'n' little hunks, and feed data into each of the pieces - how do you compare under those conditions the monitor system to the time sharing system? It is obvious, under those conditions, that you probably do not want to use a time sharing system, as you never make any mistakes - you always put the right information in, and you get the answer back. Then we looked at what happens when you have some errors; the assumption was that in a monitor system, you will get the deck back, you will correct one error and resubmit it. It is a very stringent assumption, because normally people correct two or three errors and make one or two errors, but we did not want to take that into account, so we stipulated that each time you can correct only one error. The performance factor we defined as the elapsed time to do it with a monitor, to the elapsed time to do it with the time sharing system. Then we calculated, given a probability rate (perhaps 1 out of a 100 times, or 1 out of 5 times, or 1 out of 20 times) of making an error in a piece of data (that is, a character or a numeric value or an instruction - as long as it is uniform it doesn't matter), and said, "How many times do you have to re-run the deck under those conditions? How many times to resubmit the job?" Our results show if you have 100 pieces of data and you make 1 error in 100, on the average you will resubmit 1.7 times. If you correct only one error at a time, according to our correlations, in the worst case you will submit it 37 times. If you have ever had a learner who was wholly incapable of learning to program, that is the 37 times case, I am sure. But the thing that is surprising, is how many times you have to resubmit the deck if you start making errors, under these very stringent conditions.

Now in time sharing, because you are breaking the program up into little slots, what happens is that the number of data items you are submitting at any time is small. You expect a recurrent factor in time sharing just because of a few pieces of data, rather than what happens in batch processing when you have a large number of pieces of data; we have correlated this somewhat with the work that has been done in an assembly language-programming, when we talk about statements as being pieces of data, and get reasonable correlation, even though it is an over-simplified view. What it amounts to is that because the job is broken up for time sharing, you are re-running only a fraction of the program at a time.

Let us take some experimental data for this whole study. A, B, C, and D are four different modes of using the console, which is otherwise titled "From Failure to Success". 'A' is using a written three dimensional descriptive geometry language with the ability to do full debugging at the console of the alphanumeric characters, the character statements. You can have displays and drawings generated to look at, you can see output, but you really cannot do anything with it, and every design task is essentially specialised for that job when at the console. The set of operations is not usable from the console alone. You have no real graphic control over the screen and performance is not significantly (economically) better than using a monitor system. In the second mode 'B', all the operations that are in the descriptive geometry language are available at the screen, but you cannot build up any programs from them, all you can do is apply the operations sequentially. You get something like 2 to 5 times the performance obtainable with monitor systems, and this is starting to get in the payoff region for the price of the graphics systems. In the third mode 'C', you can start using both the written geometry and the display mode on the screen and they will interface. You still cannot remember sequences that you do at the screen, however, but you can separately write descriptive geometry language statements and have them interface at the screen and switch to mode of applying operations. Here you get a 4 to 6 factor in performance range. List processing techniques came in, and we started using the screen with no program specialization. In the final mode, 'D', this shows 8 to 10 times the performance of monitor systems - this is where you can apply operation sequences at the screen and recover, essentially remember them, so that you can come back in the future and apply that same sequence to a new set of data. So now you really have all the technologies working for you, and then you get performance in the 8 to 10 times the performance of doing the job with batch processing. I have compared this to batch processing because at the time I did it I could not compare it to manual techniques. It turns out that for the type of work we are discussing here, there is about equivalence between the time it takes to do it for batch processing, and the time it takes to do it for monitoring techniques using a computer - that is, you just cannot run these jobs economically on a computer without graphics. So you get about a 10 to 1 gain over manual techniques, including all the overhead you have working with this system. I think that summarizes the situation.

4. DISCUSSION

<u>Bobrow</u>: In your windshield design graphics system is there any constraint satisfaction built in so that if you make a change, if you move things around, it would be satisfied?

Jacks: No, this is all direct construction techniques. We check to see that requirements are met, but there are no constraint satisfactions. In all our work, the construction approach is used rather than the constraint approach on the design.

Overheu: Could you build your own software?

Jacks: Yes, these designs are working within a descriptive geometry language in which points, lines, circles, surfaces, vectors, vector functions, scalar functions, display statements, exist, and all operations are carried out in three dimensions. When I say a line, that is any curved line in space, essentially. The language has also been used for a lot of the construction work that was done by the designers: they can go back and forth from the descriptive geometry language in its written form, to a mode of operation in which the operations are available at the screen with no written form seen by it. Essentially they pick a line off the screen, pick another line off the screen, and call the intersection operation to be invoked until the intersection of the two lines can be found. Now the display appears to be two dimensional, but the data that is represented is treated as, and is basically the three dimensional information projected to it. So what you see on the screen are the variables essentially, in an analogue fashion, being displayed so that the designers can get their hands on them. There is a 'choice' package in the descriptive geometry language by which you can have a branch in the language, which has a different construction from what you find in normal programming languages: it says, 'display inbranch on a selection', and the display comes up with six or eight options, and then based on the man's selection, that branch is carried. It is a fairly simple way of doing it.

Overheu: How many man years did it take to develop the system?

Jacks: Behind the system there is something like 120 man years of effort, but that number contains a lot of experimental work - it contains the effort for development of a multi-programming operation for the 7094 which we had running in early 1963. It contains the effort to do numerical control machining of three dimensional surfaces, and its output from this operation. Prior to Chevrolet's looking at that data, Styling had worked on the data, generated the surfaces, Fisher Body had picked up the data and machined wood models from it. The effort included work in pattern recognition and character recognition, and the development of all the descriptive geometry we had to have behind it and the techniques for the iterations - all in three dimensions. The iterations are carried out to three significant decimal digits (for most of our information, including our final answers for x y z positions, lines moving, and so on). There is a lot of work to get the accuracy required. In that man-effort we also had another system which was strictly experimental and which used list processing techniques, and the experimental paging systems on the 7094. There was a compiler modification also. The program to do the windshields, itself took six man months to develop.

Clowes: Do we take it that you now have abandoned the normal projection methods?

Jacks: To give you information bearing on that, General Motors (GM) has on one floor in one building, a thousand people doing drafting, and on the floor above that, they have a thousand people doing drafting, and a floor above that they have another thousand - that is in Fisher Body alone, which is responsible essentially only for the firewall back and the sheet metal, not the chassis, not the wheels or anything else. So with one console, we have hardly made a dent in that problem. To give you a feeling for it, the console is running from 7 in the morning until about 3 at night.

Bobrow: Do you do anything more than windshield design?

Jacks: All the exterior surfaces on the XP 887 (which is the next American small car from GM, scheduled out in 1970) went through the system DAC, picking up the raw data from Styling, smoothing it out, making it consistent so that in spite of the manufacturing operations, it is theoretically smooth, and it fits together properly; and then doing model machining from that - all of that went through this equipment: all of the machining and the model smoothing for the small car.

Bobrow: Does that mean you have completely circumvented the drafting phase?

Jacks: No. This is exactly the point that you have to watch out for. Someday you are going to, but the tail end of the movie shows them making drawings and the reasons those are being made, are so you can plunge it back into the existing system. So that information could be taken out of the computer and put back to the hordes of draftsmen and eventually put into the shop in one way or another for different purposes. In the windshield case, the information is also used as evidence for the government that the windshields have been checked for visability, which is a legal requirement now - all the 1969 GM cars have had their windshields checked in this way, and are now all designed in this way as well.

Kovarik: It seems from the film that it must be very much easier to create a small change in the existing design or a small variation in the current concept which you provide: this may discourage any radical change in the overall effect. It may very well be that technology will be frozen.

Jacks: You have just missed the key nature of human beings, who will change everything if they get a chance; we just had one case where we really lost control. On the XP 887 windshield, this component is the governing shape on the front end of the car in terms of slowing down development because of manufacturing factors. That piece of glass is extremely expensive. So the process is to find a design that is satisfactory for the windshield, one that can be manufactured, and meets a certain cost: then they proceed with the rest of the process. On this small car, Styling played around with their renderings and their clay models and came up with a design for the windshield which they thought matched the rest of the car very well. They released it, and the people who checked the windshield to make sure they can manufacture it went to the DAC equipment, checked it, and did the surfacing of it. (Surface on a windshield is a developable surface by definition, so you can roll it flat without local distortion of the glass - it may not appear this way when you see it, but it is designed in this way; the manufacturing process puts some additional curvature in, but basically it must be a developable surface). So the people who use the console to develop the surface and in fact make a wood model after Styling, proceeded to do this, then came back to the Stylist and said "well, we've got that done, what next?" Nine complete windshields later, they found out that they will never go back to the Stylist and tell him how quickly they can do the job, because in the past, that job took three weeks on the drafting board and it was too late for the Stylers to go back and make changes. In this case the people were turning out in about an hour and a half to two hours, all the development work, including the drawings; so they turned them out to let the Stylers know; and the Stylers went nuts - "Change it!" So it has just the opposite effect. In fact one of the things we have to learn is how to control the process. The designer is now going to have a lot more power than he had before - tremendous addition to his ability to get from the design stage through the effort of the mechanical techniques that were there before.

Kaneff: Perhaps the point that is being made is that you have certain inbuilt concepts, such as windscreen wipers of a certain type that move over a certain area in a particular way, and so on; certain boundary conditions have to be satisfied.

Although you can alter the shape and dimensions in some way, it does restrict you to certain concepts, whereas if you did not have this built into the program, is not the implication that you would be much less restricted?

Jacks: You are asking, "what is the flexibility given these approaches?" The graphics system is no good unless the flexibility is greater than you have on the drafting board today. One of the things you will find, is that there are a lot of small operations, and you can start at any place in the process: this essentially gives more flexibility, not less, and it is a necessary requirement for graphics to have that flexibility. For instance, the system is such that, as I mentioned, all the 1969 surfaces were done. At the console they can do every surface that is on a GM car now, as far as sheet metal surfaces are concerned. They have carried out a definition of them and proceeded to do the machining of them. They have done just about every conceivable body panel you can find on an automobile - so we do not restrict them. This means that the geometry techniques and the surfacing techniques have to be very flexible. There is a whole new type of functional approximation that has come out of graphics, this represents essentially a new development in the question of surface interpolation. There is a paper to be published in the American Mathematical Journal by Bill Gordon from our organization, which explores the meaning of the interpolation techniques. This has changed the technology for interpolation, and the reason is fairly simple - normal three dimensional interpolation has been built around the concept of a mesh of points in space; this is the normal technology, and somewhere or other you interpolate a surface between the points. It has been pointed out that this is not what you are doing; in fact what you are doing is a functional - you have a grid of surfaces where you are interpolating between the functions and the only requirement is that a set of functions in a certain direction intercepts the set of functions in another direction at these points; in fact the whole interpolation formula can meet completely arbitrary boundary conditions, and it has been extended to the point where any order partial derivative across those boundaries can be satisfied. So from graphics has come some interesting new mathematics. If you are familiar with spline interpolation in two dimensions, which essentially gives you a continuous first derivative and continuous curvature, and gives an optimum in a sense over a series of data points, the paper to be published extends that to the three dimensional situation, and does the interpolation in space between these arbitrary functions. So you don't see the restriction if you are careful, if you work at it; you find just the opposite is going to happen. It is true that through programming techniques you can end up with restrictions, but the real challenge is not to.

Clowes: How much understanding of format does the program for producing final drawings through the drawing equipment, have? If you start placing symbols and lines all over the place, does it not get rather messy at times?

Jacks: That program has a fair amount of understanding of format. For instance, if you have a set of lines (and this is fairly common), and you want to label each line, the program automatically will put in the next label. If you have a set of lines and you want to get the labels on, it stacks up the dimensions so that they are clear.

Clowes: Does that imply that the software that does this stacking function of the pictorial organization, can analyze any graphic or is it really tied to the knowledge of what that graphic is?

Jacks: It is tied to the knowledge of dimensioning groups, not the image. There are rules for how you make dimensions, but it doesn't concern itself with what the objects are.

Bennett: Would you have to do these in a certain order to get the best effect?

Jacks: Yes, in this particular phase of the program. We had an experimental program that did shuffle them around, but we have not used it.

Clowes: Does not this ability imply that the program must be able to recover the fact that lines are parallel, for example?

Jacks: No; it does recover information about data, but usually it is just dimensional data. It does at times have to dimension 'parallel' to the given lines, for instance - it has to know that. The program, because of some of the conventions, quite often has to be aware of certain horizontal and vertical lines - it has to dimension to the nearest five inch grid line, rather than to anything else. This is because the drawings are usually 50 inches by 50 inches, and the easiest thing to do is to dimension to the grid, - so it is picking up that sort of information. This program has probably more list processing in it than most of the other programs.

Bobrow: You mention the bit rate of a human being is around 10^7 bits/second. Where do you get that number?

Jacks: It came out of psychological experiments. It's an eyeball rate and it's a burst rate. What I'm saying is that given the situation of a million white balls and one black one, you can find the black one in a second. Or given the situation of the Southern Cross, how long does it take you to find that in the sky? Again you find that very rapidly, or given the classical pictures of President Kennedy a few years ago in a crowd, and it contains 50 to 75 people, and they show the picture to people, and immediately Kennedy pops out; it is this capacity of human beings that is why you use graphics. The same thing happens on displays. All those lines are very complex and the drawings are very dense on the display, but if there is one slight error, the designer picks it out rapidly on the display.

Anonymous: It is not very useful though. After all he has got to interact with the machine; the operator has to interact with the light pen for example, and that is not at 10^7 bits/second.

Jacks: No, his response rate is somewhat slower. The data off the DAC CONSOLE (which was shown in the movie), the normal rate there is 10 times a minute - the operators will make operations on the graphic console screen at this rate; where an operation is one such as 'select a line', at that rate they are doing ten a minute, or 600 an hour. I would rather use 600 an hour because there is a rather skewed distribution on it, and at times they do a lot of very quick pokes and then there are long delays. There is no indication that this rate will go much over a thousand in the environment we are in, because the time sharing system takes some time in coming back to the individual.

One other comment that is relative to the individual here is that we have some programs which are very erratic in their response. Sometimes they take one second, and sometimes fifteen seconds. In those cases we find the individuals are pretty unhappy. They are much happier when the system is consistent in its response, and when we are designing a time sharing system we feel this is important. They would rather know that it is going to be a two second delay and have consistency, than being able, say 80% of the time, to have one second, and every once in a while to get fifteen seconds delay.

Stanton: Do I take your remarks to mean it might be worth putting in forced delays?

Jacks: We haven't seen if we can.

Bobrow: This is done currently in a number of commercial time sharing systems in the U. S. A. They put in forced delays, just so that the difference that the people see between their response time with nobody else on the system and with everybody else on the system, is no more in fact than 2 to 1; they find people are much happier about that.

Lance: Surely this is really rather psychological. If a person knows there are a lot of calculations going on, he doesn't mind waiting say fifteen seconds.

Jacks: No, here we are talking about one type of response. For a given response, 'is it consistent?', not across all different responses. If the job is big you won't argue with fifteen seconds. The mode that most of our designers work is, 'scratch their heads for a period of time', and then they sit down at the console and work very rapidly; so there appears to be a burst mode operation, not a continuous steady pace; one in which they analyze, figure out what direction to go on, and try that direction, which may result in five to ten minutes of operation.

Bobrow: Do you think operators can get up to 60 interactions a minute in burst mode?

Jacks: I would not expect that for any substantial period of time.

Speight: This burst mode seems to be very important. If you have something automatic like typing, the more rhythmic it is the better, but if you are thinking of the machine that is doing the work, then there is virtually nothing rhythmic.

Jacks: That is true. It does not operate at a continuous rate, and the time sharing system we are working on now, gives first priority to a response from the man when he hits the screen. We actually put that job at the head of the time slice queue, so it will be the next one that is operated on, so we can try to get him quicker response. Now if he needs special attention, of if he immediately generates a request for data from outside of the main memory or program, then he gets in a normal queuing operation, but in most of the operations when he is doing the fast burst rate of communication, it is on a series of decisions which you can plan ahead of time. You construct your header time; it is a particular mode, I don't know if I want to call it a sentence mode, but there is a collection of information he wants to give the machine, and there is a high probability that the control codes would be in the machine for that, so we give first priority just to give him rapid response when he is in that mode.

Bobrow: Have you tried alternative input devices such as using the teletype at the point where he can go into these high burst data rates:

Jacks: This high burst mode when he is selecting variables off the screen - such as points and lines, in which he does not know the name of the variable - if you put them on a typewriter, you slow him down. I agree that if you have sequential information to enter in, then you want to do typing, but in the burst mode he wants, for example, to give a program a series of points for curve fitting, then he will just pick the points sequentially off the screen as fast as we can recognize which ones he is picking.

One additional problem we have with these rates, the man does sometimes make errors, and so, on the light pen, for example, the normal mode of operation is when his light pen hits the screen, the object hit is brightened. This is done with either intensification control or hardware, or by just duplicating the display - so he gets immediate feedback. On the 7094 display, the device we have is a voltage pencil and when it comes off the screen we say that is what he wanted. If he is taking an option which he has thought too quickly, and he realizes that he has already done something he didn't want to do, he can just slide the pen off and down to a 'no-op', or he can slide the pen off and on to any other object, to select what he wants.

Bobrow: You do a lot of sequences to be recorded. Have you developed a programming language in which there are variables that you specify essentially? Having got a sequence, can you parametrize the sequence and make a program out of it?

Jacks: I discuss what we are trying to do there in my second paper. "Designer's Choice at the Console".

Bobrow: When do you predict that most of the 3000 draftsmen will be starting to be put out of work?

Jacks: I don't know. Investment is needed for large systems, the evolution has to take place - it is just so large that it is very hard to predict. The answer usually to that question is that you are always pessimistic - you are always wrong when you try to predict. We are running 6 graphic consoles now; on the /360, we will soon have more - we are developing a time sharing system which will handle between 12 and 16 terminals.

Bobrow: That replaces 120 draftsmen, is that right?

Jacks: I don't know what the number is. I don't think it replaces any, because the whole technology is going to change. GM has about doubled their staff of draftsmen in the last 5 years.

Bobrow: Because they have doubled the number of products?

Jacks: Yes, the product line is about double now. There is technology coming along to make changes, there is the increased volume of design coming along also, so the individual doesn't seem to be in much trouble. The approach that we are using in GM to keep away from that problem is by education. All the draftsmen in the corporation should know what is being done, and have been given liberal opportunities. It is an interesting problem the way it has been approached in Fisher Body, which has the very largest drafting rooms. The numerical control drafting table was put in the middle of the room, so there are a thousand people who understand, "O boy, automatic drafting is coming, and I'm in trouble" and right in the middle of that, you put this beast that is called an automatic drafting machine, and after two weeks of watching the electrical engineers and maintenance people try to get the thing to run, they stop even looking at it.

Macleod: How long do you think it will be before we see the end of the hard copy drawings?

Jacks: We have it in principle on the 1969 windshields, for instance. The release of the information from Styling to Fisher Body was in the form of a letter which stated "Use design record number so and so in the research laboratory's computer and in that you will find the windshield for a 1969 model A, model B, model C car, plus these other cars", and the Fisher Body people then picked the information up from there and did all their work from that. About two weeks later, the part copy drawing did get sent to Fisher Body, but it is a clear-cut case where the drawing was not used as a transmittal technique. We find other cases like that, but across the board, it is a long way off.

Overheu: You mention the production of blackboard drawings, the very large ones, which give an entire picture of the car. Have you investigated these very large displays?

Jacks: Yes. There is nothing economically feasible now. Technically in terms of accuracy that you want, as the size goes up, the absolute dimensional accuracy goes down. The closest thing we see to anything we can use is to take a drafting machine and put it on line with the console and use that as a feedback mechanism, so you quickly draw out full scale drawings from what is being controlled from the console. That work is getting second level priority for solving the problem at the next level, where all the work is done. In the design area, Chevrolet has perhaps six designers at that level, and each one of our car divisions has probably six designers at that level - there are very few of them, otherwise you would get a hotch-potch for designs, so they keep the number of people that are working at that original conceptual level of sheet metal forming, way down. The pay-off is still low - we are doing that later.

Bobrow: Are you looking at any kinds of three dimensional output displays?

Jacks: We are studying holography, for a lot of different reasons, but that doesn't look too hopeful now. The best thing we have seen for the problem of visualization, and it is not for the Stylist - good heavens they can make a picture manually in a few seconds that's amazingly good - it is for the engineer trying to visualize this thing, and the best thing we found in our console work is just to go from the XYZ space to the U, V and intensity space, as in the ADAGE terminal. That is the best one we have found - you make the intensity such that the objects that are near to you, the lines that are near to you, very bright; those in the distance are faded out, this gives you very good spatial vizualization on the screen.

Bobrow: Have you tried any stereo displays,

Jacks: Yes. You cut your screen in half and the guys' eyeballs are glued to the machine. It doesn't really help the problem as far as we can see. We played around with stereo, the two image problems, the hidden line problem, the rotating object problem games, and at one time we had stereo rotating objects, hidden line program, and intensity control programs. It was very obvious when you had those four programs with one complex object to play with, that the intensity control fed you very quickly the information you wanted - which was depth information. The other thing that's nice about intensity control is that the object on the screen is not a display - it is your source of variables for operations. That is what the object is there for, so you can do an operation on it to construct something. If you use intensity control, your object is still there. Both the lines in the background and those in front, and those that would be considered hidden, they are all right in front of you, and if you want to construct a line between points, you merely point to the points. Doing the same thing in stereo, you have to point at one or two images - that's what it amounts to. You have to get a cursor on that image, into both images, to lock on. If you are using rotation, you catch it on the flyback. Using the hidden line problem, you cannot get to those back lines, but if you use intensity control, you can get at them. You don't have to rotate it or anything else, and if anybody here is an artist, and you start making line drawings, you realise you would use the intensity control on your lines. One reason that intensity control works on displays, is automatic width control - as the intensity goes up, you get secondary emission and so these lines are wider, not only are they brighter but wider.

Pratt: What is the engineering solution to fore-shortening? You have variable intensity along the line, do you do this in increments or gradually?

Jacks: The ADAGE equipment actually takes a U, V, I input at the end of the lines and interpolates as a function of I. On the ADAGE equipment, it is all done by hardware. They do rotation too, at the same time, if you require it. Another trick is attractive - since they have variable intensity control, they can say, for 'I' greater than some value, 'don't display it', or for 'I' less than some value, 'don't display it', and that gives you an automatic cut through at a plane; you can get the effect of an object piercing a plane in space. I should note that we have no ADAGE equipment as yet.

Stanton: Do you feel any gain would be made if, given simple drawings, like the mouldings and so on, rough sketches could be entered through a scanner and propped up in front of the descriptions virtually to save you doing the construction.

Jacks: It's an interesting area. We wanted to do that at one time but just did not get enough time to do it. What I call 'syntax driven scanning' is really what you are doing there, because you are asking at any point, "what's my situation, what are the possible alternatives, what shall I look at?"

Stanton: Do you think the gain could be great if you could do it?

Jacks: I don't know. We know that the console communication does go very fast, and the price for it does look pretty good, so the gain may not be very large at this point.

Capon: Can you give an insight as to why the fact you can call a sequence of operations increases your efficiency, because I think you said the sequence was particularly variable to your work output, and this means presumably not a great deal of reason for doing it again and again, because you are still operating on the same variables?

Jacks: No, if you record the sequence, you can get new variables and so on.

Capon: I thought I heard you say you could not turn this into a program.

Jacks: No you cannot turn it into a program with all we have, but you can force in new variables if you apply it early in a crude sense, not that it does make any difference.

Capon: That being the case then, I've not really seen where your difference comes in. Why do you put that sequence in as part of your model rather than as part of your current operation for instance? In what sense is this a specific model?

Jacks: In the sense that if you want to reapply it, you have to go in and change the data in that model. That's how you reapply it. You cannot isolate it, you cannot cut it out and say "this is now input data for it, here is now a new interface for that sequence of operations". What you have to do is go right back into that model and change the model essentially - form a series of operations on the model and put new data in the model.

LINGUISTIC DESCRIPTIONS

M.B. CLOWES

D.J. LANGRIDGE

Division of Computing Research
C.S.I.R.O., Canberra

R.J. ZATORSKI

Department of Science Languages
University of Melbourne

1. INTRODUCTION

The idea that the sentences of one's own language have a complex structure is acquired, in school, long after the ability to create and understand those sentences. For most if not all of us, the study of grammar in school has only a marginal effect on our ability to communicate in our native language. This skill is so painlessly acquired and so readily deployed that it comes as a surprise to find evidence that a knowledge of sentence structure _is_ continuously utilised in communication. The familiar situations for this discovery include learning a foreign language, or interacting with someone learning our own language, e.g. a child or a foreigner.

A new situation of this type has arisen with the development of computer programs with which we wish to communicate in natural language. Here the need to build into the program a knowledge of the language requires us not only to 'dig out' what it is we know about the language, but also to work out a suitable notation for writing it down. This notation will have to express rigorously, the fundamental generalisations (e.g. that all sentences have parts) as well as the idiosyncrasies of particular words (e.g. 'sheep' has the same form in singular and plural).

The most highly-developed account of natural language aiming to give this degree of rigour is that associated primarily with the linguist N. Chomsky. In this paper we survey the characterisation of English sentence structure developed by his school with particular emphasis on the fundamental generalisations about sentence structure which have been postulated. Our motivation for doing this is two-fold.

Several workers in the area of natural language communication with computers (Bobrow 1967, Fraser 1967, Simmons et al. 1966, Rosenbaum 1967) have invoked some if not all of the Chomskian apparatus to characterise the English part of their programs. However, a succinct up-to-date account of this apparatus in a form intelligible to the non-linguist has not been published. Thus it is hard to know just how much of the apparatus has been taken over by these workers. To the extent that computer programs constitute a working test of the apparatus, this could be regarded as adding to the empirical support for the apparatus. Conversely, those parts of the apparatus which have not found application in these fields may be regarded as called in to question by their omission. Our first aim then is to provide a summary of that account of English sentence structure known as Transformational Generative Grammar.

In recent years we have seen a marked growth of interest in the use of structure manipulation (list structure and ring structure) by programs. In its most common form this appears as the utilisation of program structure by compilers and translators (i. e. syntax directed techniques). More dramatically however has been its appearance in the areas of graphics and picture interpretation (Clowes 1968). It is of theoretical and practical importance to know whether the structures identified in these latter cases are syntactic in nature. That is, are they characterisable in the metalanguage developed by Chomsky, or in some extension of it? What is in prospect here is the unification of these areas of interest, by the development of a single all-embracing 'theory of structure'. Such a theory is clearly beyond our ability to formulate at this time: an essential step in its development is the identification of the fundamental concepts of syntax. This then is our second goal in this paper.

To characterise our knowledge of the language - specifically our grasp of sentence structure - Chomsky has developed the concept of a linguistic rule. This specifies that grouping and interrelation of linguistic elements which mediates (according to the theory) our grasp of sentence structure. A rule should not be thought of as 'instructions to make something' - an interpretation often placed upon it. Two types of rule have been developed - to describe the structure of sentences and the interrelation of words within a sentence.

2. RULES DETERMINING PHRASE STRUCTURE

The words of a sentence are readily grouped to form intra-sentential units or constituents. Thus "the boy kicked the ball" contains the <u>phrases</u> "the boy" and "the ball". Chomsky has developed his theory of language around the idea that two distinct types of rule are required to characterise the constituent structure of sentences. We call these branching rules and transformational rules.

2.1. Branching Rules

1 (i) $$S \rightarrow NP \frown \text{Pred Phrase}$$

(ii) $$\text{Pred Phrase} \rightarrow Aux \frown VP$$

(iii) $$VP \rightarrow \left\{ \begin{matrix} \text{Copula} \frown Adj \\ V \frown \begin{pmatrix} (NP) \\ S \\ Adj \end{pmatrix} \end{matrix} \right\}$$

(iv) $$NP \rightarrow (Det) \frown N \frown (S)$$

(v) $$Aux \rightarrow \text{Tense} \frown (M) \frown (\text{Aspect})$$

(vi) $$\text{Tense} \rightarrow \begin{pmatrix} \text{Present} \\ \text{Past} \\ \text{Future} \end{pmatrix}$$

(vii) $$\text{Aspect} \rightarrow \begin{pmatrix} \text{Perfect} \\ \text{Progressive} \end{pmatrix}$$

These rules are simple, i.e. context-free, phrase-structure rules. Rule 1(i) is to be read as "S (sentence) is rewritten as an NP (noun phrase) followed by a Pred Phrase (predicate phrase)". We can illustrate this rule by the tree-diagram:-

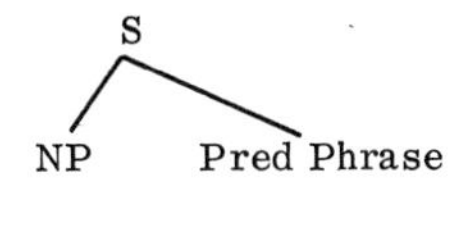

Fig. 1

Rule 1(ii) continues the specification of the structure, by rewriting Pred Phrase into Aux (auxiliary verb) followed by VP (verb phrase). All sentences would thus begin their derivations as:-

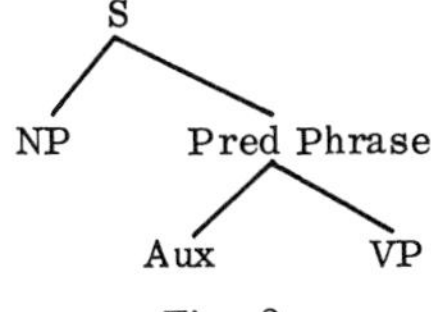

Fig. 2

Rule 1(iii) is a compacted form of five alternative rules describing the structure of VP. The round brackets surrounding NP, also Det and S in 1(iv) and M and Aspect in 1(v), indicate options. The braces enclose alternatives. The five alternatives expressed in 1(iii) are,

VP → Copula ⌢ Adj

VP → V

VP → V ⌢ NP

VP → V ⌢ S

VP → V ⌢ Adj

Fig. 3

where V is verb, Copula is the verb "be" and Adj is adjective. In rules 1(iv) and (v), Det is determiner and includes articles, demonstrative adjectives and numerals etc., N is noun and M is modal, i.e. "can", "will", "must", etc.

The application of all seven rules could give the structure

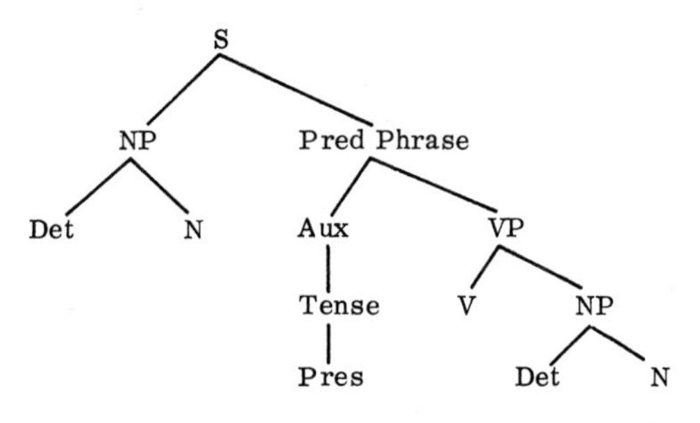

Fig. 4

In earlier versions of generative grammar (Chomsky 1957) the development of this structure would be completed by rules such as,

2 (i) N → boy, ball, Jim, week, pipe, dirt

(ii) V → kick, hit, elapse, solve

(iii) Det → the, some, six

(iv) M → may, can, will

(v) Adj → big, long, happy

Thus an approximation to the sentence

3 The boy hits the ball;

could be obtained by application of these rules.

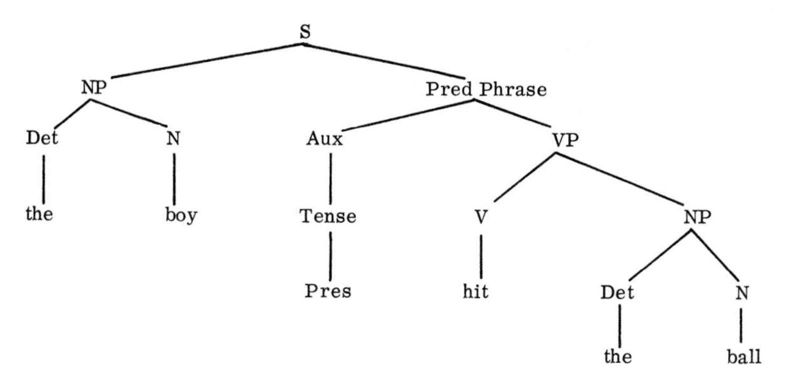

Fig. 5

In this structure the pairs of words "the boy" and "the ball" are dominated by two nodes which dominate no other words. These nodes then 'represent' the two phrases and categorise them as Noun Phrases. Similarly the node 'Pred Phrase' dominates the items which constitute the phrase "hit the ball". Thus the structure illustrated in Fig. 5 marks and categorises the constituent phrases of 3. For this reason it is usually referred to as a phrase marker.

2.2. Transformational Rules

The branching rules are adequate to describe the structure of rather simple sentences. In order to represent the structure of more complex sentences and to provide an account of certain sorts of paraphrase, it is necessary to postulate more complex rules. For example, the sentences

4 (i) The ball is hit by the boy;
(ii) Is the ball hit by the boy?
(iii) The boy does hit the ball;

are closely related in meaning to the sentence whose phrase marker is illustrated in Fig. 5.

The path from 3 to any of the sentences in 4 can be viewed in terms of operations in which whole segments of phrase markers are subject to re-arrangement according to rules such as deletion (a + b → b), reduction (a + b → c) and permutation (a + b → b + a). Rules of this kind have been developed by Chomsky and designated as transformational; they allow mapping of phrase markers into other, derived, phrase markers. An example of a transformational mapping is the following re-arrangement of an active sentence into its derived passive equivalent.

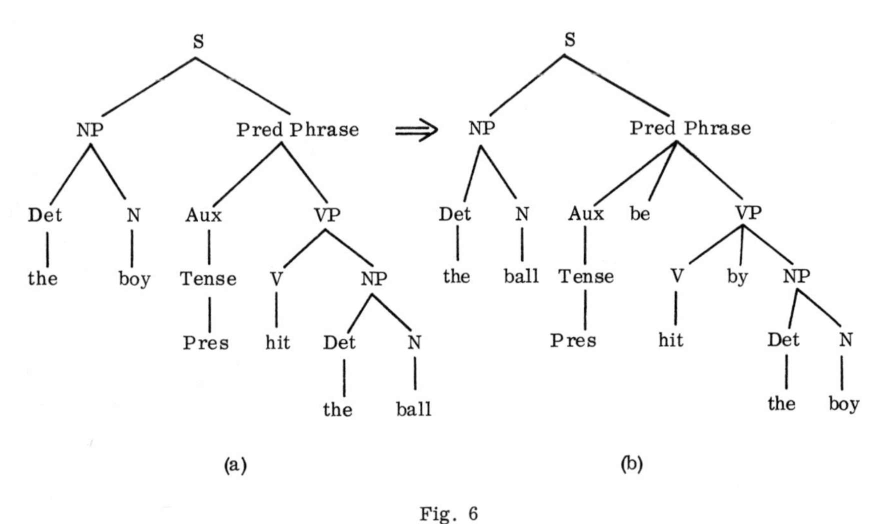

Fig. 6

Achieved by the rule:

5 $\underset{1}{NP} \frown \underset{2}{Aux} \frown \underset{3}{V} \frown \underset{4}{NP} \Rightarrow 4 \frown 2 \frown be \frown 3 \frown by \frown 1$

which acts on the phrase marker illustrated in Fig. 6(a) to give that illustrated in Fig. 6(b).

The passive transformation permits us to regard both 3 and 4(i) as originating from the same phrase marker (Fig. 5). This is the structure which underlies the surface structure of 4(i), of which Fig. 6(b) is an approximate representation.

The rule 5 provides an example of the so-called singulary transformation; other examples of singulary transformation are the interrogative, negation, etc., transformations which yield respectively:

6 (i) Does the boy hit the ball?

(ii) The boy doesn't hit the ball; etc.

In transformational rules, the double arrow '$\Rightarrow$' does not read "is rewritten as" but constitutes an instruction to map the sequence of phrases specified to the left of the arrow into the sequence specified on its right. Transformational rules are applied in ordered sequence.

The branching rules 1 specify that a sentential structure (S) may recur within a sentential structure. Such constructions evidently underlie complex sentences such as

7 (i) The boy who wears a red jersey hits the ball.

This may be regarded as implying the three simple sentences

7 (ii) The boy hits the ball;

(iii) The boy wears a jersey; and

(iv) The jersey is red;

where "the boy" is one and the same person in 7(ii) and 7(iii), and where the same "jersey" is simultaneously the object of 7(iii) and the subject of 7(iv). The branching rules (1, 2) describe this coordination of sentences as the phrase marker illustrated in Fig. 7.

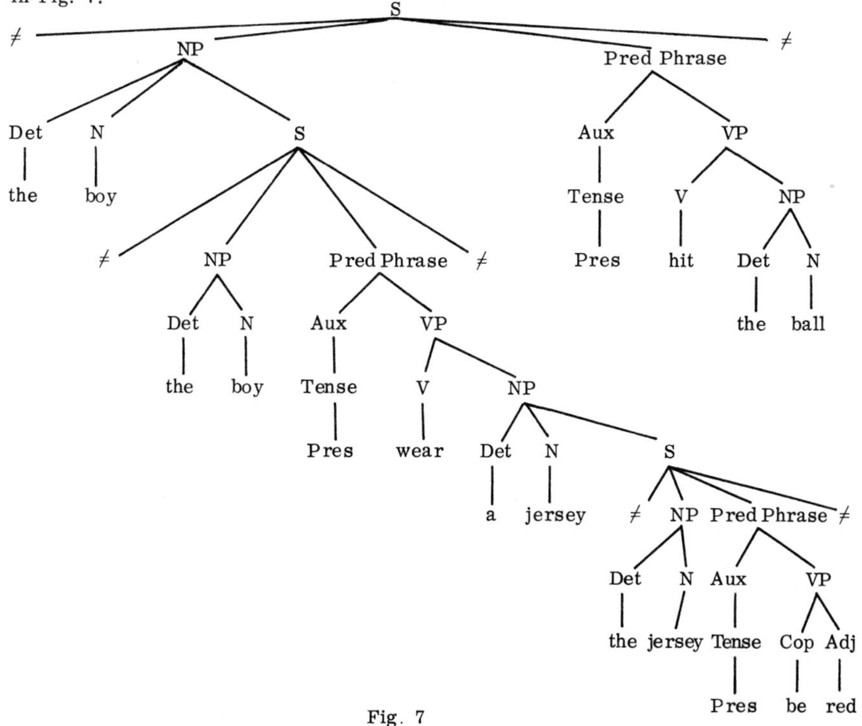

Fig. 7

which is subject to the operation of the following transformational rules:

9 (i) $\underset{1}{\text{Det}} \frown \underset{2}{\text{N}} \frown \underset{3}{\neq} \frown \underset{4}{\text{Det}} \frown \underset{5}{\text{N}} \frown \underset{6}{\text{Aux}} \frown \underset{7}{\text{be}} \frown \underset{8}{\text{Adj}} \frown \underset{9}{\neq} \Rightarrow 1 \frown 8 \frown 2$

where 2 = 5

9 (ii) $\underset{1}{\text{X}} \frown \underset{2}{\text{Det}} \frown \underset{3}{\text{N}} \frown \underset{4}{\neq} \frown \underset{5}{\text{Det}} \frown \underset{6}{\text{N}} \frown \underset{7}{\text{y}} \frown \underset{8}{\neq} \Rightarrow 1 \frown 2 \frown 3 \frown \text{wh-} \frown 7$

where 3 = 6; X and y = sequences of elements of unspecified length.

Restating the generalized phrase marker in Fig. 8 in a linear sequence, we have:

10 (i) ≠ The boy ≠ the boy Pres wear a jersey ≠ the jersey Pres be red ≠ ≠ Pres hit the ball ≠

The rule 9(i) applies to the most embedded sentential string in 10(i) to produce:

10 (ii) ≠ The boy ≠ the boy Pres wear a red jersey ≠ Pres hit the ball ≠

Finally, the application of 9(ii) yields

10 (iii) ≠ The boy who Pres wear a red jersey Pres hit the ball ≠

which phonetic interpretation rules (not discussed here) convert to 7(i).

Despite the apparent power of transformations to assemble word strings of any pattern, certain types of complex sentences either remain wholly outside the scope of the transformational mechanism or are not adequately handled by it. Specifically, the coordination of simple sentences into complex ones, as in

11 (i) John went home but Mary stayed;
(ii) John will drive Mary or Harry will;
(iii) John left for it was late;

cannot be accounted for by the present model since it involves combining sentences related only by referential meaning and lacks a formalism such as the simple identity condition of rule 9(iv), capable of filtering out the referentially incompatible sentences. Thus 11(i) might be generated by the rules of 1 in the form:

12 (i) $\neq$ John $\neq$ Mary Past stay $\neq$ Past go home $\neq$

and could perhaps be transformed into 11(i) by a rule

13 $$\underset{1}{\neq} \frown \underset{2}{X} \frown \underset{3}{\neq} \frown \underset{4}{Y} \frown \underset{5}{\neq} \frown \underset{6}{Z} \frown \underset{7}{\neq} \Rightarrow 1 \frown 2 \frown 6 \frown \text{but} \frown 4 \frown 7$$

this however would also permit the derivation of ill-formed sentences:

14 (i) *John stayed home but John stayed home;

(ii) *Mary went home but John went home;

(iii) *Mary stayed home but Mary went home; etc.

The rule 13 is obviously not capable of handling the coordination of simple sentences since it lacks the specification of some "coordinatory compatibility" which defies syntactic formulation.

This situation is not improved in the case of iteration of phrases within sentences by means of conjunctions. Consider the sentences:

15 (i) John and Mary sold the house.

Following Chomsky (1957) and disregarding the inherent ambiguity we analyse this sentence into two deep structures:

15 (ii) John Past sell the house;

(iii) Mary Past sell the house.

An early version of the branching rules 1 which did not employ the recursive device, and therefore constructed one simple 'kernel' sentence at a time, permitted the assembly of such kernels as 15(ii) and (iii) by means of so-called "generalized transformations", e.g.

16 $$\left.\begin{array}{ll} S_1 & \underset{1}{X} \frown \underset{2}{Z} \frown \underset{3}{Y} \\ & \\ S_2 & \underset{4}{X} \frown \underset{5}{Z} \frown \underset{6}{Y} \end{array}\right\} \Rightarrow 1 \frown 2 \frown \text{and} \frown 5 \frown 3$$

where Z is a "major" category (e.g. N, V), 1=4, 3=6 (Chomsky 1965, p. 212 n. 9).

Since the element Z specifies a category, rule 16 involves the identity of the categorial nodes in both phrase markers and not the identity of the lexical entries suspended from these nodes, as was the case with 9(iv). Accordingly, rule 16 permits the derivation of both well-formed and ill-formed strings:

17 (i) John and Harry sold the house;

(ii) *John and John sold the house;

(iii) *John sold the house and the house.

Equivalently, the branching rules 1 would assign to 17(i) the underlying structure:

18 $\neq$ John $\neq$ Harry Past sell the house $\neq$ Past sell the house $\neq$

and the conjunction rule would assume the form:

19
$$\underset{1}{\neq} \frown \underset{2}{Z} \frown \underset{3}{\neq} \frown \underset{4}{Z} \frown \underset{5}{Y} \frown \underset{6}{\neq} \frown \underset{7}{Y} \frown \underset{8}{\neq} \Rightarrow 1 \frown 2 \frown \text{and} \frown 4 \frown 5 \frown 8$$

where Z = "major" category and 5=7;

which, like the mechanism of 16 would fail to block the construction of sentences such as:

20 (i) *John and salesmanship sold the house; or

(ii) *Sadness and oysters upset John.

Likewise, both rules would fail to account for well-formed sentences such as:

20 (iii) John and Mary are engaged;

John and Mary are a nice couple;

John and Mary love each other;

Mary sat between John and Bill;

John combined oxygen and hydrogen;

John, Mary and Bill weigh 8, 7 and 9 stone respectively.

The derivation of surface structures of the above sentences involves the problems of reference, value-computation, determination of spatial relations, etc., i.e. problems in extra-syntactic domains whose status in language has not as yet been satisfactorily defined.

3. SYNTACTIC FUNCTIONS

The branching rules of the base component together with the transformational rules provide a characterisation of the form of a sentence in terms of the arrangement of admissible phrases and words which comprise it. In discussing the content of a sentence it is customary to break it down into Subject, Predicate, Direct Object, etc. What is the relation of these groupings to the phrases defined by the generative grammar? In the sentence "the boy hits the ball", the Subject is "the boy" which coincides with the first NP of the sentence as diagrammed in Fig. 5. Similarly, the Direct Object of the Main Verb "hits" is another NP, "the ball". It is tempting to equate these syntactic categories NP, V with the functional notions Subject, Main Verb, etc. Chomsky rejects this in favour of a relational definition. Thus Subject is identified with that NP immediately dominated by S, Object with that NP immediately dominated by VP and so on, yielding the set:

21 (i) Subject - of : [NP, S]
(ii) Predicate - of : [Pred Phrase, S]
(iii) Direct - Object - of : [NP, VP]
(iv) Main Verb - of : [V, VP]

Thus Chomsky distinguishes between grammatical function, e.g. Subject - of, and grammatical category, e.g. Noun Phrase. To ascertain the grammatical <u>function</u> of an element we require a phrase marker, specifically the underlying phrase marker.

Grammatical relations of the sort that hold between Subject and Verb can be defined <u>in terms of these functions</u>. Thus grammatical relations also require that the underlying phrase marker be employed to mark them.

4. RULES PERTAINING TO LEXICAL ITEMS

When the set of branching rules 1, 2 (section 2. 1) is used as a device to "construct" English sentences, it becomes immediately apparent that additional constraints are required if its output is not to contain unacceptable strings such as

22 (i) *The Jim kicked the ball;
(ii) *The week elapsed the boy;
(iii) *John solved the pipe;
(iv) *The dirt was happy.

In general, the deviant nature of sentences in 22 concerns the mutual associations among words, which linguists designate as "cooccurrence restrictions". Traditionally the grammarian has discussed this in terms of subcategories of Noun and of Verb (Animate Nouns, Transitive Verbs) and grammatical relations between Subject and Verb, etc.

In "Aspects of the Theory of Syntax" Chomsky (1965) introduced a number of devices to characterise this aspect of well-formedness. They are presented as branching rules, replacing those indicated in 2 above.

The devices centre upon the suggestion that a word (morpheme) is an entry in a lexicon, where each entry has a list of features or properties. A word may co-occur with others in a sentence if the feature list of its lexical entry is the same as a list of features appended to a terminal node of the underlying phrase marker, and derived by rule from the phrase marker of the sentence. This second feature list is the Complex Symbol: the rules we now present specify how it is to be 'developed' from the phrase marker.

Where a lexical entry is the same as a Complex Symbol, that word may be substituted into the phrase marker. This concept of lexical substitution suggests that view of grammatical rules which regard them as constructional, and we will employ it here.

4.1. Strict Subcategorisation

In 22(i), we have ignored the common/proper noun distinction and in 22(ii) the transitive/intransitive verb distinction.

We add the rules

1 (cont.) (viii) N → CS

(ix) V → CS

and interpret them to mean "the first two syntactic features of the complex symbol appended to a node labelled N will be + N, and a representation of the immediate environment of this node". (Equivalently for V). Then 1(viii) develops each of the four types of phrase specified by 1(iv) to give the phrase markers illustrated in Fig. 8.

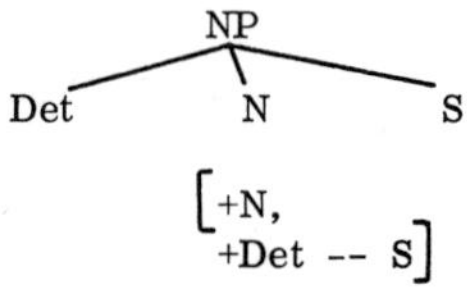

(a)

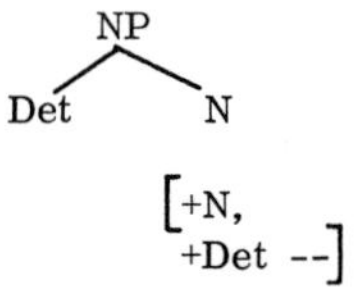

(c)

NP
N S

[+N,
+ -- S,]

(b)

NP
N

[+N,
+ --,]

(d)

Fig. 8

The term 'immediate environment' is taken to mean those categories immediately dominated by the same node (in this case NP) that dominates the node in question. The representation of this environment has the form often employed in Contex Sensitive grammars, namely a string in which -- replaces the node in question.

Each noun in the lexicon will carry a feature list which commenced with one of the four alternatives diagrammed in Fig. 8 and listed in 23

23 (i) [+ N, + Det -- S,]
(ii) [+ N, + Det -- ,]
(iii) [+ N, + -- S,]
(iv) [+ N, + -- ,]

23(i) is the category of nouns with sentential complements ("the fact that he was guilty"). 23(ii) is the category of common nouns, whilst 23(iv) is the category of proper nouns. 23(iv) is used to account for "it" in such sentences as "it strikes me that he is a fool".

"Jim" would appear in the lexicon together with the feature list 23(iv). It could not therefore be validly 'substituted' into the phrase marker Fig. 8(d) and accordingly 22(i) is shown to be ill formed.

Similarly, rule 1(ix) specifies the subcategories of verbs in terms of the type of verb complement derived by rule 1(iii). We can now obtain any of the following:

[+ V, + --], for intransitive verbs;
[+ V, + -- NP], for transitive verbs;
[+ V, + -- S], for verbs which take sentential complements;
(e.g. "I believe it to be unlikely")

and $\left[+ \text{V}, + \text{--} \text{ Adj}\right]$, for verbs which assume adjectival complements; (e.g. "I feel sad").

We can see that some aspects of traditional grammar have been formalised in a very natural way by the use of strict subcategorisation. (Note - The interested reader is referred to pp. 102-106 of "Aspects" for more subtle uses of strict subcategorisation, where the non-passivization of "middle" verbs is discussed).

4.2. Context-Free Subcategorial Rules

1 (cont.) (x) $\left[+\text{N}, +\text{Det -- (S)}\right] \rightarrow \left[\pm \text{Count}\right]$

(xi) $\left[+\text{Count}\right] \rightarrow \left[\pm \text{Animate}\right]$

(xii) $\left[+\text{N}, + \text{-- (S)}\right] \rightarrow \left[\pm \text{Animate}\right]$

(xiii) $\left[+\text{Animate}\right] \rightarrow \left[\pm \text{Human}\right]$

(xiv) $\left[-\text{Count}\right] \rightarrow \left[\pm \text{Abstract}\right]$

(xv) $\text{Det} \rightarrow \left[\pm \text{Definite}\right]$

All the rules, except 1(xv), continue the specification of the syntactic features of the CS of a noun. They are not to be interpreted as rewriting rules, but rather as specifying an addition of a feature at each application of a rule. Thus applying 1(x), (xi) and (xiii), we could derive the following CS for a noun

24 $\left[+\text{N}, +\text{Det -- S}, +\text{Count}, +\text{Animate}, +\text{Human}\right]$

These rules complete the CS development of nouns, and 'substitution' of lexical formatives of nouns into the phrase-marker can now be attempted. Assume for example that the following fragment of phrase-marker had been developed,

```
        NP
       /  \
    Det    N
           [+ N,
            + Det --,
            + Count,
            + Animate,
            + Human ]
```

The lexical formative 'boy' whose lexical entry reads

25 ("boy", [+ N, + Det --, + Count, + Animate, + Human]).

can now be substituted into the marker.

Although the appearance of features such as ± Human, ± Abstract, etc., have semantic overtones it is difficult to argue against their inclusion in a syntactic description. Knowledge of the count/mass (i.e. ± Count) distinction is required in the formation of plurals and the selection of relative pronouns depends on the Human/non-Human distinction. The other feature distinctions, i.e. ± Abstract and ± Animate are included for the purpose of restricting the selection of verbs and adjectives.

4.3. Selectional Rules

With the apparatus so far provided it is still possible to generate 22(iii), (iv). A traditional grammar employs the notion of grammatical relation, e.g. the relationship between Subject-Verb, Verb-Object and Noun-Qualifier to state that "solve" normally takes Abstract objects and "happy" normally qualifies Animate nouns.

Thus the remaining rules of the base-component are the selectional rules or "restrictions on cooccurrence" for verbs and adjectives

1 (cont.) (xvi) [+ V] → CS/ [+ N] .. -- ([+ N] ...)
(xvii) Adj → CS/ [+ N] ... --

Rule 1(xvi) completes the specification of the subcategories of verbs. The rule is read: "in the context of Subject-noun and an optional Object-noun, rewrite the verb as a complex symbol having the features already assigned to the Subject and Object nouns". Rule 18(xvii) specifies the complete CS of adjectives and is interpreted in the same manner as verbs, but here there is no Object-noun.

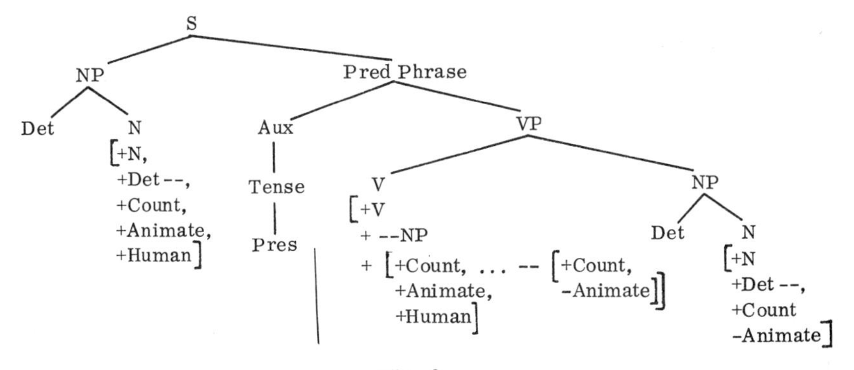

Fig. 9

Fig. 9 illustrates the development of the phrase marker of Fig. 4 by these additional rules 1(viii) - (xvii).

Since the verb in our example is transitive, both collections of noun features are transferred. We conclude this account of the base-component with a few examples of lexical formatives:

24 ("the", [+Definite])
("ball", [+N, +Det--, +Count, -Animate])
("dirt", [+N, +Det--, -Count, -Abstract])
("hit", [+V, + --NP, +[+Count, +Animate, +Human] . . . --[+Count, -Animate]])
("elapse", [+V, + --, +[+Count, -Animate . . . --]])
("may", [+M])
("hairy", [+Adj, +[+Count, +Animate, +Human . . . --]])

It will be seen that the well formedness of 3 is treated much more comprehensively by lexical substitution of these entries (as appropriate) into the phrase marker of Fig. 9 than by the simple rewriting rules of 2.

4.4. Lexical Substitution and the Senses of a Word

The complex symbol has been introduced to act as a 'local repository' for information about features of the structure and of features of items in the structure. Thus information which is distributed over the phrase marker may be localised at any point according to rules which identify that point and the distribution. The rules presented above identify these points as the terminal nodes of the phrase marker.

Lexical items are associated (in generative grammars) with terminal nodes of the phrase marker. Thus these rules provide the means whereby any information associated with a lexical item may be 'coordinated' with information drawn from the phrase marker and concentrated at a terminal node. The particular manner of co-ordination chosen by Chomsky - 'lexical substitution' - requires a one-to-one correspondence between information associated with the lexical item and that occurring in the complex symbol. Given some set of rules (as above) which develop complex symbols, this definition of lexical substitution specifies the composition of the information associated with a lexical item, i. e. specifies the structure of a lexical entry.

[Note that the lexical substitution rule provides for the development (= re-writing) of the phrase marker conditional upon a 'match' between two structures. The one (a complex symbol) derived from the phrase marker, the other (a lexical entry drawn from a (dictionary) list. To the extent that the Complex symbol is descriptive of the structure of the phrase marker, this condition is of the same form as that obtaining in transformational rules. Consequently lexical substitution is a variety of transformation - Chomsky suggests the term 'local transformation' to characterise it.]

To the extent that the complex symbol reflects non-structural information (i. e. features such as $\pm$ Animate, developed independently) we are free (in principle) to supplement the lexical entry with any information thought pertinent. The restrictions we would place upon such supplements would be the requirement that they provide an account of some example(s) of paraphrase, ambiguity or anomaly, according to the methodological principles illustrated in §2. Thus we have seen (§4. 3) that features such as $\pm$ Abstract can be deployed to account for the anomaly inherent in "John solved the pipe". The ambiguity of "the watch was unbroken" ("watch" having the senses "they kept watch" and "they smashed the watch") might similarly be traced to such a feature of the noun "watch". These varieties of ambiguity (and anomaly) devolve upon the multiplicity of senses which a word may have over and above the multiplicity of syntactic categories to which a word may belong. The latter manifests itself in multiple phrase structures for the sentence (e. g. as in "flying planes can be dangerous"), the former in the several modes of agreement possible between words embedded in the same phrase marker, where these words have multiple senses.

An alternative approach to this co-restriction of word-senses, has been advanced by Katz and Fodor (1963) and elaborated in Katz and Postal (1964). The account presented here will be based upon the earlier version (Katz and Fodor, 1963). While the subsequent development (Katz and Postal, 1964) contains a good deal which is new, the central idea of interpretation is unchanged. It is this idea which we summarise below.

4. 5. Interpretive Rules

As before we assume a lexicon, with highly structured lexical entries. The co-relation of word senses afforded by their occurrence in a phrase marker is achieved by the so-called 'projection rules'.

4.5.1. Lexical Entries

These take the form of list structures having four types of list element distinguished by brackets. Thus unbracketed elements denote syntactic categories, items enclosed in round brackets "(" ")" are semantic markers; items in square brackets "[" "]" are called distinguishers; items in angle brackets* "<" ">" comprise Boolean functions of semantic markers. Illustrative lexical entries for three words are given below. (*or "⟨" "⟩")

ball, Noun

(Social activity) (Large) (Assembly) [For the purpose of social dancing]
(Physical Object) [Having globular shape]
(Physical Object) [Solid missile for projection by an engine of war]

In diagrammatic tree form this lexical entry takes the form:

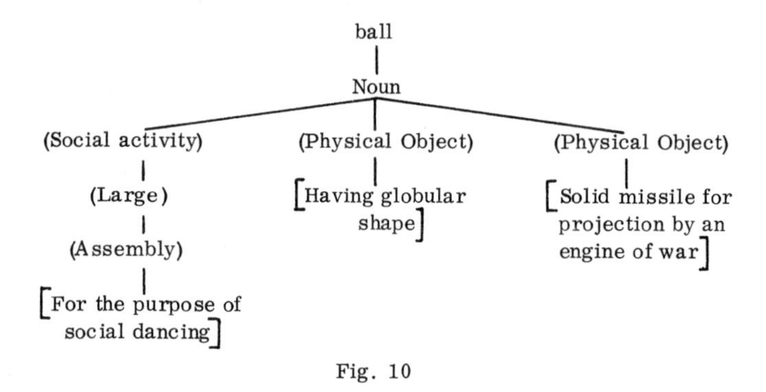

Fig. 10

---

the, Noun Phrase, Definite Article
[some contextually definite]

---

hits, Verb, Verb transitive, (Action)(Instancy) (Intensity)
[collides with an impact] < Subject: (Higher Animal) v (Improper Part) v (Physical Object), Object: (Physical Object) >

[strikes with a blow or missile] <Subject: (Human) v (Higher Animal), Object: (Physical Object), Instrumental: (Physical Object)>

The number of distinct paths through each entry reflects the number of senses which each word can assume: these paths are based upon the entries in the Oxford English Dictionary. Thus the noun "ball" has several senses, ranging over social activity and physical objects, as illustrated. The verb "hits" can either mean 'collides with' or 'strikes' in the sense "he struck the ball with the bat".

Where "ball" and "hits" occur in a phrase, e.g. "hit the ball" it is intuitively clear that there is a mutual restriction of senses. Thus the sense of "ball" meaning 'social activity' would not be something we could strike or collide with. It is therefore eliminated. What is wanted then is some algorithm which exhibits this elimination. This is the role of the 'projection rules'.

4.5.2. Projection Rules

"There is a distinct projection rule for each distinct grammatical relation" (Katz, 1966, p.165). Since grammatical relations are specified in terms of the structure of the underlying phrase marker (specified, that is, in terms of Syntactic Functions), the operation of projection rules inevitably reflects that structure. The formulation of the projection rules is in fact in terms of operations upon the base phrase marker. If an independent notation had been introduced in which to express grammatical relations, e.g. a string notation similar to that introduced for Syntactic Functions (§2.3), projection rules could have been formulated on expressions in that notation. In the absence of such a notation, we treat the underlying phrase marker as pairs of constituents dominated by a constituent node, e.g. NP dominating Det, N (in Fig. 5); between each pair a distinct grammatical relation holds. Thus in the fragment VP of the phrase marker in Fig. 5, we have the grammatical relations

(i) Article - Noun

(ii) Main Verb - Direct Object

Prior to the application of the projection rules, lexical entries are appended to the phrase marker to give the initial situation depicted in Fig. 11.

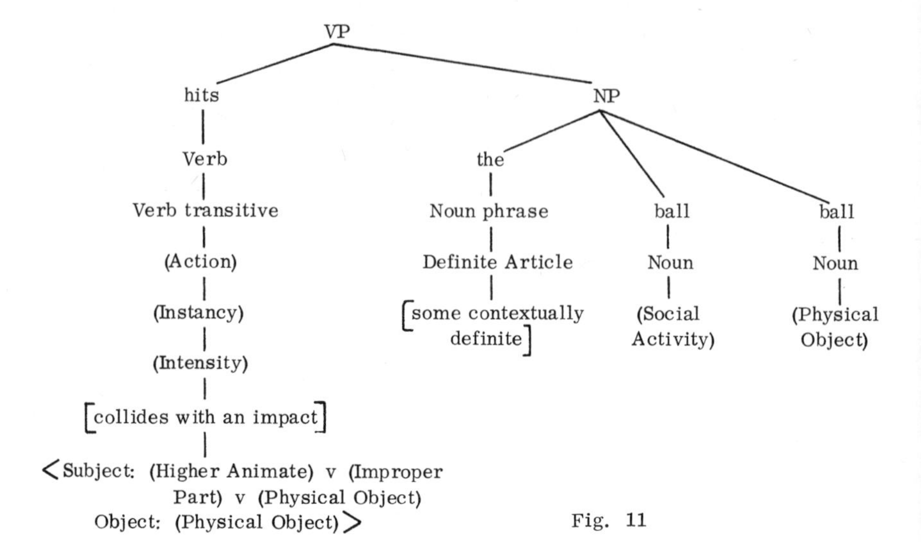

Fig. 11

In Fig. 11 we have omitted that sense of 'ball' which identifies it as a 'cannon ball'; and that sense of 'hits' which identifies it with 'strikes', for simplicity. We will pass over the formulation of the projection rule applying to case (i) since it has no selective effect here. The result of applying it is to replace the node NP in Fig. 11 by the unlabelled node in Fig. 12 dominating the two 'readings' shown. Notice that these two readings have the same general form as a lexical entry. This is essential to the recursive application of projection rules.

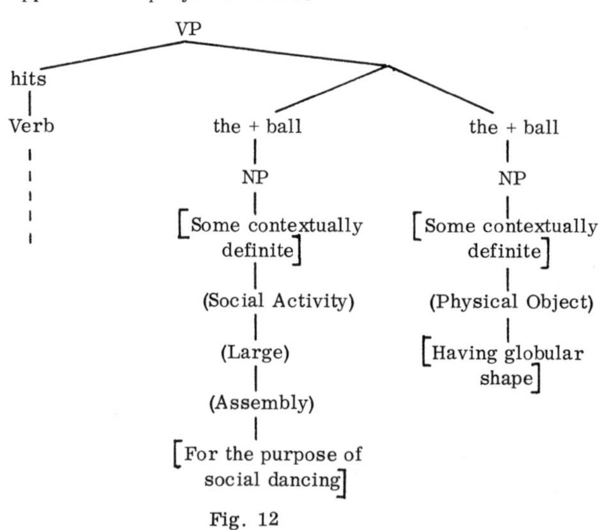

Fig. 12

The projection rule for Main Verb - Direct Object takes the form:-

Given two readings associated with nodes branching from the same node labelled SM, of the form

Lexical String$_1$ - syntactic markers of main verb - $(a_1) \ldots (a_n)$ - $[A] \langle \Omega \rangle$

and the other of the form

Lexical String$_2$ - syntactic markers of object of main verb - remainder of path

such that the string of syntactic or semantic markers of the object path satisfies the Object restriction in Ω, then there is a derived reading of the form

Lexical String$_2$ + Lexical String$_1$ - SM - $(a_1) \ldots (a_n)$ $[A]$ remainder of object path $\langle \Omega' \rangle$ where Ω' is Ω less the Object: restriction.

This rule applies to Fig. 12, since 'hits' is Main Verb, and each occurrence of 'the + ball' is Direct Object. The Boolean conditions applicable to Object, originating in the lexical entry for 'hits' permit a retention of the (Physical Object) sense of 'the + ball' because that semantic marker satisfies the restriction demanded by the rule. The reading of 'the + ball' identifying it as (Social Activity) is however rejected. Fig. 12 becomes accordingly

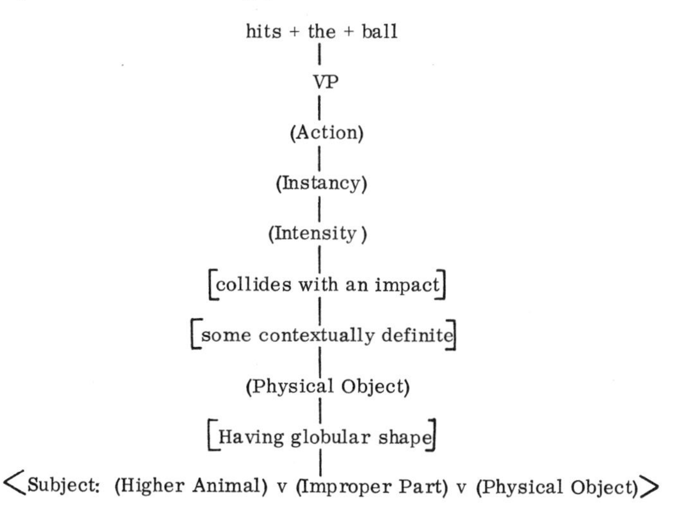

Fig. 13

This reading is then in turn subjected to the next applicable projection rule which for the phrase marker of Fig. 5 would be that appropriate to the grammatical relation between Subject and Main Verb.

The result of applying these rules is to show that (according to Katz and Fodor) the sentence in Fig. 5 is four ways ambiguous due to the congruence of the two (physical object) senses of "ball" with both senses of "hits". The third sense of "ball" (Social Activity) having been eliminated when the projection rule for Main Verb : Object, juxtaposes "hits" with that sense of "the + ball" which identifies it as a social occasion.

Thus this interpretive system provides us with an alternative way of 'rejecting' lexical entries to that outlined earlier. In operational terms we might say that Chomsky's generative approach attempts to prevent words being substituted into the phrase marker while the approach due to Katz and Fodor would allow any syntactically plausible substitution, subsequently rejecting it when it proved incompatible with the sense of the whole sentence or the sense of any part of the sentence. Taking that view of lexical substitution rules in which they are transformational in character, we see that projection rules are in a sense 'inverted' transformational rules. Overtly, Chomsky's formulation makes no appeal to the sense of the utterance - he is apparently concerned with syntactic not semantic problems. In practice however the means he uses to accomplish this, 'syntactic' features like +Animate and -Abstract, and selectional rules relating Verb to Subject and Object express the same kind of knowledge about sentence structure as do semantic markers and projection rules in the interpretive approach. The principal difference lies in the elaborateness of the lexically-encoded knowledge of words in the interpretive system. This knowledge reflects a great deal of what we know about the events and objects to which these words refer. Thus "hit" has a semantic marker (Action). This is due of course to the fact that Katz and Fodor formulated their interpretive system as a semantic model: the distinction between syntax and semantics is not one that we wish to explore at this point. Our concern is simply to give an account of the technical apparatus assembled by the linguist to characterise our knowledge of a natural language.

5. DISCUSSION

As we have seen, Chomsky has suggested that our knowledge of English concerns not only the structure of sentences (§2 above) but a great deal of information about individual words expressed through a lexicon (§4.1 - 4.3). One interpretation which can be placed upon the introduction of such lexical knowledge is that it suggests for example that we 'know' that "dirt" is -Animate because it cannot act as Subject to verbs like 'solve' which take +Animate Subjects. That is the distinction $\pm$Animate between "dirt" and "man" is to be understood in terms 'cooccurrence restrictions'. This seems counter intuitive: we would rather say that these cooccurrence restrictions and other differences in the usage of the words "dirt" and "solve" reflect what we know about the <u>reference</u> of these tokens. Linguistics, narrowly conceived as the study of language is powerless to treat this problem. Katz and Fodor (1964), try to resolve it by a so-called semantic theory conceived as a part of linguistics distinct from syntax. As we have seen, their formulation adds little to the characterisation of English not already expressed in Chomsky's Selectional rules.

Chomsky (1965) is alive to these issues which he considers as concerning the boundary between syntax and semantics (Ch. 4 op cit). Thus the ill formedness of

25 (i) *The cut has a finger;
(ii) *The arm has a man;

"cannot in any natural way be described within the framework of independent lexical entries". If we take this dismissal to include both Chomsky's theory of the lexicon <u>and</u> that of Katz and Fodor, we are still left with the need to formulate some account of the notion 'semantics'. It is our view that such an account is to be found in a relatively informal embryonic state in the development of computer-based question-answering (QA) systems.

The question raised in the Introduction: as to the extent to which these systems employ Transformational Grammars is more often than not dependent upon what is meant by "employ". Thus while several papers avow their dependence upon Transformation Grammar, there are others, e.g. Baseball (Green et al, 1963) where it is implicit in the parsing procedures without overt acknowledgement. The live issues in QA would appear to concern

(1) A construction of a model of the problem domain, i.e. some characterisation of the universe of discourse.

(2) The formulation of algorithms which will translate expressions or structures in this model into English sentence structures and vice versa.

It would be highly satisfactory if our model of the problem could be expressed in some formal language which directly reflects our knowledge of the problem domain in the same sense (and for the same motives) as the language of generative grammar reflects (allegedly) our knowledge of a natural language. It is difficult to see how our knowledge would be explicated in any other way than in terms of objects, relations and attributes of objects (see Clowes, 1969; Narasimhan, 1968). At present QA programs express whatever knowledge of the problem they have (often quite scanty) in the organisation of the data base and in subroutines which manipulate this data base and queries directed to it. This is not the place to embark upon a discussion of the formalisation of QA programs. We would merely draw the reader's attention to the papers on this topic by Fraser (1967) and Barter (1969).

The second question raised in the Introduction concerned the extent to which Chomsky's formal apparatus might provide the basis of a universal 'theory of structure'. We have already indicated our doubts about the lexicon as a representation of our knowledge of the language. The other principal devices postulated by Chomsky are the Branching Rules (§2.1) and Transformational Rules (§2.2).

Branching Rules: There are reasons for thinking that Phrase Structure Rules (PSR's) are of extremely limited utility. We can best illustrate this by examining, briefly, their application to the characterisation of picture structure.

Evans (1968) has developed a descriptive language for geometric drawings in which he characterises the structure of a triangle as

26 $$\text{TRIANGLE}\langle \underset{1}{\text{PT}}, \underset{2}{\text{PT}}, \underset{3}{\text{PT}} \rangle \left[\text{line}\langle 1, 2\rangle, \text{line}\langle 2, 3\rangle, \text{line}\langle 3, 1\rangle, \text{noncollinear}\langle 1, 2, 3\rangle\right]$$

where PT stands for point and 'line<PT, PT>' denotes the predicate "there exists a line segment whose end points are the arguments of this predicate". The predicate 'noncollinear' has its obvious interpretation. Within such a language we can readily express what is expressed by a PSR. Thus 1(i) would read:

27 $$\text{S}\langle \underset{1}{\text{NP}}, \underset{2}{\text{Pred Phrase}} \rangle \left[\text{followed by}\langle 1, 2\rangle\right]$$

This identifies the symbol "⌢" used by Chomsky to denote the relation of 'followed by' as a particular (primitive) predicate, appropriate to certain data domains, i.e. string data. Domains where the relationships between the parts of an object (treating S as an object) are more numerous and not necessarily primitive can be expressed in a language like Evans' which explicitly denotes relations but cannot be expressed in PSR's. Attempts that have been made to characterise structure in such complex domains, e.g. picture structure, with PSR's - notably Narasimhan, 1966; Shaw, 1968 - require the introduction of various interpretive devices into the meta-system. Using Shaw's formalism, a characterisation of polygons by the Phrase Structure Grammar

28
$$\begin{aligned} \text{polygon} &\rightarrow \text{piece} * \text{line} \\ \text{piece} &\rightarrow \left\{ \begin{matrix} \text{piece} \\ \text{line} \end{matrix} \right\} + \text{line} \\ \text{line} &\rightarrow \text{pt pt} \end{aligned}$$

assigns the following structural description to a triangle

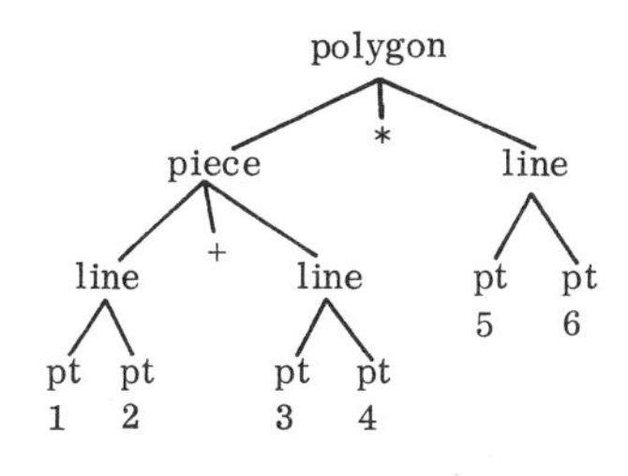

Fig. 14

The operators '*' and '+' describe certain types of relationship between the items which they separate. Thus the expression

29 $$S_1 + S_2$$

has the interpretation

30 $$\text{head}\ (S_1)\ \underline{\text{cat}}\ \text{tail}\ (S_2)$$

where <u>cat</u> specifies the spatial coincidence of those positions on S_1 and S_2 called head and tail. Identifying the points labelled 1 and 3 in Fig. 14 as tails, makes 2 and 3 coincident.

The expression

31 $$S_1 * S_2$$

has the interpretation

32 $$\text{tail}(S_1)\ \underline{\text{cat}}\ \text{tail}\ (S_2)\ \&\ \text{head}\ (S_1)\ \underline{\text{cat}}\ \text{head}\ (S_2)$$

In Fig. 14, S_1 is 'piece': 32 requires that 'piece' then should have a pair of positions, head and tail. To do this it is necessary that operators not only specify coincidence relationships but also specify the formation of compound objects having a single head/tail pair at designated positions. For the rule

33 $$S \rightarrow S_1 \text{ op } S_2$$

in which op stands for any of the (four) operators employed by Shaw, the interpretation which specifies this notion of 'formation' is

34 $$\text{tail}(S) = \text{tail}(S_1)$$
$$\text{head}(S) = \text{head}(S_2)$$

There is thus a heavy burden of interpretation placed upon the meta system and this formation rule 34 is clearly alien to any interpretation normally invoked in PSR's. Even this extension of the notion of a PSR fails to do justice to the problem of picture description as Shaw and Mill (1968) remark: "The chief limitations of the descriptive scheme are the restricted set of relations that may be expressed, the practical constraints resulting from only two points of concatenation for a primitive, and the absence of a general mechanism for hierarchic semantics."

The extent to which PSR's describe strings is the extent to which these strings have parts between which there is a single relationship 'followed by'. The additional devices introduced by Chomsky § 4.1 - 4.3 clearly hypothesise other relations between parts of a string, despite their statement in PSR form. The lengthy and ill formalised interpretation placed upon them is not unlike that resorted to by Shaw and we conclude that PSR's have little or no place in a characterisation of natural language. The specification of a base phrase marker upon which Transformational rules 'operate' is more likely to be tied up with the procedures necessary for mapping from an object or problem domain as outlined earlier. Chomsky's so-called 'base-component' (§2.1 and 4.1, 4.2, 4.3) is a convenient fiction which replaces these procedures. Whether the base phrase marker would still retain the form assigned by the branching rules (§2.1) also seems questionable. In the absence of a formal characterisation of the object domain and of mapping procedures, it is however pointless to speculate further.

Regarding the transformational component of the grammar there are many rules which appear to be very plausible, for instance those for negation and interrogatives and in fact any rules that introduce purely linguistic tokens, i.e. that have no problem reference, and hopefully rules of this type are now beyond dispute. The status of Transformational Rules which involve problem reference, e.g. reference to temporal relations in the universe as in 11 and 14, is intimately bound up with the form assumed by the base phrase marker. The uncertainties about the form of the latter outline above, make further comment pointless.

In conclusion we believe that more insight would be gained into natural language if a model of an object domain was formally characterised in a computational language and that the rules which map from expressions within this language into English and vice-versa were similarly formalised. A formalisation of our knowledge of the problem would replace much of the lexicon as formalised by Chomsky and the mapping rules would remove the need for a phrase-structure characterisation of sentence structure. These together with a 'purely' linguistic transformational component would constitute a viable model of natural language competence in this domain.

6. REFERENCES

BARTER, C.J. (1969), "Data structure and question answering". This conference.

BOBROW, D.G. (1967), "Problems in natural language communication with computers"; IEEE Transactions on Human Factors in Electronics, H FE-8, pp. 52-55.

CHOMSKY, N. (1957), "Syntactic structures"; The Hague : Mouton & Co.

CHOMSKY, N. (1965), "Aspects of the theory of syntax"; Cambridge, Mass. : The MIT Press.

CLOWES, M.B. (1968), "Transformation grammars and the organisation of pictures"; Seminar Paper No. 11, C.S.I.R.O., Division of Computing Research, Canberra. Presented at the Summer School on 'Automatic Interpretation and Classification of Images', NATO Advanced Study Institute Program, Pisa, Italy, 1968.

CLOWES, M.B. (1969), "Picture syntax"; Seminar Paper No. 18, C.S.I.R.O., Division of Computing Research, Canberra.

EVANS, T.G. (1968), "A grammar-controlled pattern analyzer"; Proceedings of the IFIP Congress, Edinburgh.

FRASER, S.B. (1967), "On communicating with machines in natural language"; Computer and Information Sciences - 11. Academic Press, New York.

GREEN, B.F. et al., (1963), "Baseball : An automatic question-answerer"; E.A. Feigenbaum and J. Feldman (Eds.), Computers and Thought. New York: McGraw Hill, pp. 207-216.

KATZ, J.J. and FODOR, J.A. (1964), "The structure of a semantic theory"; Language, 39, pp. 170-210.

KATZ, J.J. and POSTAL, P. (1964), "An integrated theory of linguistic descriptions"; Cambridge, Mass. : The MIT Press.

KATZ, J.J. (1966), "The philosophy of language"; New York, Harper & Row.

NARASIMHAN, R. (1966), "Syntax-directed interpretation of classes of pictures"; Comm. ACM. 9, pp. 166-173.

NARASIMHAN, R. (1968), "On the description, generation and recognition of classes of pictures"; Computer Group, Tata Inst. of Fundamental Research, Colaba, Bombay 5. Presented at the Summer School on 'Automatic Interpretation and Classification of Images', NATO Advanced Study Institute Program, Pisa, Italy, 1968.

ROSENBAUM, P.S. (1967), "A grammar base question-answering procedure"; Comm. ACM. 10, pp. 630-635.

SHAW, A.C. (1968), "The formal description and parsing of pictures"; SLAC Report No. 84, Stanford Linear Accelerator Centre, Stanford, California.

SHAW, A.C. and MILLER, W.F. (1968), "Linguistic methods in picture processing"; SLAC PUB-429, Stanford Linear Accelerator Centre, Stanford, California.

SIMMONS, R.F. et al. (1966), "An approach toward answering English questions from text"; AFIPS FJCC 29, pp. 357-364, Washington, Spartan Books.

7. DISCUSSION

Overheu: You say in your conclusion in the very last paragraph, you are seeking a model of an object formally characterized in a computational language. What do you mean by computational language?

Clowes: I think the situation that we are trying to point to in the last paragraph (it was also probably Note 28 in the paper), was depicted very clearly to my taste in Bobrow's presentation of the TLC system of Quillion. Essentially what Quillion has is mainly a more or less adequate (I don't know how to evaluate it), characterization of what we know about the world, what we do know about clients, lawyers, etc. In that structure there is no necessary reference to English. It's explained to us in terms of English, but of course it doesn't matter which structure is used. But essentially, that is what we would think of as being an object of characterization. You notice that Quillion tries to get into that structure as fast as he possibly can. He doesn't really want to know about English syntax at all. When he has got into that structure, he essentially says, "well, supposing that I have correctly identified this fragment, this is what is being talked about, how will that be expressed in English?". These of course are his form tests. Now I think that one way of rehashing Chomsky is to say, "well , supposing that instead of a base phrase marker, one of these simple trees (we have the kind that Quillion has), then what kinds of rules would be right to turn those structures into the normal string expressions in which we find English expressed?". Those rules, we would think of as being mapping rules, and Quillion's form tests are very simple versions of those rules. For example, he looks at the apostrophe 's' between the words, in the correct order - a very simple form test. This, in other words, is the way English allows us to express a relationship in a problem domain, in an object domain, and I think this is what we are driving at. When we say a formal context of language, what we mean at least is, let's say a list structure language in which you can address things, define things, and so forth; you can construct lists, you can do some computation - that's all we mean by a formal computational language. The essential point I think that has been said, is that the base phrase marker that Chomsky throws up, at great expense, in terms of parsing and so forth, really isn't what you want anyway. We have that, and still have to get to the referential domain, and therefore the real question is, "why do you need that base phrase marker at all?". A lot of the selectional stuff that he is doing trying to match, trying to say that 'solve must take a certain kind of subject matter and object and so forth'. What this is doing is expressing knowledge that oughtn't really be linked to linguistics at all, but ought to be in the object domain, ought to be how you can tie objects together, not how you can tie the words together, and this I think is the crux of the whole paper.

Overheu: As I understood from Bobrow's talk on Natural Language Interaction Systems, they found in the TLC a greater need for looking at deeper structure anyway.

Clowes: What that means is that the notion form test as an expression of a mapping between an object complex and an English string, those form tests are much too simple. The relationship between an object structure and a way of expressing it in English is clearly much more complete, and I think that's where we will, as it were, hang on to certain aspects of the Chomsky model.

Narasimhan: I would like to add a few comments. It seems to me that the form tests do use grammatical notions. You still have to give some explication about how you are going to arrive at these form tests, and then it seems to me a much more difficult problem; does the form test only take you one way, from the given string - how to go inside? It does not tell you how to come out and produce a string. It seems to me that is the crucial problem and the linguists have never solved it. I don't think the apparatus that they have would ever take them there.

Clowes: I think that's a fact. The linguist has drawn the line on talking about what the philosopher calls ontology. I mean this notion of objects having a structure and this domain that TLC uses, this I think is the ontological structure of reality. Now the linguist has, I think for a very good reason, said "well that's just not part of my province". I think he is mistaken, but nevertheless that is what I think he has said.

Narasimhan: Yes, but why spend about 27 pages in talking about what the linguists are doing?

Clowes: Because I do think that when it comes to expressing this idea of form test, in particular to actually enriching the idea of form tests that Bobrow presented, and indicated he would have to enrich, then this is where we will find the answers we want. I think the notion that we haven't discussed very much here, grammatical relationships, is in fact one of the places where we will find the relevant powerful explication of form tests.

Overheu: What you are saying is that we ought to take a more existentional view in the semantics and pragmatics of this type of problem, rather than a more classical method of approach.

Clowes: Narasimhan drew my attention to a book which we have currently seized on; it has a particular message in it. The book is "Ontology and the Logistic Analysis of Language" - its a book on Philosophy. It has a key phrase present: the Relation of of Representation. I think that is exactly what we are talking about: the form test expressed by the book - the relation of representation. The idea that you can represent something in English. A good deal of what happens in things like STUDENT and so forth is really concerned with how you can represent something say in mathematics, in an algebraic expression, and how you can represent the same thing in English. Bobrow, in STUDENT, tries to get between those two representations, the English representation and the Algebraic representation, without ever getting to the object domain, you know, like ladders and boats, etc., and I think that it is possible to get a tremendous amount of mileage out of this particular idea.

Simmons: How do you handle a sentence such as "Time flies like an arrow" where each of the first three words could be the verb? How do you sort out which is the most likely verb, or which is the most likely word to be the verb, and how are you going to view it in the complete sentence?

Zatorski: This word obviously has a number of analyses - the sentence has a number of deep structures. It is as about as much as you can say.

Simmons: How would you choose which one you are going to select?

Zatorski: This is not an aspect of your knowledge of language, 'how would you choose it?'. The point is that a number of structures could be assigned to such a sentence, and they would all be grammatical. The choice itself is beyond the system.

Simmons: So you are just concerned with the actual analysis of the sentence rather than trying to find its meaning?

Zatorski: You are concerned with knowing which of these sentences are unambiguous, and if you have three on which you can build on a given deep structure, then you have three unambiguous sentences, that's what you have, and the question of selection of these for any specific purpose in the given context, in the given conversational situation is what we consider a matter of performance. Its a matter of actual specific usage at a given time. Your knowledge of language enables you to assign a deep structure to that.

Kovarik: I wonder if you could clarify if the study is addressed to some semi artificial computational language, or if it attempts to cope with the living natural language. I would say that the living natural language is at times supporting co-existence for conflicting rules. Would you consider that there are contradictions, that the living language hasn't really got relational structure completely?

Zatorski: No what baffles me is what you said is a notion of conflicting rules. You mean perhaps that there are rules and that there are exceptions from rules, in the way in a traditional grammar, in the way we learnt traditional grammar we are loaded with exceptions from rules, but that doesn't mean that there is any conflict between rules. That simply means that our notion of rules for that particular area is not sufficiently well specified. There is basically no conflict between rules.

Narasimhan: Could you also try to counter that by saying that you have a different set of rules for Bible English, say, would that be acceptable?

Zatorski: No.

Narasimhan: It seems to me to be plausible that one does in fact function in that manner. It seems to me that you talk about behavioural repertoire for example, then for classes of situations, I could operate on the basis of one system of rules and in another class of situations I could in fact operate in terms of another system of rules. An extremely simple obvious example is the way one functions with different kinds of people, or in different situations, say in the classroom or on the football field.

Zatorski: I am baffled by your notion of how we use the language. We only have this one language to use.

Narasimhan: That's essentially what I'm questioning.

Zatorski: The point is that we may be using different vocabulary in given situations, but the rules by which your sentences are assembled, are identical in all situations, and in any case the model itself postulates an ideal, is built in terms of an ideal speaker who handles all these differences of vocabularies that we may be employing in any particular situation. The construction of your sentence does not differ from one area to another.

Narasimhan: I don't know if that's true.

Zatorski: Perhaps psychologists approve of this, but Chomsky's model predictates an ideal speaker with an ideal knowledge of the language.

Overheu: (to Dr M. B. Clowes) I seem to recall having a discussion some time ago in which you assured me that all relationships between objects could be explained on the basis of a syntax alone. Do I take it now that your view is that one has to go into a higher level of semantics to achieve advancement?

Clowes: No. What we are saying here, and I say in a few cases this has happened, is that what people have talked about as being semantics, will have to be given a structural characterization, and that structural characterization will use a language, and will also as a matter of fact express the syntax of, say, English or of graphics or pictures or what have you. In fact the sort of arguments that we are putting up in the discussion (Notes (26), (27) of the paper) is that there is almost certainly a universal syntactic notation which Narasimhan has also talked about, namely that we talk about objects, attributes and relations, and within that notation you can express the idea of structure, the structure of language and what have you. One of the open questions is whether you can also express what I call the relations of representation, that's what I talk about in my paper on Picture Syntax.

PICTURE SYNTAX

M. B. CLOWES

Division of Computing Research
C.S.I.R.O., Canberra

1. INTRODUCTION

From a number of different points of view of the characterisation of a complex behaviour, be it of man or machine, there has emerged the idea that representation is a crucial component of that behaviour (Newell 1965, Newell & Ernst 1965, Amarel 1968). When we represent a pattern as a point in an n-dimensional property space or a game in terms of a moves tree, we are setting up a representation which may well be the only view of the problem that a computer program that is designed to 'solve' that problem will have. One of the central issues in this representational view is that of deciding what it is about the problem or task which is to be represented. Among the most forceful answers to this question is that given in the context of linguistic behaviour by Chomsky (see Clowes, Langridge & Zatorski, 1969). We might briefly illustrate his viewpoint by considering what it is about the second sentence of this paper which would need to be represented in order to paraphrase it into a more compact form. That is, what representation of the sentence would a program need in order to accomplish this task? We might for example paraphrase the second half of the sentence (beginning "we are setting up . . . ") as "we are setting up a representation which may well be a computer program's only view of the problem it is designed to solve". The principal change here has been the introduction of the possessive particle "'s" to replace a construction involving "has". That is, we have re-expressed the fact that 'a computer program will have a view'. We might say then that in tidying up this sentence, we had to understand it, and understanding it certainly involves the realisation that the concluding Verb Phrase "will have" relates "program" and "view". It is this kind of information about sentences which Chomsky is primarily concerned to represent and he does so by showing how a complex sentence may be diagrammed as a set of simple sentences related one to another in a manner which indicates how these simple sentences qualify nouns, verbs, etc. This diagram - the base phrase marker - he identifies as the underlying structure of the sentence. The correspondence between this underlying structure and the linear arrangement of words in a conventional sentence is stated in terms of transformational rules.

We might summarise the above as indicating that one aspect of complex objects (or problems, tasks) that might be represented is their articulation into simpler objects. We are thus representing something about the structure of complex objects.

There seems no doubt that aspects of the structure of sentences is crucial to their understanding, and it is inconceivable that an adequate program able to converse in English would not need to grasp a good deal of the structure of the sentences input to it. It is also clear that this program would need to have a representation of what is being talked about, (see for example Barter 1969).

In the field of Picture Interpretation, specifically that area of it known as Pattern Recognition, the problem of representation has received relatively little formal study, the emphasis being upon the characterisation of decision.

In general the decision models employed deal with the assignment of patterns to classes and require that the pattern be described in terms of the values of a defined set of measures or parameters. For the purposes of such decision makers, the whole pattern is characterised by these measures (thus it says nothing about the problem of, say, segmenting text into characters) and measures differ only in their reliability or information content (Kamentsky & Liu, 1963).

Picture Interpretation abounds with tasks which cannot be tackled by representing the picture as values of a set of independent parameters. Many of these tasks have arisen in the area of graphical communication with computers (Stanton, 1969), e.g. the recovery of an electrical circuit from its diagram. In general these tasks require the identification of relationships between pictorial entities. That is they require an articulation of the structure of the pattern. Thus a text reading machine, prior to the identification of characters must articulate a page of text into lines - and 'feed' these lines to the next stage in the order given by their pictorial (or spatial) relationships. (Note that a longhand manuscript frequently contains insertions which seriously complicate this relation between serial ordering and pictorial relationship.) Each line must be further articulated into characters, provision being made for the distinction between successive characters (as in a word) and super- and sub-scripts.

Where the text contains mathematical expressions which must be understood (as in compiling a program from its flowchart) the relations to be recovered may involve groups of characters (sub-expressions) as well as individual characters.

In commercial text readers this process of articulation is expressed in part by paper-handling mechanisms and is drastically simplified by minimising the variety of permitted pictorial relationships and specifying them as more or less precise spatial relationships. Thus characters must be aligned vertically to within some specified tolerance, (given in inches) and separated horizontally with a corresponding precision. In the absence of precise specifications of this type, the computation of relations such as alongside, above, etc. may be complex and subtle (see Anderson, 1968).

We shall assume that the notion of articulation outlined above is typical of that encountered in Picture Interpretation. It resembles the concept of linguistic structure in two ways. Firstly it requires that the pattern, like the sentence, be thought of in terms of pictorial entities such as lines and characters which may be compared to phrases and words. Secondly that a relationship 'followed by' is essential to the articulation of linguistic structure : we have seen that there appear to be several relationships ('above', 'alongside', etc.) in the case of pictorial structure. Over a period of years several workers have pursued this analogy as a basis for the formalisation of Picture Interpretation (for a review see Clowes, 1968a). The purpose of this paper is to outline some tentative conclusions which have emerged in our 'pursuit' of this analogy. As a vehicle for this exposition we will attempt the characterisation of letter recognition as a problem in articulation.

2. CYCLOPS-1 AND LETER : TWO APPROACHES TO THE ARTICULATION OF CHARACTERS

2.1. CYCLOPS-1 is a program (Marill et al., 1963) which recognizes hand-drawn characters from the set of capital letters (A-Z) and numerals (0-9). Characters are drawn with a light pen. When drawing, a program records the location of the tip of the pen every 20 millisec. Thus a picture or 'scene' which may contain several overlapping characters is specified as a table of (x, y) pairs. Interpretation of the picture is carried out in two steps (1) formation of line segments, (2) identification of groups of line segments as characters.

2.1.1. Formation of Line Segments

The table is searched to find triplets of points such that one of the points is between the other two, near to the other two and all three are approximately collinear. Pairs of triplets are chained to form segments where they have two points in common and triplets are also chained with segments to extend segments. In certain cases, e.g. intersections of lines, chaining may break down : segments are therefore examined in the vicinity of their endpoints to see whether pairs of segments might if extended form a single segment.

2.1.2. Identification

The recognition system has definitions of characters which are treated as hypotheses to be tested on the line segments recovered from the picture. An hypothesis is a series of tests which must be satisfied if the segments are to be identified with the character which the hypothesis represents. Hypotheses are tested in a fixed predetermined order. Cyclops-1 has forty-odd segment and pattern characteristics in terms of which hypotheses are stated. They include:

1 (i) <u>Predominance coefficient</u> of a line segment in the pattern (arc length of segment/total arc length of all segments).

(ii) <u>Straightness coefficient</u> of a segment. (Distance between end points of segment/arc length of segment.)

(iii) <u>Orientation</u> of the segment. (The angle between the x-axis and the straight line connecting the ends of the segment.)

(iv) <u>Intersections</u> between segments including a segment and itself.

(v) <u>Inflection points</u> of a segment.

Thus an hypothesis descriptive of an 'A' would take the form:

2 (a) An approximately straight segment (α) slanting up to the left.

(b) An approximately straight segment (β) slanting up to the right.

(c) α is nearly the same length as β, and the topmost points of α and β are close together.

(d) An approximately straight and horizontal segment (γ).

(e) The end points of γ are reasonably close to the midpoints of α and β.

Cyclops-1 also had the capacity to evaluate super-hypotheses which catered for the presence of other (possibly overlapping) characters in the picture.

2.2. LETER is the initial symbol of a 'generative grammar' for the capital letters A-Z due to Narasimhan (Narasimhan, 1966; Narasimhan & Reddy, 1967). Each letter is described as a composition of 'phrases' or 'words'. A 'fragment' of this, describing some varieties of 'A' is reproduced below:

3 (i) LETER → A | B | C | .... | Z

(ii) A(1) → INVE.h(11,23;;2)

(iii) INVE(1,2) → r.1 (11;2;2) | r.v(11;2;2) | v.1(11;2;2)

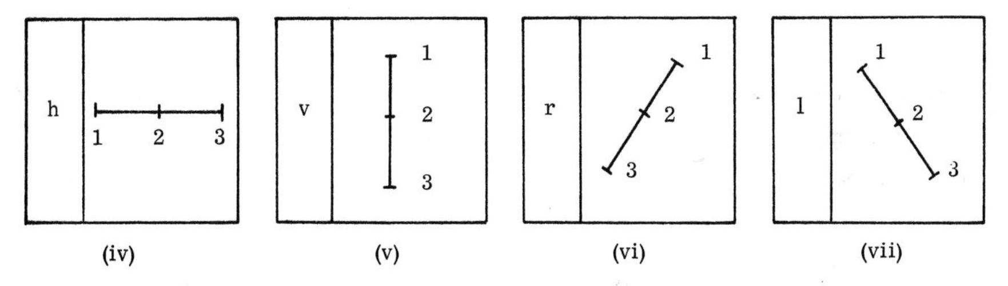

The 'words' of the grammar (e.g. v, r, l, h) are exhibited pictorially as lines having distinguished vertices labelled 1, 2, 3. INVE is a phrase having two vertices (1, 2); it has three structural compositions, the first of which is r and l spatially juxtaposed so that vertex 1 of r is coincident with vertex 1 of l. This version of INVE may be exhibited pictorially as:-

4

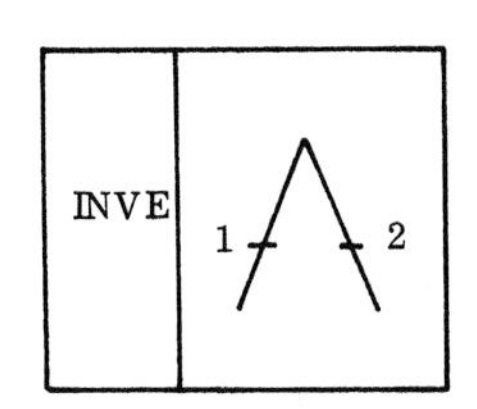

A is composed of INVE and h juxtaposed so as to make vertex 1 of INVE coincident with vertex 1 of h, and vertex 2 of INVE coincident with vertex 3 of h. To facilitate our subsequent exposition of an implementation of LETER a vertex (1) has been assigned to A, identified with vertex 2 of h. Associated with each word name (h, v, etc.) is a pair of attributes : Length and Thickness which admit of a range of values. Attributes are utilised in the computer implementation of this grammar but were omitted from its formal specification "for ease of presentation" (Narasimhan & Reddy, 1968).

In the implementation of this grammar, routines could create strokes of varying length and thickness in response to specified attribute values. The input to the program is the letter name, e.g. A, the page coordinates of the vertex of the letter, e.g. vertex 1 of A, and the required attributes. The program uses the grammar to generate the phrase INVE and the word h, assigning the coordinates of vertex 1 of A to vertex 2 of h as specified by the rule 3(ii). The value of the length attribute for A determines the version of h to be used and accordingly determines the coordinates of vertices 1 and 3 of h. The relation between INVE and h specified in 3(ii) assigns these coordinates to vertices 1, 2 of INVE. This assignment of coordinates is now repeated for rule 3(iii) thus determining the positions of all three 'words' r, l and h.

Thus the grammar provides a scheme for generating characters in a way which articulates them into parts, these parts being either phrases or 'words'. Narasimhan remarks that "the immediate relevance of such descriptions to recognition of patterns should be obvious". What is implied here is that given a description of a scene in terms of the word names of the grammar, the latter could be used 'in reverse' to assign phrase names, e.g. 'A' to groups of names (of words and/or phrases) provided that they are (spatially) juxtaposed according to the rules for 'A'. This is the so-called syntax-directed approach to picture interpretation.

2.3. 'Words', Vertices and Coincidence Computed by CYCLOPS

It will be obvious that Narasimhan and Marill et al. share a single view of the composition of an 'A'. It is of course a view held by all competent writers of the Roman alphabet; its acquisition is an essential pre-requisite to learning that alphabet. Marill's straight segments α, β, γ may be identified with the 'words' l, r, h and their end and mid points with the vertices of those words. The formation of line segments and their subsequent categorisation according to values of their attributes straightness and orientation by a CYCLOPS-like program could be likened to the formation of a description of the scene in terms of the word names l, r, h, etc. CYCLOPS' hypotheses are then an expression of the generative descriptions of LETER. Indeed if we wished to design a recognition program motivated by a grammar such as LETER there seems little doubt that much of the computation carried out by CYCLOPS would be an essential ingredient. There are however many things left unsaid in LETER which are crucial to a successful CYCLOPS. We will single out two: finding 'words' and their vertices, and the computation of vertex coincidences.

(i) Finding 'words' and their vertices. Given the power of the chaining scheme in CYCLOPS (2.1.1 above) to form segments which are not uniformly composed of local triplets, the line-forming program analysing a picture containing an 'A' would not necessarily find line segments which could be categorised as the 'words' l and r. Where the two diagonal line segments were touching, a single line segment would result. It is in the identification phase that CYCLOPS breaks up this complex line segment by looking for changes of direction.* It is only at this stage that the 'words' l and r could be identified. Identification of vertices 1 and 3 of these words is implicit in CYCLOPS' computation of straightness 1(ii) and orientation 1(iii). They are the 'end points' of the segments. Similarly the mid point identified in 2(e) may be equated to vertex 2 of these 'words'. Thus the picture analyses leading to the identification of words and their vertices may be quite complex. This complexity is not apparent in LETER nor in the 'stroke generators' of the computer implementation of LETER.

(ii) Computation of vertex coincidence. Where two 'words' have been found by breaking a line segment, CYCLOPS has in some sense 'found' a coincidence of vertices. Generally however this will not be the case and CYCLOPS' definition 2(c) for an A requiring that the two "topmost points" be "close together" shows one actual form which coinc takes in CYCLOPS. Implicit in this calculation of "close together" is a notion of a scale of judgment. While this scale might be provided in advance as a constant, the program will be able to cope with size variations only by using some attribute of the picture itself, e.g. segment length to provide a scaling factor. This contextual view of scale and size is implicit in the definitions given above of the 'Predominance' and 'Straightness' coefficients. These facets of the formal notion 'coincidence' in LETER are not exposed or defined.

* This statement is contentious, since it is not clear as to what CYCLOPS' tolerance on 'approximately collinear' is. The remark does however apply to light pen environments where each SEG is input as an ordered string of positions, which is the situation treated in § 3.2.

CYCLOPS and LETER are in a sense complementary. The former is a rich but clumsy procedure for picture analysis whose description lacks the succinctness and formalism necessary to any generalisation of it. LETER provides an intuitively plausible and superficially precise account of the same pictures to which CYCLOPS applies, but fails to specify the physical realities of those pictures. Ideally we want a model which has the virtues of both and the defects of neither: it is the formulation of such a model which is the principal objective of this paper.

3. TOWARDS A THEORY OF ALPHABETIC CHARACTERS

The central task for a theory of alphabetic characters is to show how a scheme like LETER can serve as an abstraction for the complex 'reality' manifest in the machinations of CYCLOPS and of other comparable recognition schemes, e.g. Grimsdale et al. (1959). In § 2.3 we discussed the converse of this in indicating how abstractions like 'word' in LETER require the formulation of complex picture analyses procedures if LETER is to be used as a recogniser. More generally of course we would like to know if and how LETER could act as an abstraction for pictorial data gathered from 'hard copy' by a scanner. We can summarise the account presented in § 2.3 and generalise it as follows:

A. 'Words' (e.g. l, r) are entities realised in the CYCLOPS (light pen input) environment as SEGS. Where the recording medium is hard copy, to which a recognition program gains access via a scanner, strokes are realised as uniformly coloured regions of a characteristic shape.

B. Vertices are realised as structural attributes of SEGS (and more generally of regions). These pictorial attributes are pieces of a segment or region.

C. Coinc(idence) of vertices is an abstract relationship which may be exhibited in several ways:

   (i) As a near relationship between pieces.

   (ii) As a local angular segment (a structural attribute) of a SEG - a KNEE.

   (iii) As an intersection - which we will take to be a complex structural attribute of a scene.

The theory must indicate the nature of these realisations of abstract objects, attributes and relations in a formal manner. One consequence of being able to do so would be, in principle, the capacity to translate automatically descriptions of characters written in the abstract language, e.g. LETER, into the picture description language in which computations are expressed, e.g. CYCLOPS. At this stage we can do little more than sketch out the form that this might take.

Both CYCLOPS and LETER articulate the picture. To do so they utilise objects, attributes and relations.

An object is an entity whose parts (which are also objects) are related in a specified manner. Thus in CYCLOPS, a point (PT) is a primitive object and TRIPLE is an object formed (recovered) from three of these primitive objects.

CYCLOPS

5 $\quad \text{TRIPLE}\left\langle \text{PT}_1, \text{PT}_2, \text{PT}_3 \right\rangle \left[\underline{\text{near}}\langle 1,2\rangle, \underline{\text{near}}\langle 2,3\rangle, \underline{\text{collinear}}\langle 1,2,3\rangle, \underline{\text{between}}\langle 2,1,3\rangle \right]$

5 illustrates the notation we will use to define objects. There is an object name - TRIPLE - in capital letters. The parts of this object are listed within the angle brackets, and the relations on these parts necessary if they are to form the object appear as lower case underlined names within the square brackets. A relation is a predicate having two or more arguments. Its value is true or false. In some cases it is convenient to treat a relation as having a value other than true/false. In such cases its value will be indicated thus

6 $\quad \underline{\text{relation}}\langle \text{OBJ}, \text{OBJ}\rangle \left[\text{value}\right]$

In general relations are computable, thus the value of $\underline{\text{near}}\langle \text{A}, \text{B}\rangle$ is determined by comparing the distance between A and B, with some criterion. If this comparison discloses some specified relationship, e.g. that the distance is less than the criterion, $\underline{\text{near}}\langle \text{A}, \text{B}\rangle$ is true. Distance may be regarded as a relation between a pair of points in which case distance is a relation having an integer (say) value. Alternatively we can regard it as an attribute of the pair A, B. We shall regard the position of a point (its x, y coordinates) as an attribute of that point. Attributes are written in the same form as relations but without underlining of the name. Thus

7 $\quad \text{position}\langle \text{PT}\rangle \left[\text{x}, \text{y}\right]$

Distance takes the usual Euclidean form

$$8 \qquad \underline{\text{distance}} \langle \underset{1}{\text{PT}}, \underset{2}{\text{PT}} \rangle \underset{3}{[\text{d}]} \subset \text{position} \langle 1 \rangle [\text{h, k}], \text{position} \langle 2 \rangle [\text{l, m}], \underline{\text{Eq}} \langle 3, \sqrt{(\text{h-l})^2 + (\text{k-m})^2} \rangle$$

and

$$9 \qquad \underline{\text{near}} \langle \underset{1}{\text{PT}}, \underset{2}{\text{PT}} \rangle \subset \underline{\text{less than}} \langle \underline{\text{distance}} \langle 1, 2 \rangle, \text{criterion} \rangle$$

The form taken by 8 and 9 shows them to be computable functions. The right hand side specifies the calculation to be effected to assign a value to the predicate (or relation) specified on the left hand side. CYCLOPS has these definitions of near and similarly of between and collinear 'built in' - probably as subroutines.

The definition of a line segment in CYCLOPS takes two forms:

$$10 \qquad \text{SEG} \langle \underset{1}{\text{TRIPLE}}, \underset{2}{\text{TRIPLE}} \rangle [\underline{\text{overlap}} \langle 1, 2 \rangle]$$

where overlap can be defined:

$$11 \qquad \underline{\text{overlap}} \langle \underset{1}{\text{TRIPLE}}, \underset{2}{\text{TRIPLE}} \rangle \subset 1 \langle \underset{3}{\text{PT}}, \underset{4}{\text{PT}}, \underset{5}{\text{PT}} \rangle, 2 \langle \underset{6}{\text{PT}}, \underset{7}{\text{PT}}, \underset{8}{\text{PT}} \rangle, \underline{\text{Eq}} \langle \text{position} \langle 4 \rangle, \text{position} \langle 6 \rangle \rangle, \underline{\text{Eq}} \langle \text{position} \langle 5 \rangle, \text{position} \langle 7 \rangle \rangle$$

where by convention we omit on the right hand side the defining relationships of the TRIPLEs 1, 2 for brevity, and Eq applies to the values of the attributes.

The second version of SEG is recursive in form:

$$12 \qquad \text{SEG} \langle \underset{1}{\text{SEG}}, \underset{2}{\text{TRIPLE}} \rangle [\underline{\text{overlap}} \langle \text{end} \langle 1 \rangle [\text{TRIPLE}], 2 \rangle]$$

Here we see a structural attribute end ⟨SEG⟩ whose value is TRIPLE. Regarding a SEG as comprising a set of ordered TRIPLEs we may define end ⟨ SEG⟩ thus

$$13 \qquad \text{end} \langle \underset{1}{\text{SEG}} \rangle \underset{2}{[\text{TRIPLE}]} \subset 1 \langle 2, \text{TRIPLE --- TRIPLE} \rangle, \text{ or } 1 \langle \text{TRIPLE, --- TRIPLE}, 2 \rangle$$

A structural attribute is one whose definition (as in 13) requires both a specification of the structure of the object it is an attribute of and is itself a structured object. This notion of structural attribute is a crucial feature of a picture articulation.

The definitions 5, 10, 11 and 12 are a formalisation of the characterisation of segments given in 2.1.1 above. We can now see that the performance of CYCLOPS involves the formation of objects (line segments) subject to the relationships between the parts of those objects being of specified varieties. It also computes attributes both structural, e.g. topmost point and non-structural, e.g. arc length. An hypothesis in CYCLOPS is a procedure establishing that objects of specified type (SEGS) and having attribute values within certain ranges, enjoy some set of relations. Thus the CYCLOPS definition of 'Λ' (2 a, b, c) computes the attributes straightness, orientation and length, and *relations* involving length and position.

LETER

By contrast, LETER describes the same organisation abstractly as

14 $\quad \text{INVE}\ \langle \underset{1}{l},\ \underset{2}{r} \rangle \left[\underline{\text{coinc}}\ \langle \text{vertex1}\ \langle 1 \rangle,\ \text{vertex1}\ \langle 2 \rangle \rangle \right]$

where vertex1 $\langle 1 \rangle$ is exhibited by the picture 3(vii). The 'words' of LETER (e.g. h, v, l, r) are geometrical lines, the vertices are positions *on* these lines. The vertex labelled 2 in all these words is in fact a position midway along the line. The notion line - specifically straight line - is of course the abstract entity presumed in geometry and manipulated algebraically in coordinate geometry. Thus the algebraic notion 'equation of a line' is a representation of this entity in a form which permits us to compute positions lying on the line. It characterises positions not points; points are *pictorial expressions* of position.* We can treat position (P) as an object whose representation is a pair of numbers.

15 $\quad \text{P}\ \langle \text{integer, integer} \rangle$

Since position is always relative to an origin, these numbers are in fact the value of a relation. In Clowes (1968b) a formal characterisation of this notion of relative position as a relation (*Relpos*) was given. The definition took the form

16 $\quad \underline{\text{Relpos}}\ \langle \text{AXIS, P} \rangle \left[\text{integer, integer} \right]$

* Conventional usage equates point with dot: we would treat dot as an abstract object whose representation is a region of characteristic size and shape.

Its principal merit lies in the capacity it gives to reference AXIS, an entity only indirectly characterised by the Cartesian notation. We shall regard a LINE as a pair of positions related to the same AXIS. This pair enters into the algebraic formulation "equation of a line".

$$17 \qquad \text{LINE}\left\langle \underset{1}{P}, \underset{2}{P}\right\rangle \left[\underline{\text{Relpos}}\left\langle \underset{3}{\text{AXIS}}, 1\right\rangle [h,k], \underline{\text{Relpos}}\langle 3,2\rangle [m,n]\right]$$

For the most part it will be convenient to abbreviate this to

$$18 \qquad \text{LINE}\langle P, P\rangle$$

The end of a LINE is an attribute of type P:

$$19 \qquad \text{end}\left\langle \underset{1}{\text{LINE}}\right\rangle \left[\underset{2}{P}\right] \supset 1\langle 2, P\rangle \quad \text{or} \quad 1\langle P, 2\rangle$$

The only other vertex employed in LETER is of course the median position (medpos):

$$20 \qquad \text{medpos}\left\langle \underset{1}{\text{LINE}}\right\rangle \left[\underset{2}{P}\right] \supset 1\left\langle \underset{3}{P}, \underset{4}{P}\right\rangle, \underline{\text{on}}\langle 2, 1\rangle, \underline{\text{Eq}}\left\langle \underline{\text{distance}}\langle 3, 2\rangle, \underline{\text{distance}}\langle 2, 4\rangle\right\rangle$$

where the relation on is given the conventional algebraic definition in terms of the equation of a line on 3, 4. LETER also deploys other 'words' including arcs. The definition of on would differ according to whether it is

$$\underline{\text{on}}\langle \text{LINE}, P\rangle \quad \text{or} \quad \underline{\text{on}}\langle \text{ARC}, P\rangle$$

We can now reformulate 14 in these terms:

$$21 \qquad \text{INVE}\left\langle \underset{1}{\text{LINE}}, \underset{2}{\text{LINE}}\right\rangle \Big[\underline{\text{Coinc}}\left\langle \text{end}\langle 1\rangle \left[\underset{3}{P}\right], \text{end}\langle 2\rangle \left[\underset{4}{P}\right]\right\rangle,$$
$$\underline{\text{above}}\left\langle 3, \text{end}\langle 1\rangle \left[\underset{5}{P}\right]\right\rangle, \underline{\text{above}}\left\langle 4, \text{end}\langle 2\rangle \left[\underset{6}{P}\right]\right\rangle,$$
$$\underline{\text{Eq}}\left\langle \underline{\text{distance}}\langle 3, 5\rangle, \underline{\text{distance}}\langle 4, 6\rangle\right\rangle\Big]$$

and the relationship above might take the form

$$22 \qquad \underline{\text{above}}\left\langle \underset{1}{P}, \underset{2}{P}\right\rangle \supset 1\left\langle \underset{3}{x}, \underset{4}{y}\right\rangle, 2\left\langle \underset{5}{x}, \underset{6}{y}\right\rangle, \underline{\text{gt}}\langle 4, 6\rangle$$

where gt is the arithmetic relation 'greater than'.

Implicitly above makes reference to an origin of coordinates since as we remarked earlier, the x, y representation of a position is a shorthand for a relation

(Relpos) between the position and an origin of coordinates.

The formulation 21 of INVE does not of course completely describe the geometry of this object. Implicit in the expression of INVE in LETER (3(iii)) is something about the relative lengths and orientations of these LINES. We might add appropriate relations on the two LINES which would begin to capture this. In fact of course some considerable latitude in the precise numerical ratios of these attributes is more appropriate. In Clowes (1968b) it was shown how a relational definition of position Relpos readily permits an expression of the limited accuracy of judgments of relative position. This limitation ~ 10% in the judgment of linear position and 20% in the judgment of position in two dimensions is a consistent feature of our ability to make perceptual judgments (Miller, 1956). It implies that judgments' of length and distance will be similarly coarse and that this is perhaps the origin of the requirement for 'latitude' in expressing these additional 'shape' relationships in 21.

The occurrence of above and distance in this geometrical domain brings out the close (and confusing) similarity between this domain and the picture domain. Indeed the difference lies only in the existence in the geometrical domain of the object LINE its structural position attributes and the relationship coinc.

Pictures provide the medium in which these abstractions may be expressed in a communicable form. It is this notion of expression which we can now attempt to characterise.

3.1. Pictorial Expression as a Mapping Between Domains

Undoubtedly the most interesting variety of expression is that which requires us to associate these geometrical abstractions with picture data obtained by scanning some 'hard copy' image. We shall not attempt a formulation of that problem here but confine ourselves to the treatment of the simpler picture data provided by the capacity to track the locus of a pen. This is the CYCLOPS' environment. Here a LINE is expressed as a straight segment. We can continue to use the formalism adopted for the specification of relations by treating 'expression' as relational and designating it by map

$$23 \qquad \underline{\text{map}} \left\langle \underset{1}{\text{LINE}}, \underset{2}{\text{SEG}} \right\rangle \supset \underline{\text{greater than}} \left\langle \text{straightness} \langle 2 \rangle, \text{criterion} \right\rangle$$

We would give straightness a definition comparable with that used in CYCLOPS and we will regard 'criterion' as an attribute of the picture whose value is a constant. For hard copy SEG is REGION and $\underline{\text{gt}} \left\langle \text{straightness} \left\langle 2 \right\rangle, \text{criterion} \right\rangle$ is replaced by a complex of relations specified on boundary $\langle \text{REGION} \rangle [\ ]$

The structural attributes of LINE are mapped into structural attributes of SEG:

$$24 \qquad \underline{\text{map}} \left\langle \text{end} \underset{1}{\langle \text{LINE} \rangle} \underset{2}{[\text{P}]}, \text{end piece} \underset{3}{\langle \text{SEG} \rangle} \underset{4}{[\{\text{PT}\}]} \right\rangle \subset$$

$$\underline{\text{map}} \langle 1, 3 \rangle, \underline{\in} \underset{5}{\langle \text{PT}, 4 \rangle}, \text{position} \langle 5 \rangle [2]$$

The braces '{ }' denote a set. In general, a piece $\langle \text{SEG} \rangle$ is some SEG-like structural attribute whose length is some fraction, say 1/5, of the total length of a SEG, i.e.

$$25 \qquad \text{piece} \underset{1}{\langle \text{SEG} \rangle} \underset{2}{[\{\text{PT}\}]} \subset \underset{3}{\text{SEG}} \langle \underset{4}{\{\text{PT}\}} \rangle, \underline{\in} \langle 4, 2 \rangle, \underline{\text{Eq}} \left\langle \text{length} \langle 3 \rangle, 1/5 \times \text{length} \langle 1 \rangle \right\rangle$$

The special case of piece $\langle \text{SEG} \rangle$ namely endpiece$\langle \text{SEG} \rangle$ has the additional constraint that it contain end$\langle \text{SEG} \rangle$ as defined by 13.

Thus 23, identifies a LINE with a straight SEG and 24 an end position on a LINE with the position of some point in a designated piece of the SEG, provided that the LINE and SEG 'already' enjoy a map relationship.

The third variety of pictorial expression (C) can now be indicated. 23 and 24 provide for the pictorial expression of the entities LINE, and upper end appearing in a geometrical expression such as 21. The abstract relationship Coinc between positions on a pair of LINEs, i.e. end $\langle \text{LINE} \rangle$ may be expressed as the pictorial relationship near between the relevant (pair of) pieces of the two SEGS, i.e.

$$26 \qquad \underline{\text{map}} \left\langle \underline{\text{Coinc}} \left\langle \underset{1}{\text{end}} \underset{2}{\langle \text{LINE} \rangle}, \underset{3}{\text{end}} \underset{4}{\langle \text{LINE} \rangle} \right\rangle, \underline{\text{near}} \left\langle \underset{5}{\text{piece}} \underset{6}{\langle \text{SEG} \rangle}, \right.\right.$$

$$\left.\left.\underset{7}{\text{piece}} \underset{8}{\langle \text{SEG} \rangle} \right\rangle\right\rangle \subset \underline{\text{map}} \langle 2, 6 \rangle, \underline{\text{map}} \langle 4, 8 \rangle, \underline{\text{map}} \langle 1, 5 \rangle, \underline{\text{map}} \langle 3, 7 \rangle$$

The pictorial relation near may be given a definition

$$27 \qquad \underline{\text{near}} \left\langle \text{piece} \langle \text{SEG} \rangle \underset{1}{[\{\text{PT}\}]}, \text{piece} \langle \text{SEG} \rangle \underset{2}{[\{\text{PT}\}]} \right\rangle \subset$$

$$\text{mindist} \langle 1, 2 \rangle \left[\underline{\text{distance}} \langle \text{PT}, \text{PT} \rangle \underset{3}{[\text{N}]} \right], \underline{\text{lt}} \langle 3, \text{threshold} \rangle$$

We can regard 27 as a procedure which computes distance between all pairs of POINTS in the two SEG's and selects the minimum distance. (i.e. we regard mindist as an attribute of this set of distance measurements.) If this minimum is less than some threshold then the pieces are *near*. Unlike the 'criterion' in 23, threshold might be simply the length of one or other SEG in 27, rather than some attribute of the picture as a whole. Notice that there is no requirement to map any of the geometrical relationships as might be included in a complete description of INVE (21) nor in general anything else there which characterises the *shape* of INVE. That is the shape of the character is an *abstract* geometrical notion since it can be computed entirely within that domain. The immense simplification of representation effected in the mapping from the picture domain into the geometrical domain makes the expression of shape much simpler. Of course we have to compute shape of SEG's in the picture domain in order to decide how to map these SEG's, e.g. whether they should be LINEs or ARCs and so on.

There are two other ways of mapping *Coinc* informally indicated in C above, the manifestation of *coinc* as a KNEE may be characterised crudely as:

$$28 \qquad \text{map}\left\langle \underset{1}{\text{LINE}},\ \underset{2}{\text{fragment}} \langle \text{SEG} \rangle \left[\{\text{PT}\}\right] \right\rangle \subset \underline{\text{gt}} \left\langle \text{straightness} \langle 2 \rangle,\ \text{criterion} \right\rangle$$

i.e. we do not as in 23 express a LINE as a separate SEG but as a fragment of a SEG. (Fragment is defined like piece but could be up to say 4/5 of the length of a SEG.) A pair of LINEs which have been mapped in this way, and which enjoy a *coinc* relationship may be mapped as a knee which we take to be a piece of a SEG having a characteristic shape:

$$29 \qquad \underline{\text{map}} \left\langle \underline{\text{coinc}} \left\langle \text{end} \left\langle \underset{1}{\text{LINE}} \right\rangle [\text{P}],\ \text{end} \left\langle \underset{2}{\text{LINE}} \right\rangle [\text{P}] \right\rangle,\ \text{knee} \left\langle \underset{3}{\text{SEG}} \right\rangle \left[\underset{4}{\{\text{PT}\}}\right] \right\rangle \subset \underline{\text{map}} \left\langle 1,\ \underset{5}{\text{piece}} \langle 3 \rangle \right\rangle,\ \underline{\text{map}} \left\langle 2,\ \underset{6}{\text{piece}} \langle 3 \rangle \right\rangle \ \underline{\in} \langle 5, 4 \rangle,\ \underline{\in} \langle 6, 3 \rangle$$

and

$$30 \qquad \text{knee} \left\langle \underset{1}{\text{SEG}} \right\rangle \left[\underset{2}{\{\text{PT}\}}\right] \subset \underset{3}{\text{piece}} \langle \text{SEG} \rangle [2],\ \underline{\text{lt}} \left\langle \text{straightness} \langle 3 \rangle,\ \text{criterion} \right\rangle$$

This characterisation is 'crude' in its description of the shape of a knee, i.e. as simply deviating from straightness. Of course, it is the fact that this deviation is localised within the piece which is characteristic of a knee. (It is this crude characterisation which has been used in the program PICNTERP - see Appendix).

The third 'realisation' of coinc, as an intersection will not be characterised here. In general it would appear to involve an articulation in the picture domain similar to that described by Stanton (Stanton, 1969) as a 'road map view'.

3.2. Parsing and Recognising VISTA* Pictures

In their discussion of CYCLOPS, Marill et al. comment that "if during the operation of the system one or more hypotheses have been rejected, a great deal of analysis will already have been done on the input. This analysis should suggest the next hypothesis to test. In short, new hypotheses should be suggested by characteristics already discovered in the input". As indicated in § 2.1 this does not happen in the existing version of CYCLOPS. We might extend this observation to suggest that between 'line forming' (§2.1.1) and 'Identification' (§2.1.2) we might interpose a process which recovers those 'characteristics' common to all letters recognisable by the program. The foregoing account of 'mapping' (§3.1) identifies what these characteristics might be. Specifically, the identification of near and points of inflection provides a level of analysis from which to attempt recovery of characters. Thus given an input of the form illustrated in Fig. 1 we would form the end pieces of each SEG (two in this case) and compute near between each end piece and all other pieces. This would recover the fact that the left hand end piece of the horizontal is near one or more pieces - centrally disposed - of the vertical.

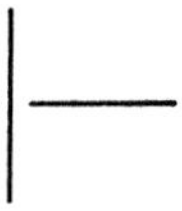

Fig. 1

Such a procedure extended to include KNEE's and intersection would yield an articulation of the scene from which a mapping into LINEs and Coinc's can be made. The number of LINEs thus identified in the scene and the variety and number of Coinc relationships between them provides a rough basis for selecting 'plausible' character hypotheses. Thus one of the defects of CYCLOPS may be immediately rectified by effecting changes in it which would make it perform the mapping process abstractly characterised above. To do this we would require a means of representing the scene in both its pictorial and abstract forms. Experimental studies now in progress suggest that a ring structure representation such as has been used by Stanton (1969) to characterise plane region maps is adequate. These studies have also shown that detection of near and of points of inflection is reasonably straightforward. This

* VISTA: Graphic Display, Time Shared with the CDC 3600 Computing System (at C.S.I.R.O. Division of Computing Research, Canberra).

experimental work has as its goals a realisation of the mapping theory developed in §3 and the demonstration of some limited recognition ability.

4. CONCLUSIONS

This account of a model or theory of alphabetic letters has been the vehicle through which we have attempted to decide 'what is it about pictures that we want to represent?'. Arguments have been presented which suggest that letters can be viewed as having a syntactic structure part abstract (geometrical) and part pictorial. We call this structure syntactic because it involves an articulation of objects into parts, necessarily related. This differs from Chomsky's syntactic theory (Clowes, Langridge & Zatorski, 1969) not only in invoking two domains and a mapping between them, but also in the variety of relations involved and the fact that these relations may have quite complex definitions. A special feature of this syntactic theory is the status of attributes. Attributes like end etc. are essential to the statement of relationships. Essential to the specification of attributes is an articulated description of objects. The description of objects, relations and attributes is crucial to adequate syntactic description. In an earlier version of Picture Syntax (Clowes, 1968b) a good deal of confusion was present over the distinction between objects, relationships and transformational rules. We would now regard the latter as presented by Chomsky, as an imperfect characterisation of the map relation. The distinction between objects and relationships we have now explicated.

The map relation introduced here may be dubbed 'the relation of representation' (Küng, 1967), as we have seen its recovery is crucial when we think of interpretation as a process of translation. Classificatory approaches to picture interpretation view the latter as concerned with the recovery of the relation of class membership. The interpretation of $\frac{a^2}{2}$ as (a**2)/2 viewed as a translation process identifies the two ways in which the abstract mathematical notion 'index' (to the power of) may be exhibited by pictorial relations of relative position and size (in a^2) or by the pictorial objects '**' (relationally) interposed between these pictorial objects. We cannot regard the interpretation of $\frac{a^2}{2}$ as one of determining class membership without invoking a non-finite set of classes, for to do so is to regard $\frac{a^2}{2}$ as a character like say E. There is an infinite number of such 'characters' obtained by replacing any character in $\frac{a^2}{2}$ by $\frac{a^2}{2}$ and so on. Each 'new' character is well formed. Thus we are unable a priori to define the classes to which these pictures are to be assigned and this view of interpretation is thus inapplicable.

In earlier papers (Clowes 1968a, Clowes 1968b) an argument was developed to suggest pictures were representational. For example that circuit diagrams had an underlying electrical reality involving relationships like phase difference, voltage drop, etc. which were clearly non pictorial. While the notion of representation was always regarded as some kind of mapping, the details were wholely obscure. Adopting the infix notation of the predicate calculus we might denote this vague idea of mapping as

31 $\quad A \; \underline{r} \; B \rightarrow A' \; \underline{r}' \; B'$

Stanton (1969) has persuasively argued that this suggests that relations ($\underline{r}$) in one domain are mapped as relations ($\underline{r}'$) in another. Similarly for objects $A \rightarrow A'$, $B \rightarrow B'$. Instead he suggests consider the possibility that relations can be mapped into objects. Thus the connected relation between positions in a circuit is exhibited pictorially as an object : SEG. It was his advocacy of this view of mapping which led to the formulation of the 'Theory of Alphabetic Characters' presented here. This view-point, and such aspects of it as two domains may have many relations in common so that only a small degree of abstraction is involved, is likely to have wide implications.

ACKNOWLEDGEMENTS

I am indebted to R.B. Stanton for the original insight into the nature of mapping rules. Discussions with him and with D.J. Langridge have been vital in many areas of this study.

5. REFERENCES

AMAREL, S. (1968), "On representations of problems of reasoning about actions"; in Machine Intelligence 3 Michie (Ed.) Edinburgh University Press.

ANDERSON, R.H. (1968), "Syntax-directed recognition of hand-printed two-dimensional mathematics"; Ph.D. Thesis, Harvard University.

BARTER, C.J. (1969), "Data structure and question answering"; This conference.

CLOWES, M.B. (1968a), "Transformational grammars and the organisation of pictures"; Seminar Paper No. 11, C.S.I.R.O. Div. Computing Research. Presented to the Nato Summer School on Automatic Interpretation and Classification of Images, Pisa, Italy.

CLOWES, M.B. (1968b), "Pictorial relationships - a syntactic approach"; Seminar Paper No. 12, C.S.I.R.O. Div. Computing Research. Presented to the Fourth Machine Intelligence Workshop, Edinburgh. To be published by Edinburgh University Press.

CLOWES, M.B., LANGRIDGE, D.J. and ZATORSKI, R.J. (1969), "Linguistic descriptions"; this conference.

GRIMSDALE, R.L., SUMNER, F.H., TUNIS, C.J. and KILBURN, T. (1959), "A system for the automatic recognition of patterns"; Proc. IEE 106, B, 26, p.215.

KAMENTSKY, L.A. and LIU, C.N. (1963), "Computer-automated design of multifont print recognition logic"; IBM J. Res. & Dev. 7, 1, p.1.

KÜNG, G. (1967), "Ontology and the logistic analysis of language"; Dordrecht : Reidel Publishing Co.

MARILL, T., HARTLEY, A.K., EVANS, T.G., BLOOM, T.H., PARK, D.M.R., HART, T.P. and DARLEY, D.L. (1963), "CYCLOPS-1 : A second-generation recognition system"; AFIPS Conference Proceedings 24, pp.27-33.

MILLER, G.A. (1956), "The magical number seven; plus or minus two : some limits on our capacity for processing information"; Psych. Rev. 63, pp.81-97.

NARASIMHAN, R. (1966), "Syntax-directed interpretation of classes of pictures"; Comm. ACM 9, 3, pp.166-173.

NARASIMHAN, R. and REDDY, V.S.N. (1967), "A generative model for hand-printed English letters and its computer implementation"; ICC Bulletin 6, 4, pp.275-287.

NEWELL, A. (1965), "Limitations of the current stock of ideas about problem solving"; in Electronic Information Handling, Kent & Taulbee (Eds.), London : Macmillan.

NEWELL, A. and ERNST, G. (1965), "The search for generality"; Proc. IFIPS Congress 65, pp.17-24.

STANTON, R.B. (1969), "Plane regions : a study in graphical communication"; this conference.

6. APPENDIX

The experimental studies referred to in Section 3.2 have taken the form of a program which attempts to recognise characters 'drawn' with a light pen on the VISTA console. The program - PICNTERP - makes use of a low-level package VISTPACK to communicate with the console user. It also makes calls to PLEXPACK (Stanton, 1969) to create and process ring structure representations of the data.

The range of recognisable pictures is restricted to single characters from the set A E F H I K L M N T V W Y Z, i.e. to characters which do not contain curved strokes. The primary question which the program should give some answers to concerns the ease and reliability with which coinc can be recovered from near and knee. The program does not recover intersections. (Hence X cannot be recognised.) PICNTERP (see Figure A1) attempts to effect this recovery through three sequential subroutines SEGMENTR, STRUCTUR and TRNSLATR. The effect of SEGMENTR is to turn all cases where coinc is manifested as a knee into cases where it is manifested as near. The reasons for doing this are partly historical and partly for efficiency. (SEGMENTR uses an array representation which is more rapidly addressed than a ring structure representation.) Among the penalties paid are (1) the need to use a rather crude characterisation of knee;

(2) the decision to translate knee into near is both premature and irreversible. That is knee should be represented as a structural attribute of a SEGMENT in the PICTURE representation.

The ring structures used in STRUCTUR to represent the PICTURE are illustrated in Figure A2. Those used in TRNSLATR to represent LETER are illustrated in Figure A3.

TRNSLATR effects the mapping described in Section 3.1. In the event that a near is encountered which cannot be translated into an end or a mid pos of a line (i.e. the stroke is incorrectly positioned) an error message is displayed by EXHIBTR together with a pictorial representation of the structure of LETER (Figure A4). TOPO effects an analysis of this LETER structure in terms of numbers of lines, end-to-end coincidences, end-to-middle coincidences and the number of mid-positions involved in these coincidences. This is used as the basis for deciding what geometry to test on the character. (It is this feature which represents our practical development from CYCLOPS) (Figure A5). The decision tree used by TOPO appears in Figure A6.

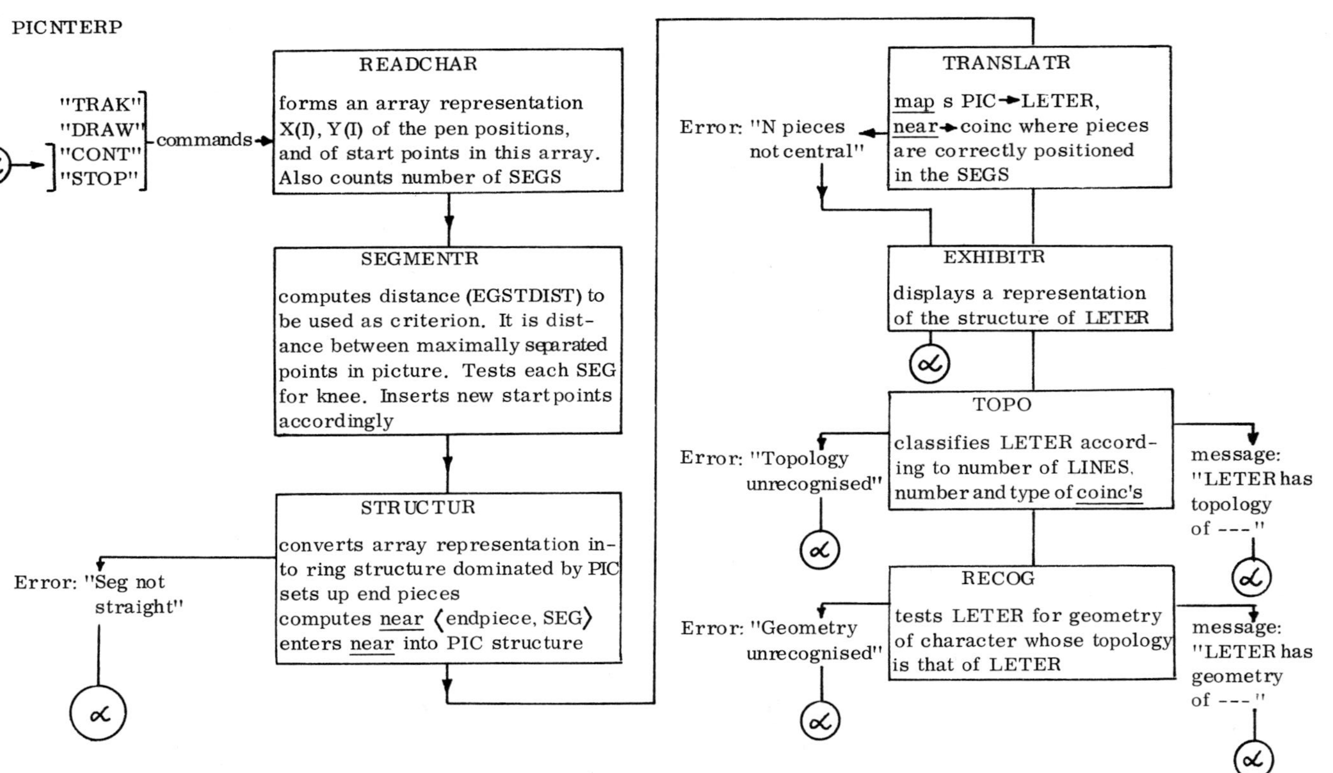

Fig. A1 - Flowchart of PICNTERP.

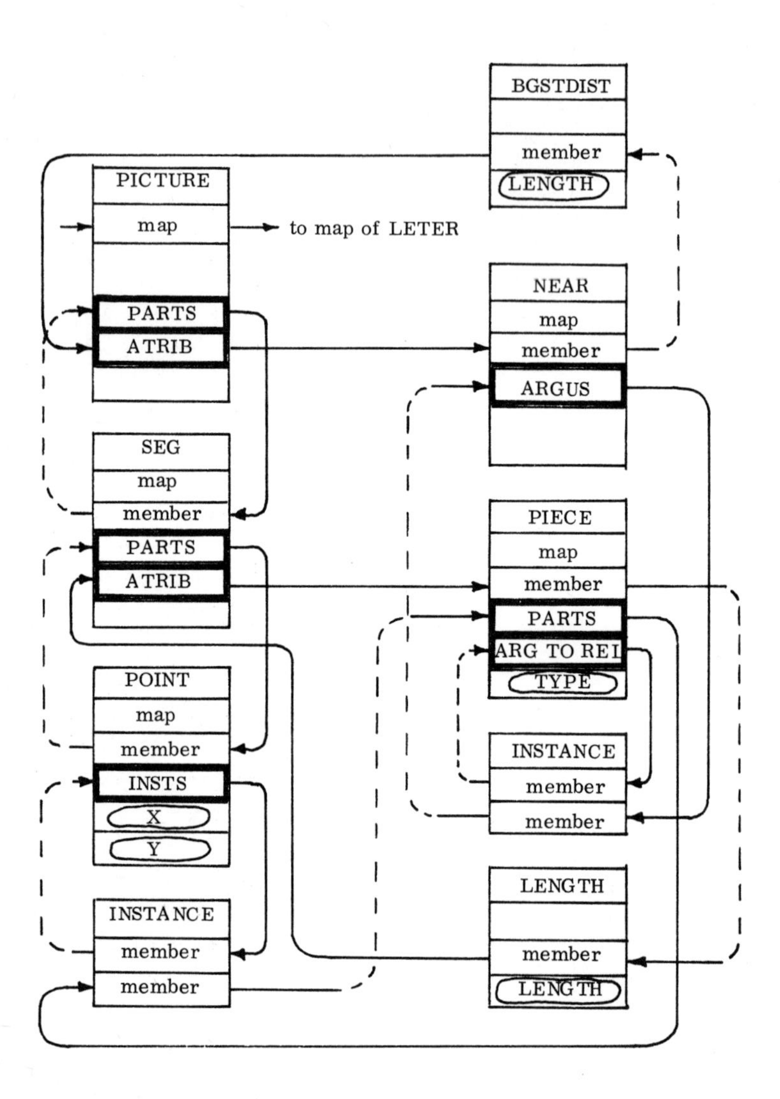

Fig. A2 - Ring structures used in the representation of the picture.

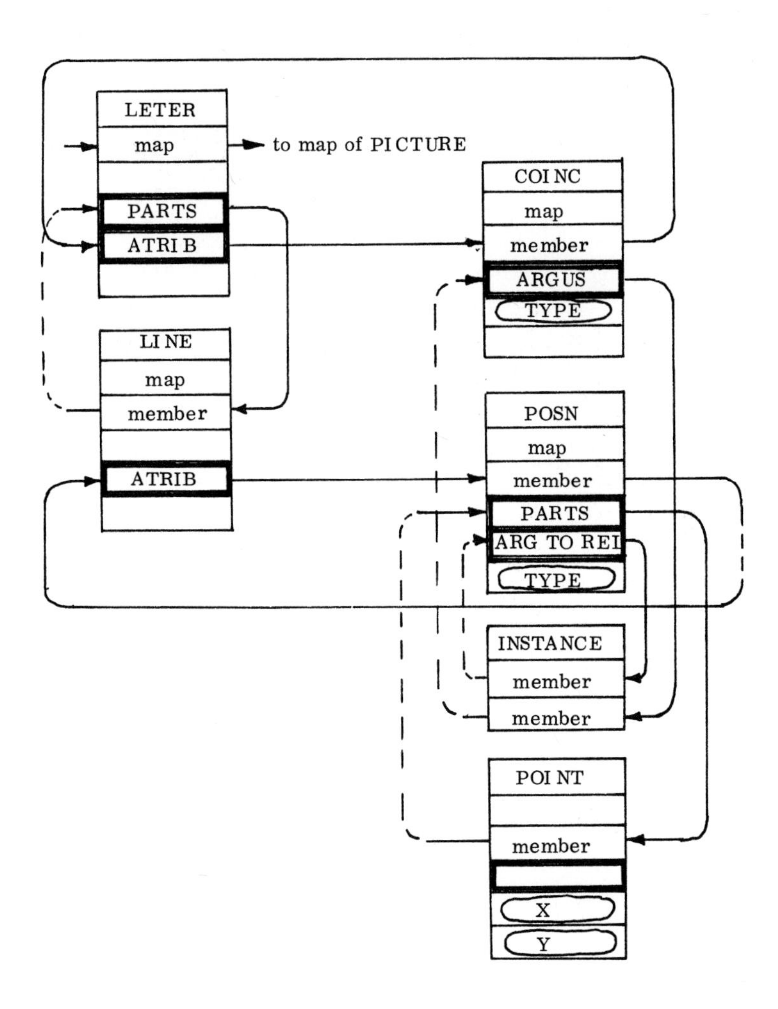

Fig. A3 - Ring structures used in the representation of the LETER structure abstracted from the picture.

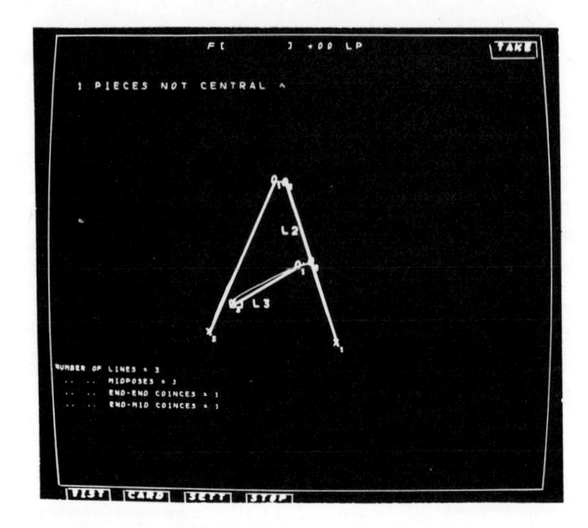

Fig. A4. LETER structure and error message for an input not accepted by TRANSLATR.

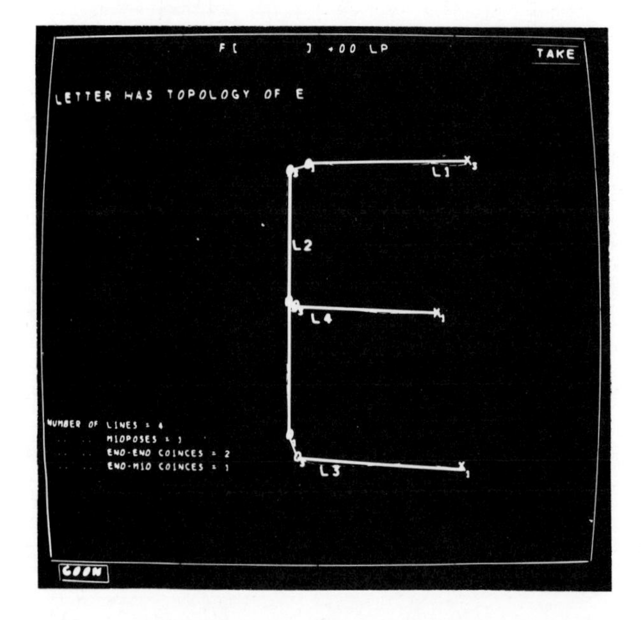

Fig. A5. Acceptable input with "topological" features as listed in lower left. Classification of its "topology" appears in upper left.

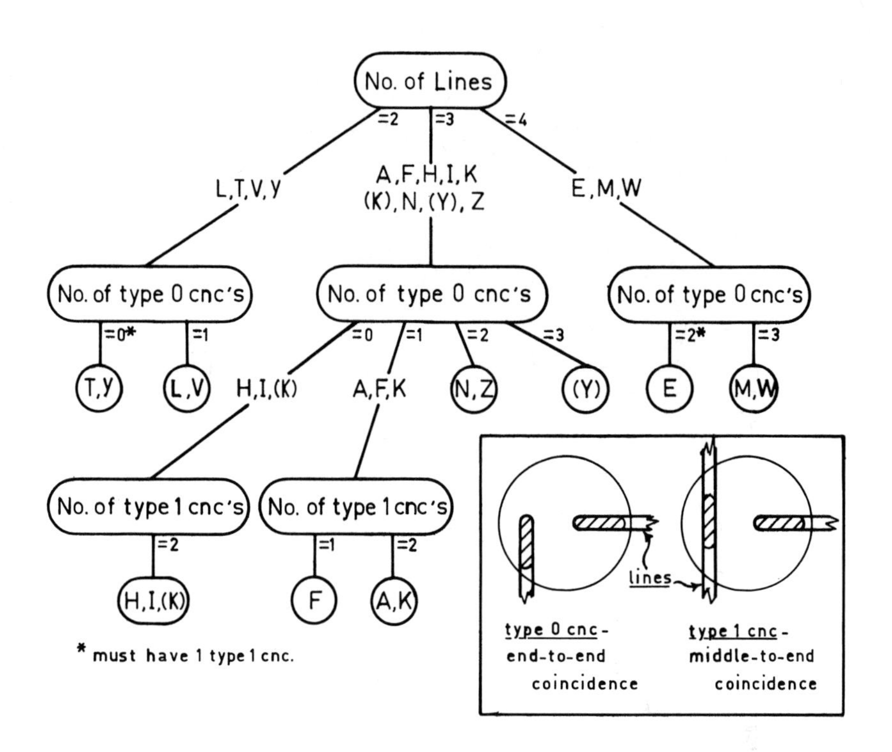

Fig. A6. Decision tree used in TOPO to sort LETERs into plausible character classes prior to making geometrical tests.

The geometrical relationships tested include:

<u>parallel</u> ⟨LINE, LINE⟩, <u>perpendicular</u> ⟨LINE, LINE⟩,
<u>same side</u> ⟨LINE, LINE⟩, <u>equal length</u> ⟨LINE, LINE⟩, etc. (Figure A7)

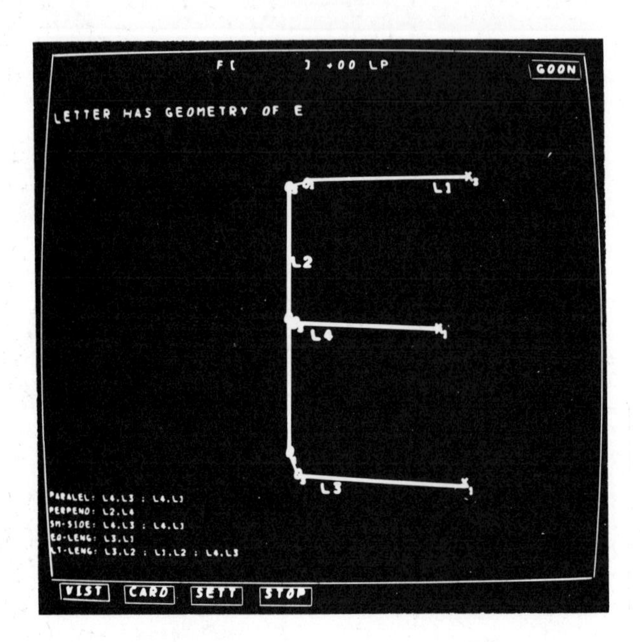

Fig. A7. 'Recognised' character with listing of relevant geometrical relationships (lower left) and classification (upper left).

In Figure A8 we illustrate some characters which were acceptable to PICNTERP and in Figure A9 some that were not. These latter are in three categories: the top row are those which fail TRNSLATR, the second those whose "topology" is unrecognised, and the bottom row those whose geometry is unrecognised.

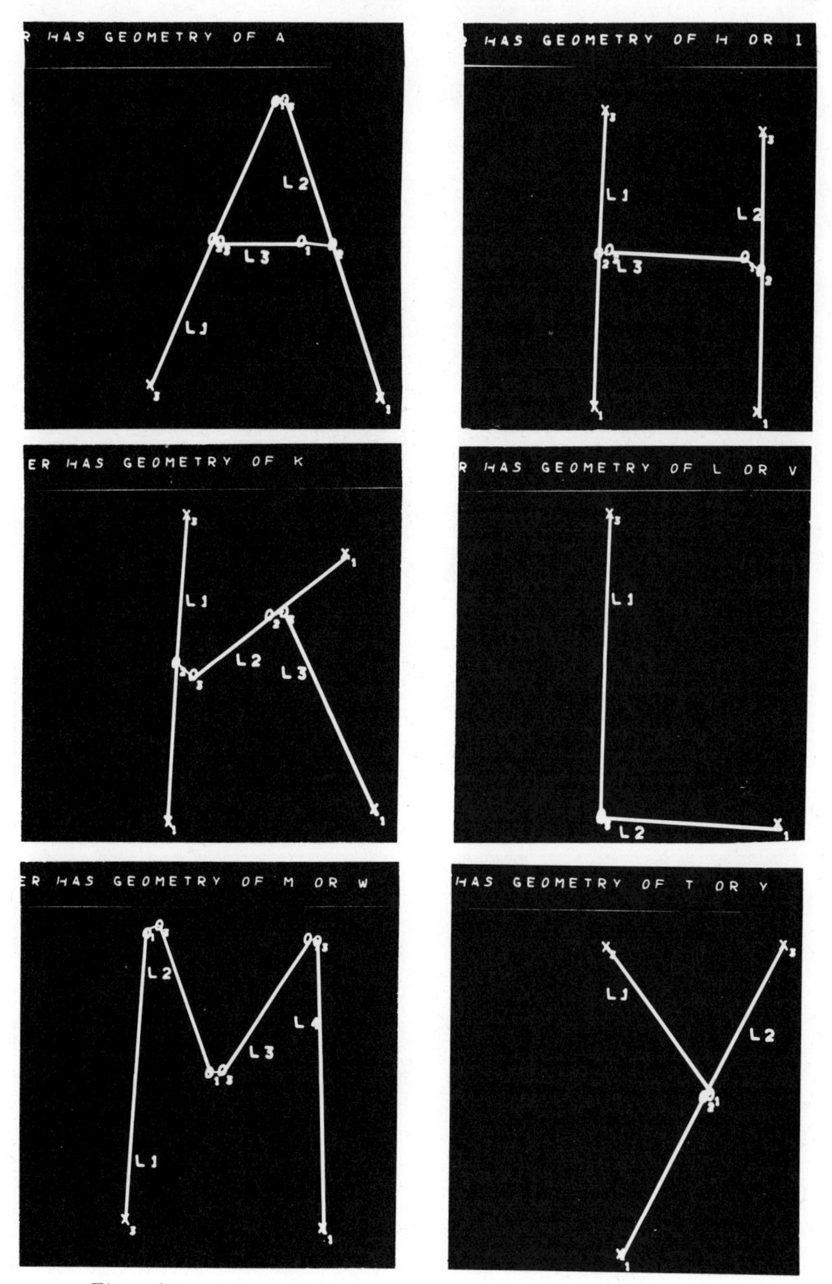

Fig. A8. Some characters out of the set of straight line characters, recognisable by PICNTERP.

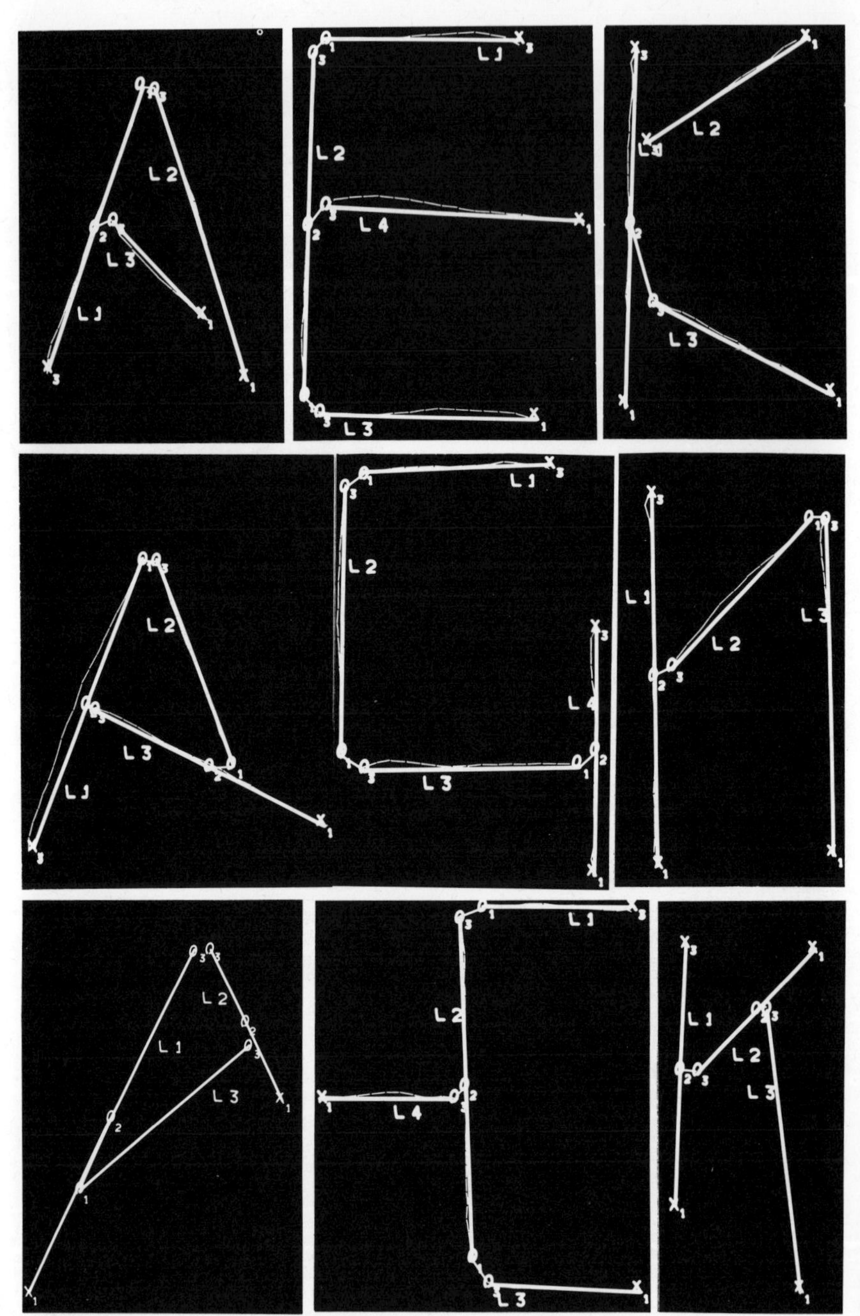

Fig. A9. Inputs which fail to satisfy the requirements of TRANSLATR (top row), TOPO (middle row) or RECOG (bottom row).

7. DISCUSSION

Moore: Recognition of two dimensional characters necessarily involves geometry, and I think we should keep to the proper terms. In geometry there are three fields - metric, projective and topological geometry. For example, a 'V' is topologically equivalent to just a line. I would like to make the point that with your system, and its a criticism that I'm making of practically all systems that have been made up to date, you are studying projective properties and metric properties but not topological properties. Your 'A' for instance, coincidence, straight lines, these are all projective properties and not topological ones. Writers like Poincaire would say, however, that most of the intuitional judgements that we make of things, are topological properties.

Clowes: Thank you for the comments. I would like to think, however, that when you talk about fields of geometry in this sense, what you are talking about is different varieties of abstraction. What I have tried to point to is to characterize the whole notion of abstraction. You see, we do form these abstract systems, and what is it to form an abstract system? I don't think mathematics tells me anything at all about that - perhaps it shouldn't. In other words if I gave you a picture and you tell me its topology is so and so, you have done a fantastic amount of abstraction, and in a sense that's what I've got to handle - that kind of abstraction. When you look at pictures, you don't see the little dots that the scanner sees, that's abundantly clear. You see a pretty abstract complex organisation, and it is essentially the abstraction that the program has to go for!

Moore: The problem with say an 'A' drawn, instead of with a vertex at the top, with just a semi-circle, would create a lot of difficulties with your system.

Clowes: What you are telling me is that the abstract relationship of coincidence between the ends of pairs of lines can be exhibited in several ways, but my view of 'knee' can actually be rounded.

Moore: Well, the 'A' is topologically equivalent to an 'O' with two pot handles on it, and my point really is that you are definitely only measuring projective properties, straight lines, intersections and so on. I think an important attribute of character recognition is to recognise topological features, which your system does not do.

Cook: I would have thought that the point you just made, indicates that topology is not the feature we are tending to operate on. You have said that a 'V' is topologically equivalent to a straight line. On the other hand we don't see the letter 'V' as the letter 'I'. This seems to indicate the eye is not simply responding just to topology.

Moore: Oh, I think that projective aspects are important too. Projective properties as aspects of your memory are important. It is also important to look at topological properties, which this system doesn't - the concept that that 'A' is topologically equivalent to an 'O' with pot handles on it.

Cook: I would have thought, since you have quoted Poincaire, that topology comes intuitively, one would have great difficulty in establishing that topological properties are recovered in a simple way by the visual system. It is characteristic of topology that it generally produces intuitively queer results, things which are topologically equivalent are very surprising, because they are not the sort of things that we regard as being equivalent very often.

Moore: Put it this way; together with Dr Clowes' projective aspects, we also have the additional information that 'A' is topologically equivalent to 'O' with pot handles on it. This would certainly aid you with classification of it, and you could use his grammar, etc., to help.

Stanton: There is no reason why the topological view of these things can't be characterised, especially as the question really is, do we need it? If we can establish that we need it, then presumably you have found out something about what we need to describe 'A', but its not been clear so far that topological properties are involved.

Bobrow: What notation do you use to write your language statements into the computer? Do you actually have some mapping of this notation so that you can write it directly in, so that you can do the things you say, namely characterise all these things and write abstract forms and change your mind? Do you only get them in by writing a program or do you have some interpreter?

Clowes: No, I don't have an interpreter, and there is one reason for this.

Bobrow: No compilers?

Clowes: No, there is one excuse for this. For a long time we have thought about it, and we could never decide just what kinds of notation in which to express the range of things we are talking about. I have the feeling that we now do have a good language in which to write these things down - it is, however, different from anybody else's.

Bobrow: You are then the compiler for this language?

Clowes: Yes.

Hext: I'd like to make one or two comments about a question which has worried me. The question in advance is - "you said you have to have a line specified by two end points and an axis? (Relation 17) How do you define the axis?". Secondly, on the general question of something is 'so and so', for instance, you said that an end point is, not just a point but an area. I'd say that is purely a matter of functions that operate on how you represent your end problem. For instance, if your end point is a pair of co-ordinates, and you define a function 'coinc' to, between two end points, it calculates the distance between, and says it is less than so and so. If you define your end points as area, you cannot say whether the areas overlap, it is purely a matter of the functions that operate on what you have put down.

Clowes: What I am saying is that in fact the definition of coincidence is given as a mapping. I don't give any other definition. We can define coincidence by mapping. That's what I'm saying. I could be wrong, and in fact what you are pushing me to do, is to add 'coincidence' to the system of notation, in what we think of really as being the ideal case. They are the same value, that's what we mean by coincidence.

Hext: Is a mapping a function?

Clowes: Yes, but I'm defining coincidence in terms of near, or knee or intersection. I'm defining it that way, not in terms saying that pairs of values are the same. So I'm tying it to the way you draw, not to the way you think. The way you think is the way you think in arithmetic, and there you require coincidence to have a definition of two values being the same.

Hext: Well, finally, if we say something is the letter 'A', either we mean, I think it looks like the letter 'A', or we mean it satisfies a certain function operating on 'it', namely this program set; this conflict between the two ways of defining 'A' has been running through a lot of what has been said - sometimes we confuse the two.

Clowes: All I can say is that this expresses in a form which is less obscure than the program, more obscure than English, what the definition of 'A' is. The definition it uses is essentially the same as definition of and in the program. The program uses this definition - this is not procedural.

Hext: Now I ask what your axis is?

Clowes: The axis taken is simply the axis that when you handle the light pen you get certain X, Y values; in other words the screen axis.

Hext: If I have a program which decides whether or not this is a letter 'A', you supply this function with parameters and you are saying you supply it with end points and the axis. Is the axis a parameter of that function, or is the axis built into the function?

Clowes: The answer to your question is to be found in the definition of 'above'; it is in Note 23. There should be a relation on 'relpos' but there is no mention of 'relpos' at all, just X, Y, and for that reason I can never change my axis. As far as pictorial judgements are concerned, as far as judgements of shape are concerned, I think it's sufficient to stop this idea of 'relpos' evolving an axis.

PLANE REGIONS : A STUDY IN GRAPHICAL COMMUNICATION

R.B. STANTON

Department of Electronic Computation
University of New South Wales

and

Division of Computing Research
C.S.I.R.O., Canberra

1. INTRODUCTION

There has been a good deal of interest over the past years in computer graphic systems and their uses. A principal motivation behind the development of computer graphics has been the hope of achieving graphical communication with machines. Literature in this area has largely been concerned with describing particular systems and with describing aids to implementation, such as general processing functions and 'flexible' data structures.

It is probably fair to say that few, if any, of the computer graphics systems described have exhibited that freedom of graphic expression which we identify with man-man graphical communication. (In fact sometimes the graphics are little more than visual confirmation that particular buttons have been pressed.) Also, very little has been done to identify just what graphical communication is (as it relates to console work). In-so-far as graphical communication is dependent on an ability to 'interpret' a graphic, many of the difficulties in producing these systems are being brought into focus through recent work in the picture interpretation field. Essentially the situation can be stated as follows: Man-machine graphical communication is possible only if the machine can see the same things in a picture that man does. Providing machines with the capacity to see has been the business of picture interpretation, so its relevance to graphical communication is not surprising.

Twelve months ago, a system for inputting plane region maps through a computer console was developed in order to explore the nature of graphical communication. The problem of formalising the system provided many insights into the communication problem, and together with correlative study of other existing systems, has resulted in the uncovering of principles which are thought to illuminate graphical communication.

The body of this paper is in three sections. The first presents a brief description of general concepts, the second gives a description of the plane regions system together with a semi-formal account relating it to the concepts expounded in the first section, and the third contains a discussion and conclusion, and gives the relationship between this and existing work.

2. GRAPHICAL COMMUNICATION : SOME PRINCIPLES

Communication is tied up with the idea that we have something to say. In particular, graphical communication can be said to have taken place only when the viewer sees the things in the graphic that he was meant to. That is, both the communicator and the communicatee understand the picture in the same way. In the case of man, this understanding would typically be manifested in the ability to answer questions relating to the understanding of the picture. In the case of a machine as a communicant, we can relax the requirement to that of being able to identify, within the machine, a representation of the information abstracted from the picture. We imagine that such representations underlie the capacity to answer questions (see Barter, 1969). (By representation we mean a description in some language - i.e. the language statements themselves.) We shall attempt to illuminate this remark by way of example and by identifying the components which seem to be essential to our understanding of graphical communication systems.

The claim is often made that graphs, perhaps the most frequently occurring example of computer graphic output, exemplify graphical communication between man and machine. Mostly, the machine representation of a graph is an array giving the positions, relative to a pair of axes, which constitute the graph. (The ordering within the array may be used to encode one of the coordinates.) This representation is interpreted as a set of points which have three positions* exhibited by the array, (either absolutely or relatively). Additionally, the ordering of the positions can be interpreted sometimes as an 'immediately to the right of' relationship between

* We regard the representation where the points are the end position of vectors as effectively being of a larger set of positions; the vector acting as a rough interpolation device.

two points (although this is usually accidental, being forced by the nature of machine storage and algorithms). This relationship however is never used as such with the possible exception of guiding a plotting device efficiently.

The machine is very competent at displaying position, particularly so when scaling and zero suppression are taken into account. However users do not often look at graphs in order to recover the positions of the graph points.* Rather, the viewer attempts to see the shape of the graph. While shape is doubtlessly predicated on position, there is no sense in which the machine has a grasp of the shape of the graph. In this case then, in so far as the machine and the man 'understand' the picture differently, i.e. use different descriptions, we can conclude that graphical communication does not occur. If we could program a machine to capture shape, we might have graphing systems which would, (1) select scaling factors according to criteria which preserve shape; (2) avoid confusions arising from the production of multigraphs; (3) go some of the way toward producing 'intelligent' legendry and labelling.

There are two observations we can make about pictures which become compelling motivation for a particular form of a theory for graphical communication. The first is the apparently central role played by pictorial relationships in descriptions of pictures. Relationships such as near, touch, inside, join, cross, overlap, adjacent, above, etc. seem to be essential to the description of a picture. The second is the fact that we see things (objects) in pictures. The objects we see range from those which are decidedly pictorial in nature (e.g. edge, lines, points, figures, etc.) to those which we regard as more 'worldly' objects (e.g. car, circuit, ball, etc.). Associated with the latter type of object are relationships which are non-pictorial, i.e. related to the particular domain of the object. Thus we have pictures displaying two circuits in series; a glass on a table, two wheels coupled together; two gears meshed, one man bigger than another; and so on. Based on these observations we will distinguish between two or more domains. One of these domains, that of pictorial objects and relationships, we will call the picture domain. The other domains, those which group objects and relationships belonging to a specific area of knowledge, we will call problem domains.

Most objects in a domain can be thought of (equiv. defined) in terms of other objects in that domain, which are considered to be parts of the defined object. The

* We are ignoring the case where the graph is used as an alteration to a table.

particular organisation of the parts of an object is identifiable with relationships which bind them together in order to create the organisation. Thus we might have the same parts forming two different objects - e.g. in Fig. 1 three lines form a triangle or a star depending on the relationships obtaining between the lines.

Fig. 1

Similarly the straight lines in Fig. 1 have as parts a set of points enjoying particular relationships viz. 'nearest neighbour' and 'collinear'. In another domain we might have a logic circuit consisting of some registers connected together in various ways where these registers themselves consist of a grouping of interrelated logic elements - and gates, or gates, flipflops, etc. In a similar but more complicated way the definition of the existence of a relationship between two objects may involve the existence of subordinate relationships between the parts of the objects. Thus two lines crossing involves the idea that two points on these lines are coincident. The process of identifying the parts and relationships of an object is called the articulation of the object and we note that there are a set of primitive objects and relationships (i.e. those which are not articulated, e.g. a resistor in a circuit; 'nearest neighbour' relationship between two coloured positions in a picture; a character in a word).

The complete description of an object in terms of its articulation, including the articulation of the parts recursively to the level of primitives, is the structural description of the object. The specification or description of a domain has to be such that it (the description) can be used to assign structural descriptions to all of the objects encompassed by the domain.

We can see then, that the information contained in a domain is the set of all possible structural descriptions (in many cases, an infinite number), which can be assigned by the description of the domain.

When we use pictures to represent information in a particular problem domain, we associate the structural description of that information with the structural description of a picture. This association, called mapping, in more detailed form is the association of problem objects and relationships with pictorial objects and relationships often in a complicated way. For example the divide relationship between two numbers may appear as a line, i.e. a pictorial object bearing pictorial relationships with the two pictorial objects into which the numbers are mapped. Similarly exhibition of the relationship between a name of a river say and the river may involve a set of complex pictorial relationships. Also the pictorial mapping of a primitive in a problem domain may result (indeed mostly does) in a complex pictorial figure, e.g. a resistor. In this case we refer to the pictorial figure as a symbol.

There is one aspect of object description which in the interests of clarity has purposely been left until now. In defining relationships (and hence also objects) we often need to refer to a specific 'part' of an object. Those objects are called attributes, and are often not parts (as defined above) in the sense that they may not have an independent existence. Rather they are thought of as objects which have intrinsic relationships to the object of which they are a 'part'. Examples are, end points of a line; side of a square; top of a table; input impedance of a circuit; colour of a book, etc. Clearly the last three do not enter into the structural description of the objects of which they are attributes and in this sense are not parts of these objects. They may, however, become parts of other objects by entering into their definition. Mostly, attributes have the status of objects and as such enter into mapping rules and relationships.

Earlier, we said the graphical communication involved a common understanding or 'seeing' of a picture. Based on the above ideas, we can equate 'common understanding' with recovery of similar structural descriptions. Thus, in the case of graphs, the machine would be required to recover a structural description which articulated the graph into its 'shape' components (which hopefully are the components which are meaningful in problem interpretation).

A further two examples will assist in clarifying these principles developed so far.

Consider a system in which engineering drawings are digitised and held in a file, indexed by a classification number. A user is allowed to view any drawing through a display or plotting device by supplying the relevant index. This type of system can be likened to a slide projector or tape recorder. Clearly there is no graphical communication taking place between man and machine. Rather we would say that the composer of the drawings or slides (man) is communicating with the viewer. The only man-machine communication is either 'yes, the drawing can be displayed' or 'no, the drawing is not filed'. The former is given implicitly and the latter presumably by a text message.

Secondly, consider systems where the machine is required to produce wholly novel pictures from a non-pictorial description of the picture. An example would be the production of flowcharts, sensibly laid out in two dimensions from listings of the code. In this case, a structural description of the code would be recovered, representing flowchart objects and relationships such as sequence of boolean decisions and blocks of assignments. This description would then be mapped into a pictorial structure from which a flowchart would be composed. *

The flowchart so produced allows a viewer to recover the organisation of the code; equivalently to recover a structural description similar to that recovered from the code by the machine, so that we can say that graphical communication has been involved. Another example would be a computer aided design system which presented the viewer with a machine designed circuit. A description presented to the machine might be 'design a three stage amplifier having an emitter follower input, gain of 100 using type X transistors, and to supply 10 watts into an 8 ohm load'. The machine might examine different configurations so as to optimise certain factors such as cost, bandwidth, etc. The result of the computation could be a novel circuit given by a structural description. The mapping would carry the circuit into a picture in which the relationships in the articulated circuit are easily recovered by the viewer, i.e. the mapping process invokes standard (albeit circuit specific) pictorial formatting to exhibit the problem structure.

The main point made in these examples is that for a machine to assume the role of a significant communicant with man using the picture medium, it must be able to <u>decide for itself</u> how to use the graphic domain in order to exhibit non-pictorial information.

An existing system which reflects this machine capacity is Martin's (1965) program which converts a mathematical expression in storing form into a two dimensional (text book style) display. Syntax rules involving relative positions are used in conjunction with the better known string syntax rules (e.g. Algol 60) for arithmetic expression. (See also Anderson's (1968) work with mathematical expressions.)

* Many flowcharts are, of course, functionally organised so as to reflect the actual problem being solved by the program. Often this organisation is given by procedural blocks of code, however it may not be. To this extent, knowledge of code alone may not produce a sensible flowchart. Overcoming this would involve the relevant problem domain and associated mapping.

Pictures have often been referred to as expressions in a language. Seen this way, the description of the picture domain becomes the syntax of the picture language. Problem domains, being the same kind of entity as the pictorial domain can be thought of as the syntax of problem descriptions; and, together with the mapping into pictures, can be thought of as a semantic component in a picture grammar. Continuing the analogy with language, recovery of a structural description is identified with parsing. This relationship with a language allows us to draw on a powerful concept in syntactic theory; that of competence, (see Chomsky, 1965). In writing a generative grammar (equiv. description of a domain) we attempt to characterise the knowledge "... in the most neutral terms possible ..." (Chomsky, 1965, p. 9). Thus we exclude from a grammar that knowledge relating to the way in which the language is used, and concentrate on the knowledge underlying such usage, i.e. competence. In this paper we adopt this concept, viewing the descriptions of domain as providing competence models on which systems can be based, and not as descriptions of the system itself. In so far as a grammar can be used to assign descriptions which apparently reflect our competence, it is said to be _descriptively_ adequate (Chomsky, 1965, p. 24).

In summary then, graphical communication implies that man and machine recover similar structural descriptions of a picture. The structural descriptions are assigned to the picture through descriptive apparatus embodying our competence in the relevant area. The description of our competence can be partitioned into domains, wherein problem specific descriptions of objects, relationships and attributes are given, (in terms of primitives) and mapping between domains (or some of them), in which descriptions in one domain are associated with description in another domain, often in complex ways.

3 PLANE REGIONS

3.1. Regions Problem

The particular system selected for study is one which provides a graphics terminal user with the facilities for inputting a number of plane regions to some problem program. Such a system would allow the user to both draw and edit (carry out modifications to) the input data.

We are, of course, considering a system where the user is inputting a 'map', not the regions by themselves. This distinction is very important. Every line drawn by a user is a division between two regions, not the boundary of a single region. This point provides us with a picture/problem distinction. (We use picture, problem as references to the respective domains as outlined in the previous section.). In a pictorial sense any line divides the space into two regions or areas; however we can imagine a map region being displayed alone, in which case the background would not be thought of as being a region in the problem sense (Fig. 2(a) and (b)).

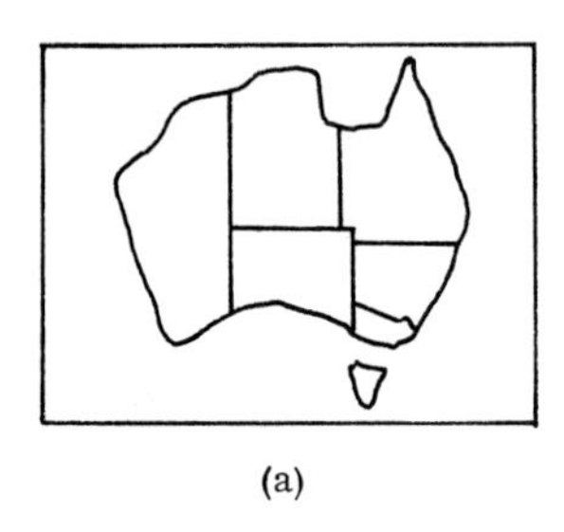

(a)

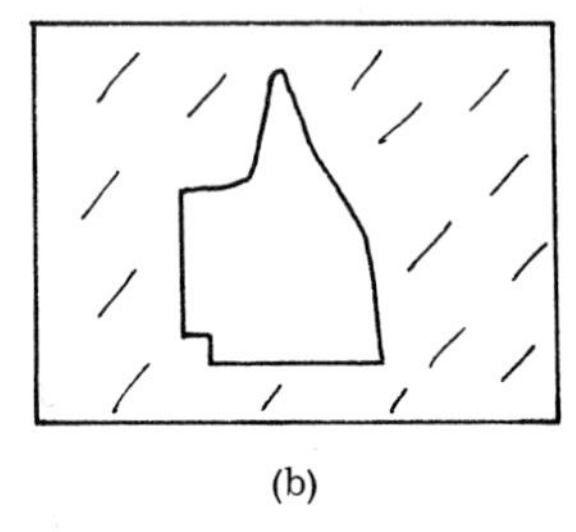

(b)

Fig. 2

This problem is of interest to some workers in the geomorphic and land systems fields, who want to study contextual properties of regions.* (See Cook, 1967 and Thomlinson, 1968). Cook has provided a definition (pictorial) of plane regions and has used a data structure which will be discussed below. Loomis, 1965, has provided an English description of plane regions. For the purposes of this study, the problem component is reduced to a minimum, consistent with maintaining significance for the study. This we do by requiring it to produce a picture of a particular region, together with a few simple characteristics such as area, perimeter or adjacent regions. The system environment is shown in Fig. 3. The regions processor is a parser which interprets the pictorial data input by the user.

The editing and construction processor permits a user to construct a map by drawing any sequence of lines and to modify the map by deleting any interjunction boundary part. (A, B, C --- in Fig. 4 are in interjunction boundary parts).

There is an important distinction we can draw between two views of the drawn, or partly drawn, map which the operator has when using this system. The

* In practice, it seems that the regions view of many graphics (independent of problem) would be the initial view to be recovered when interpreting a graphics generally (Narasimhan, 1964, 1966; Guzman, 1967; and in particular the connectivity graphs of Rosenfeld, 1966, essentially do this).

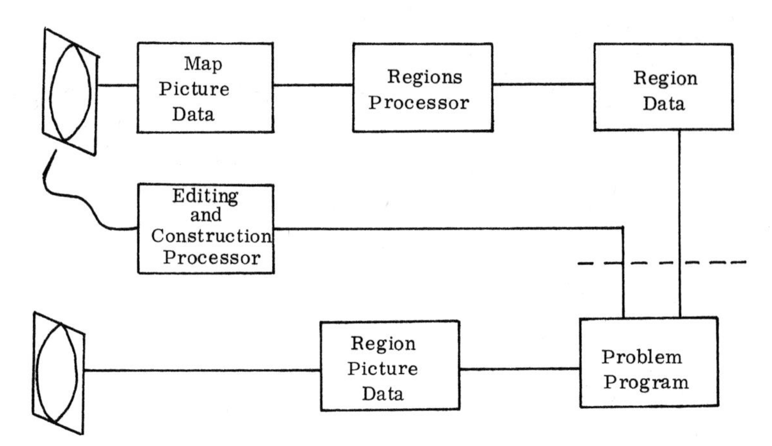

Fig. 3

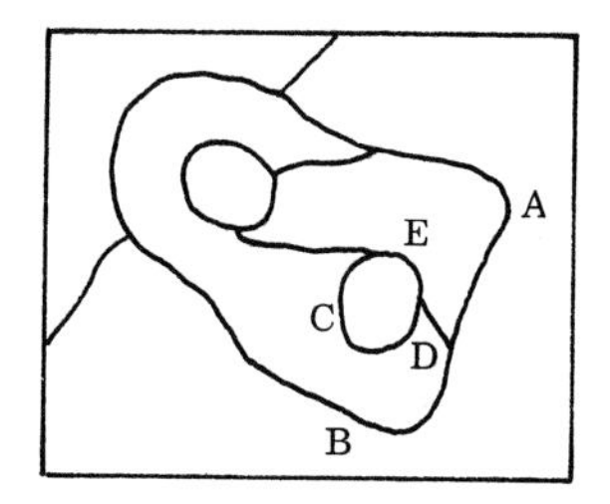

Fig. 4

first is that which the operator sees when the map is considered to be the set of regions which the problem program is to see, as in a political map, or the view which would have to be recovered in order to place names in the regions. The second is the view seen by the user when constructing and editing the map. In this second view the lines forming the boundaries are seen as figures (as opposed to the area figures seen in the regions view) and is referred to hereafter as the road-map view. The two views are shown in Fig. 5.

Consistent with the earlier discussion on graphical communication, we require that the machine be able to see both views. In the next section we will show that the roadmap view can be thought of in terms of a language structure and the regions view, the problem structure.

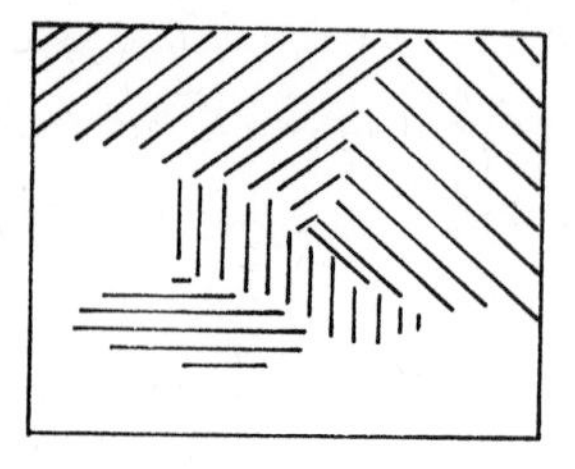

Regions View

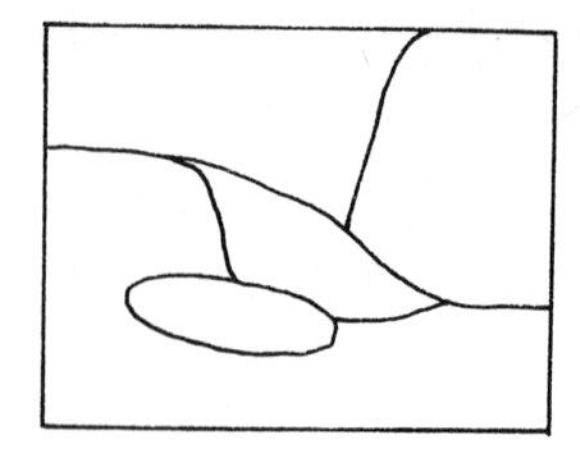

Road Map View

Fig. 5

3.2. Toward a Formal Specification of Plane Region Maps

The description of plane region maps will contain three components as outlined in Section 2. They are the picture language domain, the regions problem domain and the mapping rules. To facilitate the descriptions of domains, the following simplified notation is used. *

(a) Objects: The general form of an object description is given by 1. The NAME preceding the brackets

1 NAME ⟨OBJECTS⟩ [relationships]

is the name of the object defined by 1. The objects inside the angled brackets are the parts of the defined object and the relationships are those which bind the parts together such that they form the defined object. Inferior integers are used to relate the parts to the arguments of the relationships. For example, 2 gives a definition of an equals sign by relating two lines as being near parallel and having the same lengths. (Lengths are attributes of lines - see below.)

2 EQ $\langle \text{LINE}_1, \text{LINE}_2 \rangle$ [near ⟨1, 2⟩, parallel ⟨1, 2⟩, above ⟨1, 2⟩, same ⟨NUMBER.LENGTH ⟨1⟩, NUMBER.LENGTH ⟨2⟩⟩]

(b) Relationships: The form of a relationship description is given by 3. The name is the name

3 name ⟨OBJECTS⟩ [subordinate relations]

* This notation is for illustrative purposes only, being inadequate for a complete description of domain and mappings. Some fairly substantial ideas on notational requirements are emerging from current work and will be published in the near future.

of the relationship and the OBJECTS are the arguments. The subordinate relationships are those which must exist in order that the name relationship can exist. For example, 4 states that two lines enjoy the join relationship just in case the endpoints of the lines are coincident.

$$4 \qquad \underline{\text{join}} \left\langle \underset{1}{\text{LINE}},\ \underset{2}{\text{LINE}} \right\rangle \left[\underline{\text{coinc}} \left\langle \text{PT.ENDPT}\,\langle 1\rangle,\ \text{PT.ENDPT}\,\langle 2\rangle \right\rangle \right]$$

(c) Attributes: The general form of an attribute description is given by 5. NAME is the name of the attribute. The OBJECT

$$5 \qquad \text{OBJECT.NAME} \left\langle \text{OBJECT} \right\rangle \longrightarrow \text{r. h. s.}$$

occurring first in 5, is the OBJECT which is identified by (i.e. is the value of) the attribute (NAME), taken with respect to the OBJECT inside the angled brackets. The r.h.s. is a description of the relationship of the attribute to the structural description of the object which is being attributed. For example, 6 states that a diagonal of a square is a line such that the endpoints of this line are opposite corner points of

$$\begin{aligned} 6 \qquad & \underset{1}{\text{LINE.DIAGONAL}} \left\langle \underset{2}{\text{SQUARE}} \right\rangle \longrightarrow 2 \Big| \left\langle \underset{3}{\text{LINE}}, \underset{4}{\text{LINE}}, \underset{5}{\text{LINE}}, \underset{6}{\text{LINE}} \right\rangle \\ & \left\{ \underset{7}{\text{PT}}, \underset{8}{\text{PT}} \right\} \text{.ENDPTS}\,\langle 1\rangle \Big| \Big[\text{coinc} \left\langle \underset{9}{\text{PT.ENDPT}}\,\langle 3\rangle, \underset{10}{\text{PT.ENDPT}}\,\langle 4\rangle \right\rangle, \\ & \underline{\text{coinc}} \left\langle \underset{11}{\text{PT.ENDPT}}\,\langle 4\rangle, \underset{12}{\text{PT.ENDPT}}\,\langle 5\rangle \right\rangle, \text{coinc} \left\langle \underset{13}{\text{PT.ENDPT}}\,\langle 5\rangle, \underset{14}{\text{PT.ENDPT}}\,\langle 6\rangle \right\rangle \\ & \underline{\text{same}} \left\langle 7, 10 \right\rangle \ \underline{\text{same}} \left\langle 8, 13 \right\rangle \Big] \end{aligned}$$

the square. A special case exists where the attribute is a part of the attributed object, e.g. side of a square. In this case, rather than give the full structure of the attributed object, the attribute 'PART' is employed. 7 gives the definition of side of a square using 'PART'.

$$7 \qquad \underset{1}{\text{LINE.SIDE}} \left\langle \underset{2}{\text{SQUARE}} \right\rangle \longrightarrow 1.\text{PART}\,\langle 2\rangle$$

'PART' behaves recursively so that any point on a square is given by 8.

$$8 \qquad \underset{1}{\text{PT.POINT}} \left\langle \underset{2}{\text{SQUARE}} \right\rangle \longrightarrow 1.\text{PART}\,\langle 2\rangle$$

(d) Mapping: The mapping will be exhibited by writing a map relation between descriptions in two domains, e.g.

$$9 \qquad \underline{\text{map}}\ \langle \text{EXP}\ \langle \underset{1}{\text{EXP}}, \underset{2}{\text{DIV}}, \underset{3}{\text{EXP}} \rangle\ [\underline{\text{Follow}}\ \langle 1,2\rangle\ \underline{\text{Follow}}\ \langle 2,3\rangle]\ ,$$

$$\text{FIG}\ \langle \underset{4}{\text{FIG}}, \underset{5}{\text{LINE}}, \underset{6}{\text{FIG}} \rangle\ [\underline{\text{above}}\ \langle 4,5\rangle, \underline{\text{below}}\ \langle 6,5\rangle, \text{near}\ \langle 4,5\rangle\ \text{near}\ \langle 5,6\rangle]\rangle$$

$$[\underline{\text{map}}\ \langle 1,4\rangle\ \underline{\text{map}}\ \langle 3,6\rangle]$$

gives a mapping between an arithmetic expression, specified as an object in a string domain and its figural form in a picture domain. We imagine that the expressions which are the string arguments of the divide operator have previously been mapped into figures.

(e) Sets: A requirement for the descriptive apparatus for domain description is the ability to refer to named sets, both ordered and unordered. These sets function as objects, but may include as members objects or relationships. They are distinguished by brackets as shown

$$\text{Set Name} \qquad \{\ \text{members}\ \}$$

Regions are sets of interrelated positions on a two-dimensional surface. They can enjoy relationships which are usually thought to be pictorial in nature, e.g. adjacent, touch, overlap, etc. However, these objects and relationships are rarely represented directly in the picture. Rather a language is used to represent the regions information - i.e. the language of curves, lines and points. This is the basis for considering the regions domain to be an abstract problem domain, descriptions in which are mapped into the pictorial domain.

A region is a set of interconnected positions, P, such that any two positions can be connected by a chain of positions enjoying nearest neighbour relationships (nn). We will allow that two such positions are related by a connect relationship (CON) where CON would be expanded in terms of the nn relationships between positions which are members of the set of points forming the region. Rules 10, 11, 12 give the region definition and Fig. 6 shows the primitive nn relationships between a given position and the associated flanking positions.

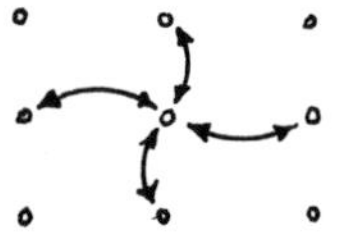

nn rels.

Fig. 6

$$10 \qquad \text{REGION}\ \langle \text{PNS}\ \{\text{P---P}\}\rangle\ [\forall(\underset{1}{\text{P}} \in \text{PNS})\ \forall(\underset{2}{\text{P}} \in \text{PNS})\ \underline{\text{CON}}\ \langle 1,2\rangle]$$

$$11 \quad \underline{\text{CON}}\left\langle \underset{1}{P}, \underset{2}{P}\right\rangle \left[\underline{\text{nn}}\langle 1, 2\rangle\right]$$

$$12 \quad \underline{\text{CON}}\left\langle \underset{1}{P}, \underset{2}{P}\right\rangle \left[\underline{\text{nn}}\langle 1, \underset{3}{P}\rangle \; \underline{\text{CON}}\langle 3, 2\rangle\right]$$

Using the characterisation of REGION, we can define the region boundary as an attribute. There is, however another approach to the description of a REGION based on considering it to be an object with a set of boundary closures as parts. The set of positions in the region are then addressed as an attribute of the boundary, i.e. the inside of the boundary. This latter approach seems to be more applicable to the computing situation and is adopted in the following discussion. A CLOSURE is defined as an ordered set of positions, where any one position enjoys a nearest neighbour relationship with just two other positions - see rule 13.

$$13 \quad \text{CLOSURE}\left\langle \underset{1}{\text{PNS}}\left\{\underset{(1}{P}, \underset{2}{P}\text{---}\underset{i}{P}, \underset{i+1}{P}\text{--}\underset{n)}{P}\right\}\right\rangle \left[\underset{2}{\forall(P_i \in 1)}\,(2 \neq P_n)\,\underline{\text{nn}}\langle 2\; \underset{3}{P_{i+1}}\rangle \right.$$

$$\left. \underline{\text{nn}}\langle P_n, P_1\rangle\right]$$

A region is composed from a closure representing the outer boundary and any number (including zero) of closures representing the inner boundary. Each of the closures of the inner boundary must be <u>inside</u> the outer boundary and <u>outside</u> every other member of the inner boundary. This is stated in 14, 15, 16.

$$14 \quad \text{REGION}\left\langle \underset{1}{\text{CLOSURE}}, \underset{2}{\text{CLS}}\left\{\underset{(1}{\text{CLOSURE}}, \text{------}\underset{n)}{\text{CLOSURE}}\right\}\right\rangle$$

$$\left[\forall(\underset{3}{\text{CL}} \in 2)\,\forall(\underset{4}{\text{CL}} \in 2)(3 \neq 4)\;\left(\underline{\text{inside}}\langle 1, 3\rangle, \underline{\text{outside}}\langle 3, 4\rangle\right)\right]$$

$$15 \quad \underset{1}{\text{CLOSURE}}.\text{OUTBNDRY}\left\langle \underset{2}{\text{REGION}}\right\rangle \rightarrow 1.\text{PART}\langle 2\rangle$$

$$16 \quad \underset{1}{\text{CL}}\left\{\text{CLOSURE----}\right\}.\text{INBNDRY}\left\langle \underset{2}{\text{REGION}}\right\rangle \rightarrow 1.\text{PART}\langle 2\rangle$$

That a position is <u>inside</u> a region is given by 17

$$17 \quad \underline{\text{inside}}\left\langle \underset{1}{P}, \underset{2}{\text{REGION}}\right\rangle \left[\underline{\text{inside}}\left\langle 1, \text{CLOSURE.OUTBNDRY}\langle 2\rangle\right\rangle, \right.$$

$$\left. \forall(\underset{3}{\text{CLOSURE}}.\text{INBNDRY}\langle 2\rangle)\;\underline{\text{outside}}\langle 1, 3\rangle\right]$$

The <u>inside</u> relationship between a closure and an arbitrary position would be given by the relative position between the specified position and a set of axes given by segments of the closure.

Although the possible relationships which can exist between two regions are many - e.g. overlap, touch, adjacent, near, inside, etc., we will consider only adjacency -

$$18 \quad \underline{\text{Adj}}\left\langle \text{REGION}_1, \text{REGION}_2 \right\rangle \left[\underline{\text{Join}} \left\langle \text{CVE}\{\text{P--P}\}.\text{SEG}\left\langle \text{CLOSURE.OUTBNDRY}\langle 1\rangle\right\rangle, \text{CVE}\{\text{P---P}\}.\text{SEG}\left\langle \text{CLOSURE.OUTBNDRY}\langle 2\rangle\right\rangle\right\rangle, \underline{\text{Outside}}\langle 1,2\rangle \right]$$

Where joined curves (CVE is a SEGment of a CLOSURE) means that the two curves are alongside one another at all positions. Another version of 18 is required to give the situation where the outer boundary of one region is adjacent to an inner boundary of the other.

The rules 10 to 18 capture a fragment of our knowledge of regions. To express this knowledge in the form of a plane regions map a description of the pictorial domain, i.e. curves language, is required. Curves are perceptually manifested as a set of points (coloured positions) related in a similar manner to those in a closure.

There are two primitive relationships in this domain. They are nearest neighbour and coincident defined on points (PT)'s. There are several relationships which two curves (CURV) can enjoy. Four of them, join, meet, cross and touch are shown in Fig. 7 a, b, c and d respectively.

Fig. 7

Where three (or more) curves join at a point we can define an order in a particular sense. This ordering we will represent by a between relationship as shown in Fig. 8 where B is between A and C (in an anticlockwise sense).

The join and between relationships are given by 19, 20 and 21. For reasons of clarity, the right hand side of 21 contains references to Fig. 8(b) rather than to the structures of the relevant curves. 21 states that for B to be between A and C, either

19 $\underline{\text{join}} \left\langle \underset{1}{\text{CURV}}, \underset{2}{\text{CURV}} \right\rangle \left[\underline{\text{Coinc}} \left\langle \text{Pt. END} \langle 1 \rangle, \text{Pt. END} \langle 2 \rangle \right\rangle \right]$

20 $\underset{1}{\text{Pt. END}} \left\langle \underset{2}{\text{CURV}} \right\rangle \rightarrow \text{P} \{1, \text{pt----pt}\}. \ \text{Part} \ \langle 2 \rangle$

$\text{or P} \{\text{pt, pt----pt}, 1\}. \ \text{Part} \ \langle 2 \rangle$

21 $\underline{\text{between}} \left\langle \underset{A}{\text{CURV}}, \underset{B}{\text{CURV}}, \underset{C}{\text{CURV}} \right\rangle \left[\underline{\text{rside}} \langle \text{P1, P2, P4} \rangle \ \underline{\text{rside}} \langle \text{P2, P3, P4} \rangle \ \underline{\text{rside}} \langle \text{P1, P2, P3} \rangle \right]$

$\text{or} \left[\underline{\text{rside}} \langle \text{P1, P2, P4} \rangle \ \underline{\text{lside}} \langle \text{P1, P2, P3} \rangle \right]$

$\text{or} \left[\underline{\text{rside}} \langle \text{P2, P3, P4} \rangle \ \underline{\text{lside}} \langle \text{P1, P2, P3} \rangle \right]$

(a) P4 and P3 have to be on the right-hand side of vector P1 to P2 and P4 has to be on the right-hand side of vector P2 to P3, or

(b) P4 has to be on the right-hand side of either vector P1 to P2 or vector P2 to P3 given that P3 is on the left of vector P1 to P2.

Fig. 8

The mapping problem is one of identifying objects and relationships in the regions domain with objects and relationships in the curves domain.

If we wished to exhibit one region, the mapping could be quite simple - i. e. each position on a boundary closure could be mapped into a point on a curve. If we wished to exhibit two overlapping regions the mapping rules would specify how the curves were ordered at each junction such that the two overlapping boundaries could be recovered - see Fig. 9. However if two regions could be either overlapping or touching, then mapping out the boundaries as curves and exhibiting the relationships by junctions leads to ambiguities. Fig. 10 shows a <u>touch</u> organisation of the Fig. 9(a) regions. In many cases, of course, the touch overlap can be distinguished by continuity of boundaries, i. e. a junction may be formed by two curves <u>crossing</u> (in the regions overlap case) or by two curves touching (in the regions touch case). However, ambiguous or not, the correspondences of regions and regions relationships with curves and curve relationships is the role of mapping.

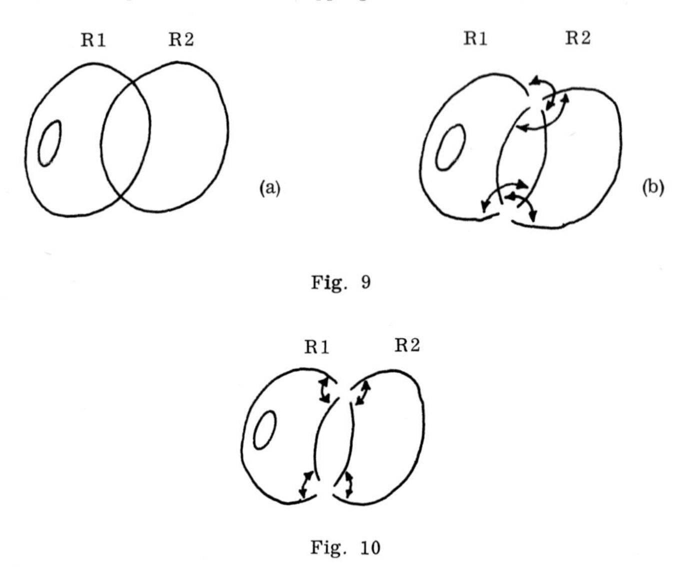

Fig. 9

Fig. 10

Plane region maps are a rather special case of the general regions problem in that every position on the two dimensional surface containing the regions, belongs to one and only one region. This fact indirectly allows us to use a curve as the pictorial representation of the primary regions relationship, adjacency, i. e. when two regions are adjacent, we produce a curve in the picture of the same shape as the segment of boundary which is involved in the adjacency, Fig. 11.

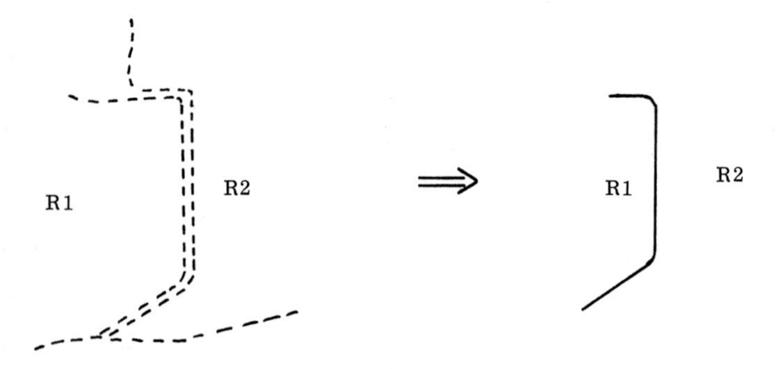

Fig. 11

This is characterised by the mapping rule 22.

$$22 \qquad \underline{\text{map}} \left\langle \underline{\text{Adj}} \left\langle \underset{1}{\text{REGION}}, \underset{2}{\text{REGION}} \right\rangle \left[\underline{\text{Join}} \left\langle \underset{3}{\text{CVE}}, \text{SEG} \langle 1 \rangle, \underset{4}{\text{CVE}}.\text{SEG} \langle 2 \rangle \right\rangle \right], \underset{5}{\text{CURV}} \left\langle \underset{6}{\text{P}} \{\text{Pt---Pt}\} \right\rangle \left[\underline{\text{Equal}} \langle 3, 5 \rangle \right] \right\rangle$$

The CURV into which the Adj is mapped has CVE_3 positions for the set of points from which it is made.

The mapping of 3 adjacent regions may give rise to ordering relationships at a junction. This is given by 23 and illustrated in Fig. 12.

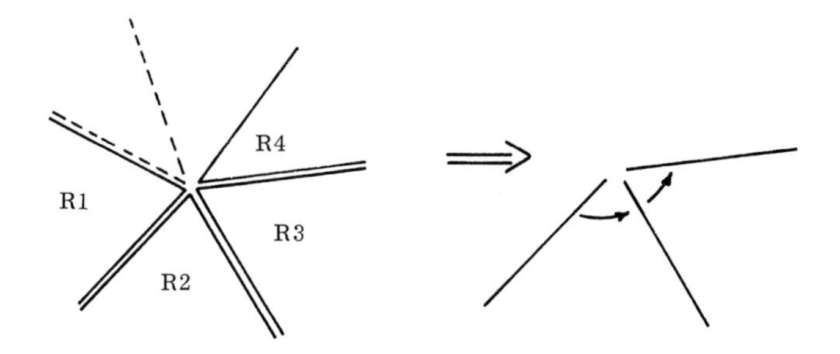

Fig. 12

$$23 \qquad \underline{\text{map}} \left\langle \left\{ \underset{1}{\underline{\text{Adj}}} \left\langle \underset{2}{\text{R}}, \underset{3}{\text{R}} \right\rangle \underset{4}{\underline{\text{Adj}}} \left\langle 3, \underset{5}{\text{R}} \right\rangle \underset{6}{\underline{\text{Adj}}} \left\langle 5, \underset{7}{\text{R}} \right\rangle \right\} \left[\underline{\text{Touch}} \langle 2, 5 \rangle, \text{Touch} \langle 3, 7 \rangle \right], \right.$$
$$\left\{ \underset{8}{\text{CURV}}, \underset{9}{\text{CURV}}, \underset{10}{\text{CURV}} \right\} \left[\underline{\text{Between}} \langle 8, 9, 10 \rangle \; \underline{\text{coinc}} \left\langle \text{PT.END} \langle 8 \rangle, \right. \right.$$
$$\left. \left. \text{PT.END} \langle 9 \rangle \right\rangle \; \underline{\text{coinc}} \left\langle 11, \text{PT.END} \langle 10 \rangle \right\rangle \right] \Big\rangle$$
$$\left[\underline{\text{map}} \langle 1, 8 \rangle \; \underline{\text{map}} \langle 4, 9 \rangle \; \text{Map} \langle 6, 10 \rangle \right]$$

The junctions will occur as it were 'accidentally' by the mapping of single adjacencies. However since junction relationships are very significant surface markers, (i.e. used in any sensible machine recovery of regions), omission would amount to failure to achieve descriptive adequacy.

3.3. Data Structures and Their Use

3.3.1. System Data Structures

The data structures used are hybrids, designed both from consideration of usage and from the forms dictated by the more formal framework. In particular, the latter was responsible for the decision to exhibit relationships explicitly in the structures, binding together the objects relevant to a particular view of the pictorial data.

Fig. 13 shows a road map data structure. The MAP is made up from four CURVES each of which is made from a set of points. In order to save memory space, the curves in the machine are articulated into straight segments. These segments can be thought of as two points (the end points) related by a line relationship. Again, to save space, the line relationships are not explicitly components in the structure but are represented implicitly by the ordering* of the points on the curve ring. (One can interpret SKETCHPAD structures (Sutherland, 1963) as explicitly exhibiting this relationship by the LINE elements. The joins and sames exhibit the relationships which exist in the inset diagram. The closed CURVE 23 is represented as a curve joined to itself. Probably this should be represented as a CLOSURE in its own right but there are no operational difficulties in not doing so.

A regions map data structure is shown in Fig. 14. It is the counterpart to the road map structure in Fig. 13. In it a region is depicted as being made from a set of closures. The inside relationship between the closures of the inner boundary and the closure which we have earlier called the outer boundary is exhibited, obviating the need for naming them as such (in outer and inner boundary). The adjacency relationships between regions are exhibited in terms of the subordinate relationships of joins between closures and in turn sames between points. (Only one adjacency is shown to avoid pictorial confusion.) The outer boundary of region 1 is the frame of the picture and is supplied by the system.

3.3.2. Parsing

The parsing process referred to here is that process which builds a region map data structure given a road map data structure and is represented by the region processor in Fig. 3. There is, of course, a similar process involved in forming the road map data structure from the graphics console input but this process is relatively trivial and is discussed in the following section.

Given the road map structure, the task of the parser is basically to recover

* The economies to be had by using the machine store structure as a meaningful component in a data structure await the development of adequate machine grammars and mapping rules carrying problem structure into machine structure. One feels that just the development of adequate problem grammar is difficult enough and a theory of data structuring based on this alone, would indicate the inclusion of the line relationship.

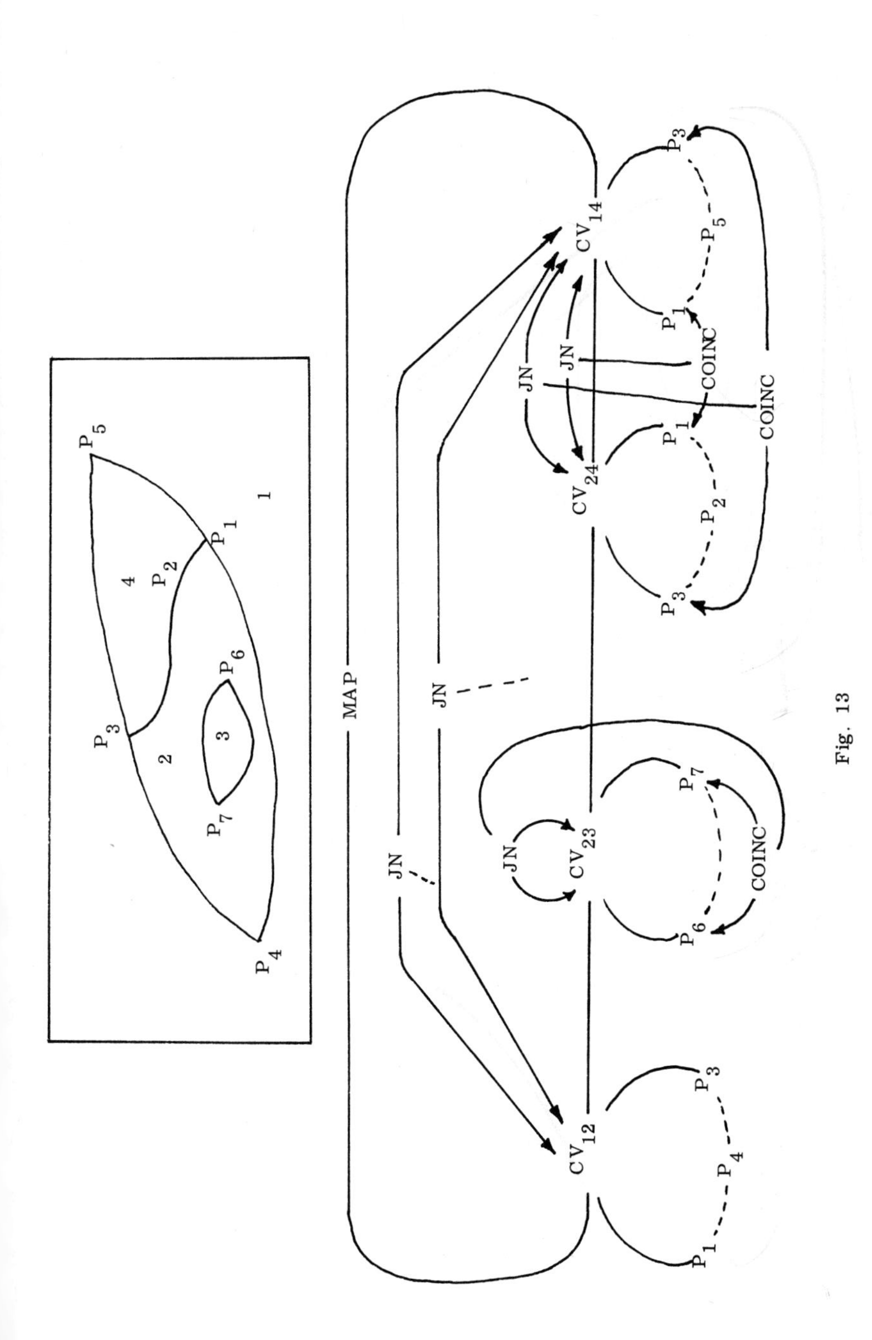
P5
P1
P2
P3
P4
P6
P7
1
2
3
4
MAP
JN
JN
JN
JN
JN
CV14
CV24
CV23
CV12
COINC
COINC
COINC
P3
P5
P1
P1
P2
P3
P7
P6
P3
P4
P1

Fig. 13

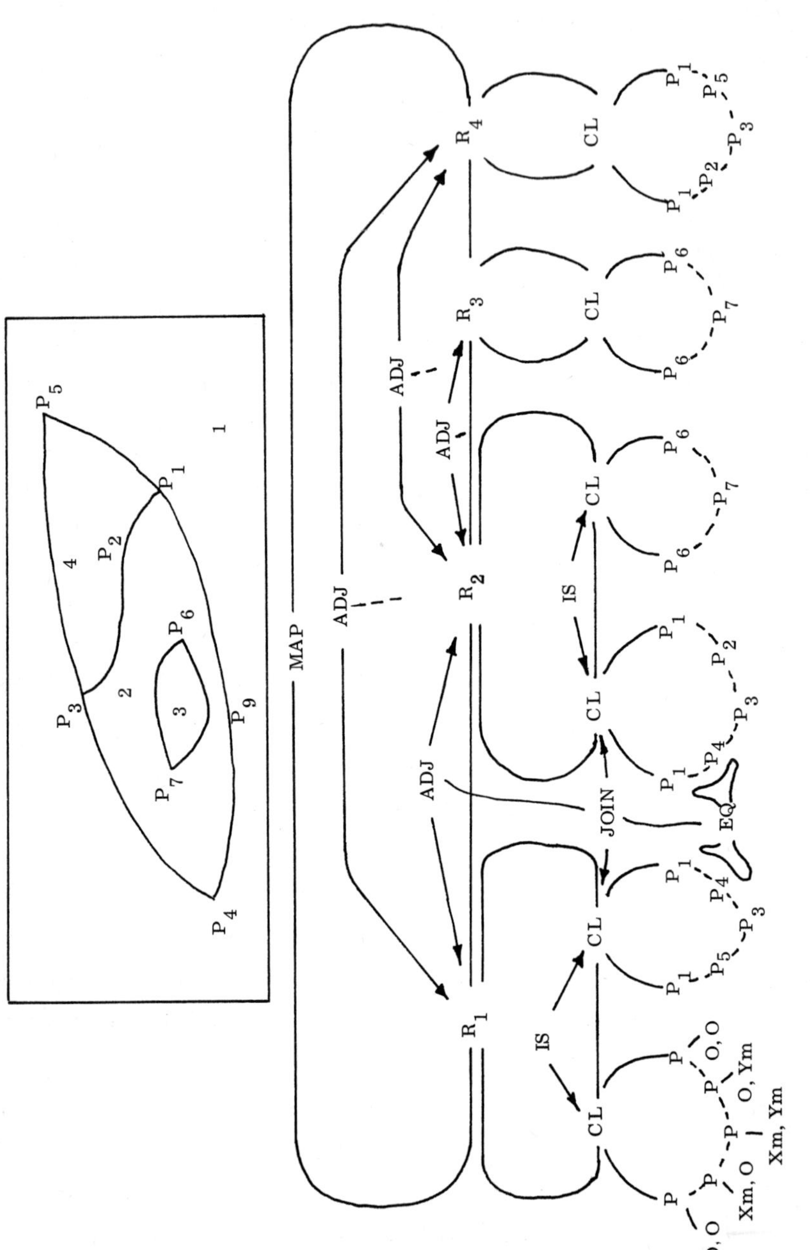
P5
P1
P2
P3
P4
P6
P7
P9
1
2
3
4
MAP
ADJ
R1
R2
R3
R4
CL
IS
JOIN
EQ
O,O
O,Ym
Xm,O
Xm,Ym

Fig. 14

closures. This it does by starting with the system-supplied frame and working inwards. The process of recovering a closure is based on a 'turn furthest to the right' procedure at junctions. Starting on any curve, the path traced out by following the curve to a junction then turning right and following the next curve, etc. until the starting point is reached, (this will always occur) will recover a closure (in the regions sense).

The closure could, however, be an outer or an inner closure depending on the 'direction' of traverse. Consider the map shown in the inset Fig. 14. If the traverse is carried out starting at P_2 in the direction P_3 to P_2, the closure of region 2 ($P_2, P_1, P_9, P_4, P_3, P_2$) will be traced out. However in the direction P_2 to P_3, the closure (P_2, P_3, P_5, P_1, P_2) of region 4 will be recovered. In either case a region closure is recovered. Consider now starting on an inner boundary closure at say P_4. Now in the direction P_4 to P_3 the region 2 closure (P_4, P_3, P_1, P_4) is recovered whilst in the other direction (P_4, P_1, P_5, P_3, P_4) is recovered, which is an inner closure for region 1. We note also that the sense is anti-clockwise when recovering an inner closure of a region and clockwise when recovering an outer closure. There is a simple test for clockwise, anti-clockwise traversal; that of calculating the area under the curve - clockwise implies a positive area and anti-clockwise a negative area. The final observation we need to make is that when all regions are recovered, each interjunction curve will have been traversed exactly twice. Using these observations, the parsing process can be stated as follows: -

(a) Start at a point which has the (or a) minimum y value (this ensures that the point is on a curve mapped from an image boundary).

(b) Trace out a closure in the direction which yields a negative area (trial and error) ensuring the recovery of an inner closure by (c) and (d).

(c) Find a region (already recovered) for which the new closure is <u>inside,</u> (i.e. inside outer closure, outside any inner closures) and assign this closure to the region as an inner closure.

(d) Starting on any curve which has been traversed only once, recover closures which have positive areas and assign these closures to new regions, making them the outer closures.

(e) Repeat (d) until no more 'once traversed' curves remain in the road map structure, then repeat from (a) until all curves are exhausted.

The above procedure assumes that the road map structure constitutes a well formed regions map. The variety of ill-formed road structures is reflected pictorially in Fig. 15. These are handled in two levels. Firstly every curve (in the road structure) is examined to ensure that it enjoys at least two join relationships, one at either end. This check would identify curves A and B in Fig. 15 as being anomalous. Curves D and C are detected by checking that no closure joints to itself. Currently, anomalous curves are deleted automatically from the road map by the system.

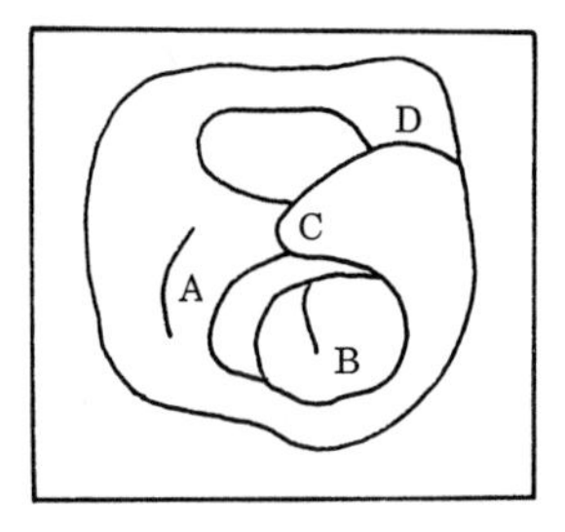

Fig. 15

It is interesting to reflect on how the parsing procedure outlined above could yield clues as to how an adequate grammar is to be constructed. For example, the between relationship at a junction is not explicitly recovered but appears in the code as the 'turn furtherest to the right' procedure. The grammar is obviously implicitly contained in the code which carries out this parsing, as is, of course, what is known as parsing strategy.

3.3.3. System Operation

When a user traces out a curve with a light pen, the points on the curve are formed into a curve structure and this sub-structure is linked into the road map structure provided that this curve is unrelated to any existing curve. Relationships between an input move and existing curves are indicated by light pen intercepts from the display of the existing curves (essentially by a touch relationship between the input and existing curves). For example, Fig. 16 shows two curves, one of which (A) is an existing curve. When B is input, the next relationship (B meets A) is momentarily exhibited and then processed by breaking A into two curves, giving the structure

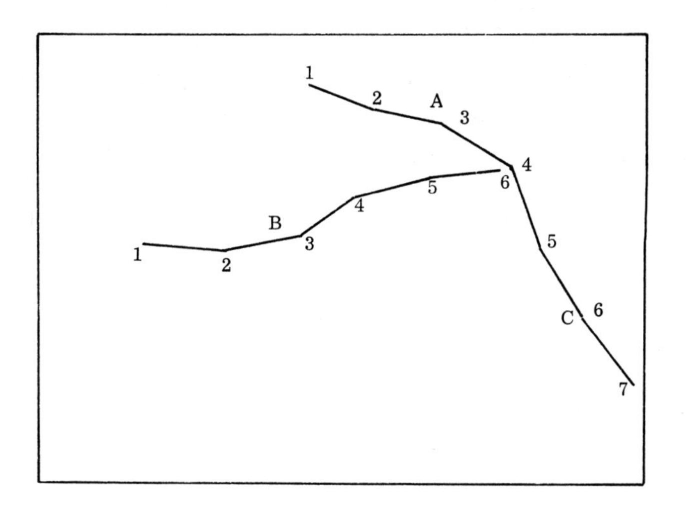
1
2
A
3
4
5
6
B
1
2
3
4
5
6
C
7

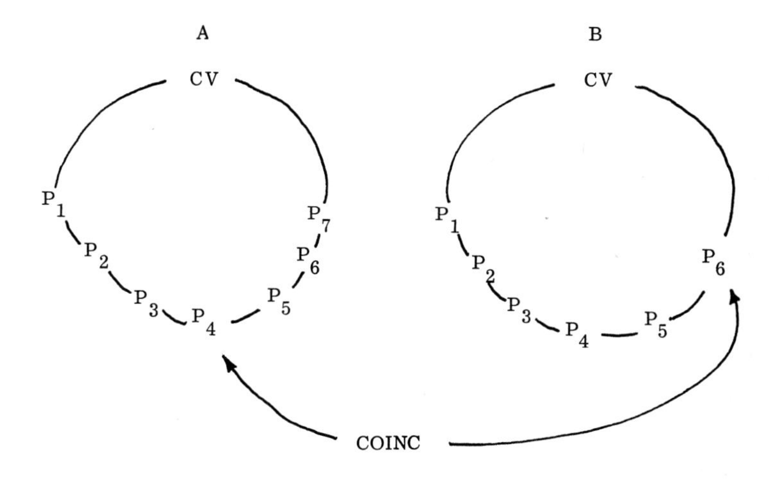
A
CV
P_1
P_2
P_3
P_4
P_5
P_6
P_7
B
CV
P_1
P_2
P_3
P_4
P_5
P_6
COINC

Fig. 16

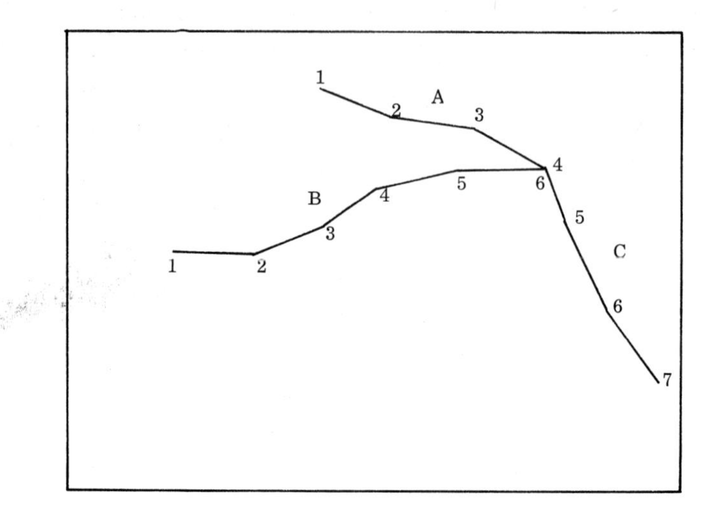
1
2
A
3
4
B
4
5
6
3
5
C
1
2
6
7

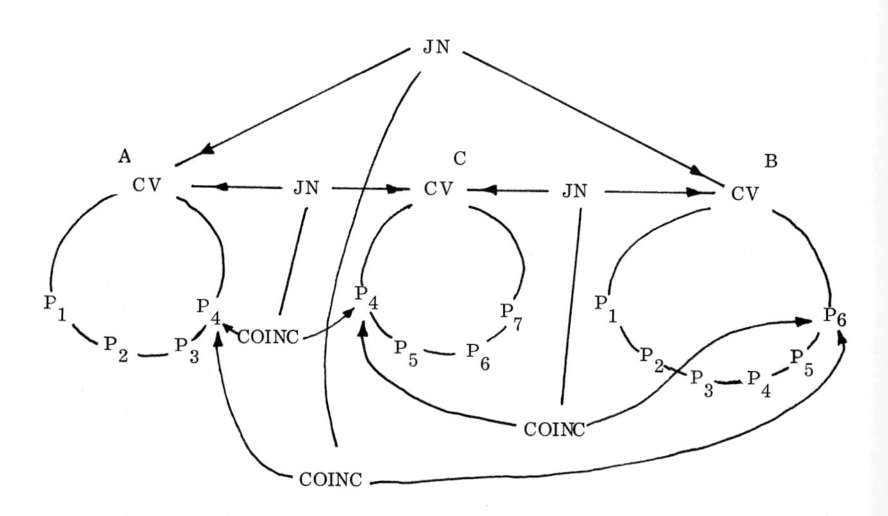
JN
A
CV
JN
C
CV
JN
B
CV
P1
P2
P3
P4
COINC
P4
P5
P6
P7
P1
P2
P3
P4
P5
P6
COINC
COINC

Fig. 17

shown in Fig. 17.* A simpler case arises when the touch indicates a join at an existing junction or (equivalently) at the end of an existing move. A touch may be indicated at the commencement of an input curve. This is handled after the curve has been finished (or when another touch is indicated). Crosses are handled as two meets. It is true that a road map view (to a user) could involve meets and crossings. To this extent, the road map structure does not achieve descriptive adequacy. After each curve is structured, any two curve joins are detected and the two curves combined into one.** Again this could cause descriptive inadequacy. (Consider Fig. 18 where two curves are joined at A. This would be represented as one move whether or not drawn as one curve.) Overcoming these inadequacies involves a grasp of shape, a problem which requires much more work, and will not be discussed here. When a request is made of the problem program, the problem program can request a regions data structure (in a look at the picture in terms of regions). Figs. 19 and 20 show the map inputted to the system and the response from the problem program when asked to display a region (pointed to) together with its area and perimeter.

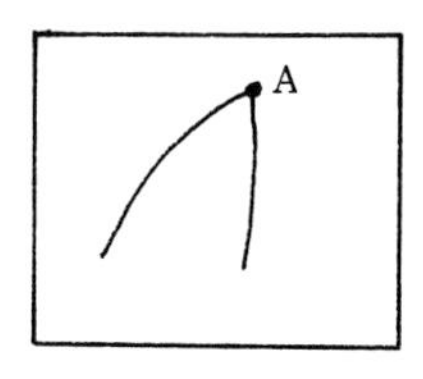

Fig. 18

At any time the user is permitted to edit the map by adding curves or by deleting them. This editing causes the road map structure to be correspondingly altered. (Deletions are carried out by simply removing a curve and all of its relationships from the structure.) This is an easy task to program for, since the road map view is the view the user has when editing.

A request for a regions structure, following any editing, causes the old regions structure to be destroyed, and a new parsing of the modified road structure to take place.

* The use of the lightpen to recover the touch relationship was investigated more or less as an experiment in lightpen mechanics. It proved to be reliable, but due to the response time of the system, it was probably not as flexible as an internal search (since tracking was interrupted with every touch).

** One of the drawbacks of using storage structure to represent the line relationship (See footnote, § 3.3.1) is evident here. When combining two curves, a special routine has to be invoked to give both curves the same 'sense of direction'. This also occurs in the regions parsing when a closure is being formed.

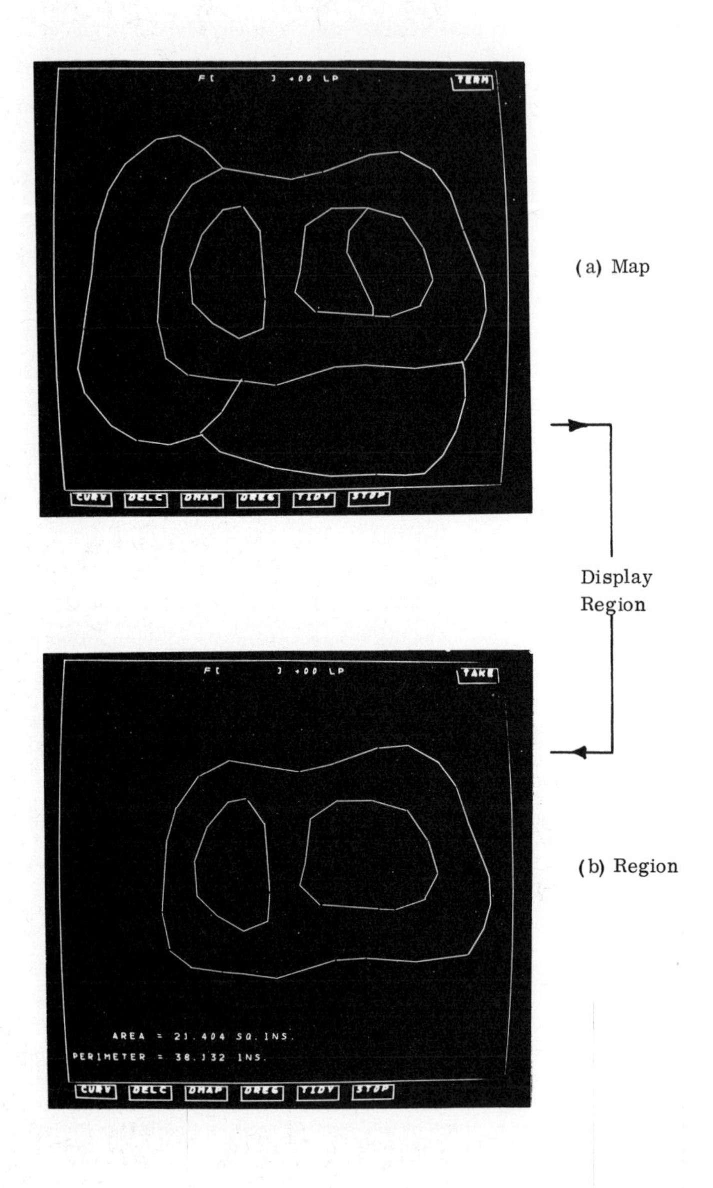

Fig. 19 - Input map and display, area and perimeter of region pointed to.

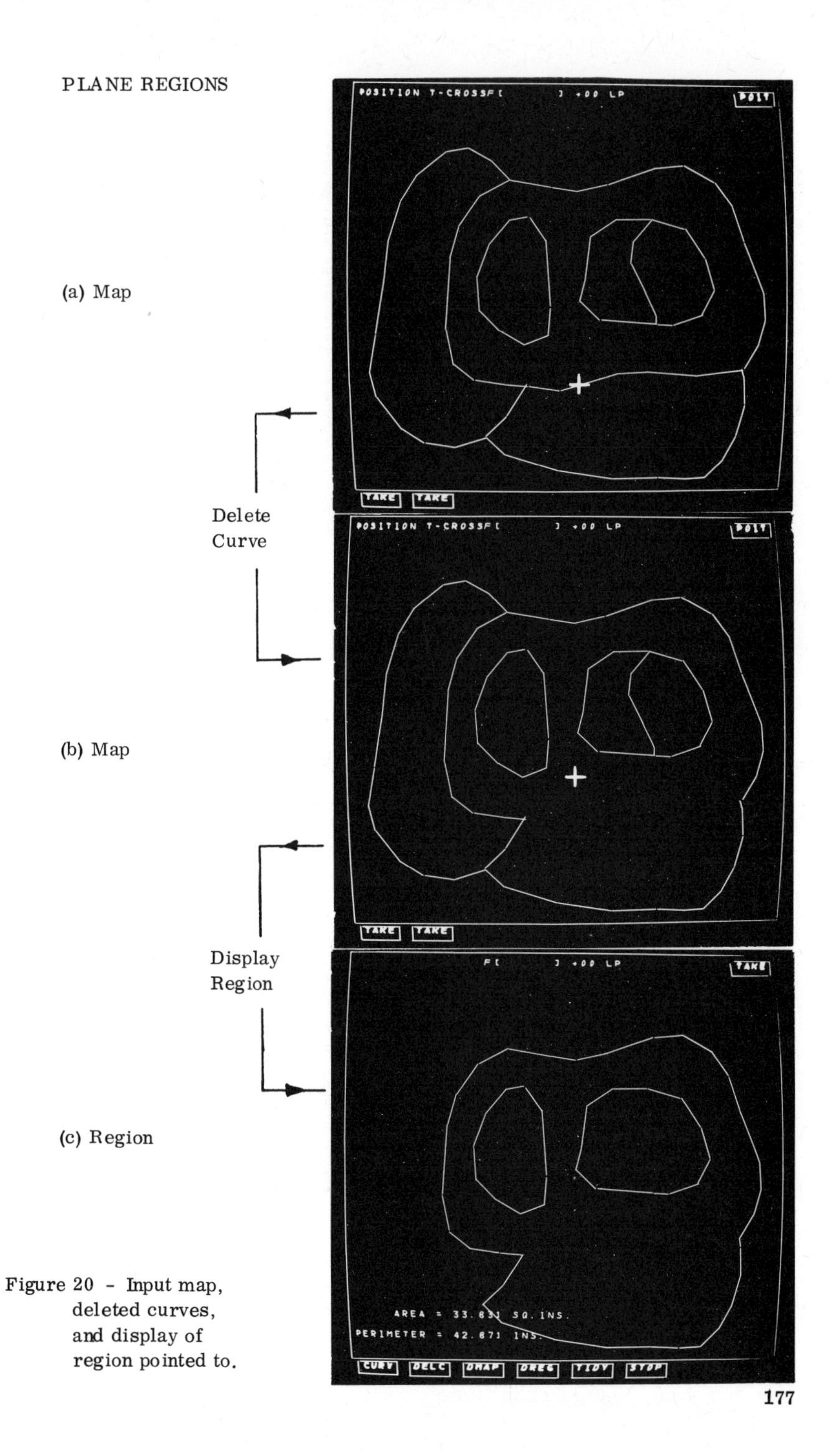

Figure 20 - Input map, deleted curves, and display of region pointed to.

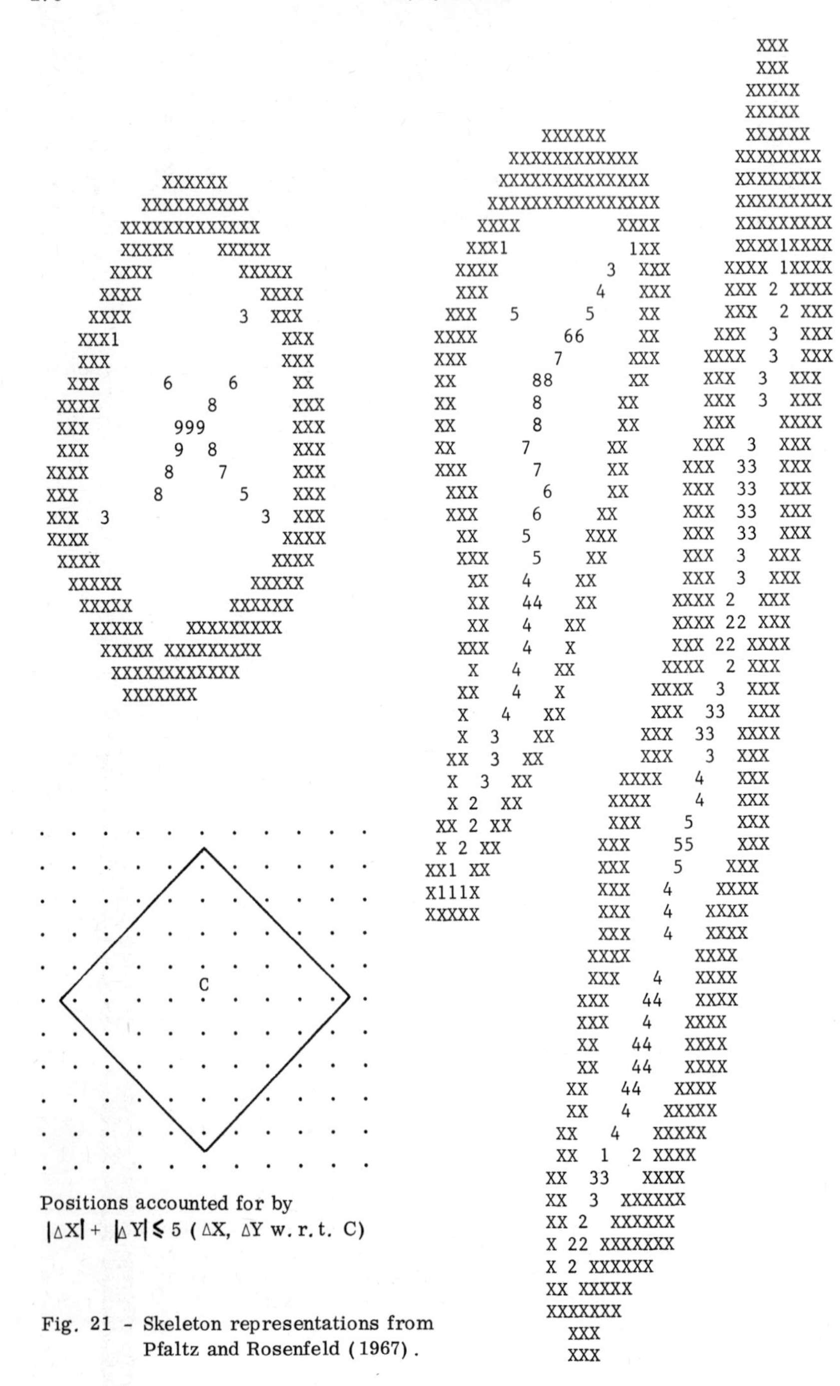

Fig. 21 - Skeleton representations from Pfaltz and Rosenfeld (1967).

3.4. Discussion of Data Structuring

Making comparative remarks about data structures, in the absence of an accepted theory (of data structuring) is a dangerous and somewhat pretentious undertaking. There are, however, certain aspects of data structuring which seem to be open to discussion in terms of their apparent suitability to a set of tasks. At worst, such comments assume the status of observations to be accounted for by a theory. We should also make it clear that we are not discussing claims made for particular structures by their inventors. Rather we are trying to say something about the extent to which such structures reflect our competence, in the belief that a competence model is the only avenue open to us through which a unified approach might be developed. And this competence is in some sense reflected in the sorts of things we want to do with a model.

The skeletal representation of planar regions by Pfaltz and Rosenfeld, (1967), is derived from fitting a series of squares to the inside of a region such that every point inside the region is contained in a square (see Fig. 21). Every position which is a centre of a square is given a value indicating the size of the square, and the resulting figure (of number/position) is the skeleton of the figure. This representation is particularly suited to recovering that a point is either inside or outside of the figure and, by extension, to the shading of the figure.* Keeping in mind that our concern is with the status of this representation as one of a characterisation of regions we can ask how it mediates a grasp of what the region represented is like (what the region is 'a definition of'). For instance, how can it be 'used' to answer questions like 'how many areas are enclosed by this region?', and 'what is the length of the perimeter?', and 'is there an enclosed area which is near the outside border?'. In this vein, we could ask a wide range of questions involving the shape, form and metrical attributes of the region and its parts. It would seem that, in order to handle most of those questions, the region boundaries at least would have to be re-recovered, and that in this sense, this representation does not mediate a grasp of regions. What then is a view of a region in terms of a set of squares as described? Tentatively we might say

* A discussion of the use made of skeleton-like representations of figures as shape descriptors by Rosenfeld and Pfaltz and in the same area of the MAF's of Blum (1964) is outside the scope of this paper. However, the relevance of these works is acknowledged, and, in so far as Blum presents his work as part of a theory of vision, the interested reader might like to compare the approach to this work with that proposed by Clowes (1968b).

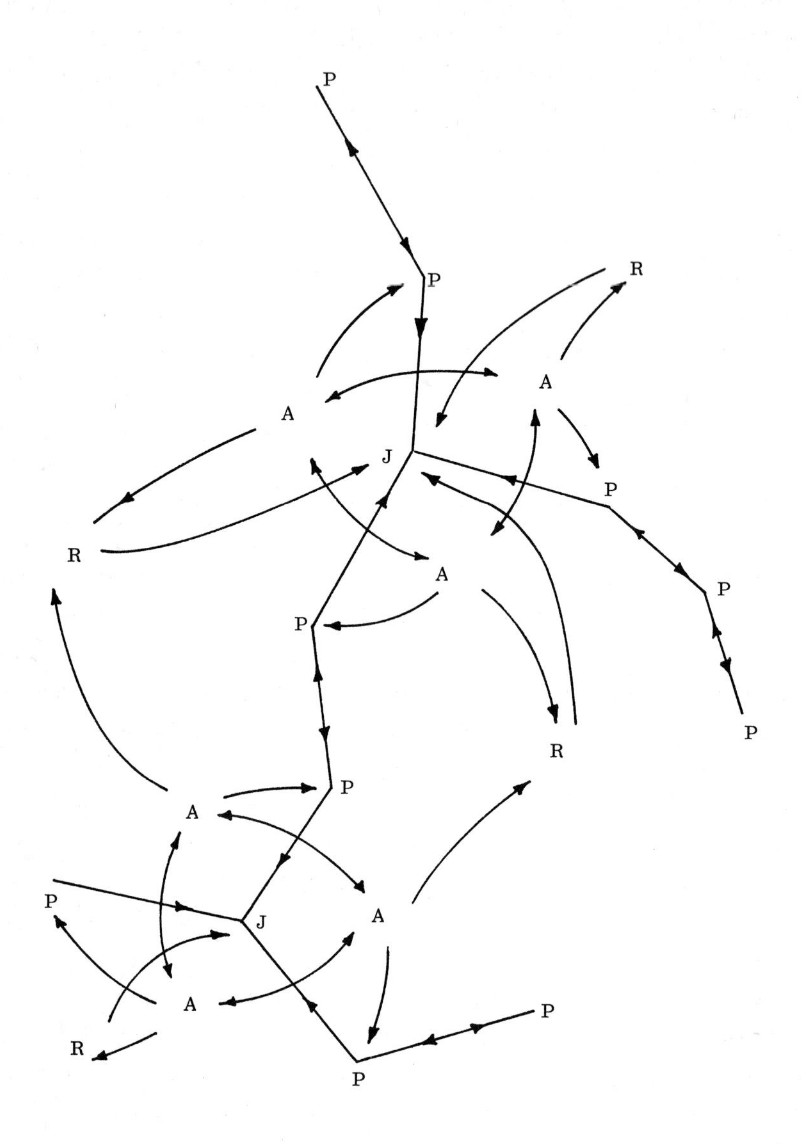

Fig. 22

that this representation grasps a particular attribute of a region: i.e. the inside of a region, rather than the region itself. We temper this remark by noting that from one point of view, the set of positions constituting the inside of a region can be regarded as the region itself (see Section 3.2).

Cook (1967) has used a structure to recover information about a regions map. The structure linkages are shown in Fig. 22; the letters represent labelled memory blocks which are linked. The disconnected boundaries (inner) are handled by 'virtual' edges which are recognised contextually by the processing routines. Cook has given routines which exploit the structure to recover the boundaries of a particular region and it is easy to see how a region's context (in terms of adjacent and touch) can be recovered. This structure can be thought of as a superposition of a road map and regions map view of the graphic where a limited number of the significant relationships are covertly characterised. Where recovery of these relationships is confused, both through the superposition of views and the covertures of relationships, a structural device (such as the 'angle' blocks - A's) is employed. We could say then, that in a significant way, the structure captures some of the 'regionness' of the graphic it characterises. However, there are penalties with adopting this structure for non-specific uses, manifested in the operational difficulties encountered when editing and extension are considered. When editing, for example, it is difficult to see how the structure could be easily transformed when two regions are 'merged'. Similarly, it would be difficult to extend the structure to include a reasonable characterisation of other relationships such as - two inner boundaries being 'near' one another. This seems to be due both to the covertness of relationship representation and to the absence of a direct 'object' representation; (e.g. in the above case, of an inner boundary). In fact, the latter implies the former. These aspects give rise to a rigidity quality of the structure, which for general usage, seems undesirable. So far, we have concentrated on the extent to which a structure reflects our competence about regions. We have said nothing about the 'suitability' of a structure to a particular application. It is clear that in many applications a particular structure, much less rich than one which captures complete competence, will give rise to time and storage space economies. For example, Pfaltz and Rosenfeld discuss compute time for insideness whilst weighing up storage requirements for skeletons versus straight line boundary segments. (We could list all of the inside points to improve processing time at the expense of storage, or store only the boundary points at the expense of processing time.)

There are two issues involved here. The first concerns limiting the characterisation (of regions in this case) to just those attributes relevant to an application, and since such simplification would seem to lead to both time and space savings, the issue is not so much a direct 'computing' issue as one about the <u>definition</u> of what we want to compute.* Such a definition necessarily involves our competence, and so we need a statement of this in a grammar.

The second issue is a space-time one which is related to structure-code trade-off. At this stage, there is very little we can say about this trade-off.** If the approach advanced in this paper is sound, the framework for discussing this issue will involve a characterisation of the machine (as indicated in the footnote of Section 3.3.1) which would satisfy the sort of adequacy requirement we have demanded of the regions characterisation.

4. CONCLUSIONS

In this paper we have attempted to show how achieving significant man-machine graphical communication rests on our giving machines the capacity to 'see'. We regard this capacity as being evidenced by machine based structural descriptions of pictorial data. The apparatus required for assigning structural descriptions consists of a set of rules, which, because of their concern with objects and relationships, are analogous with a syntactic component of a grammar.

Perhaps the most significant contribution to a syntactic theory has come from Chomsky (1957, 1958, 1965). However, Chomsky's model for a syntax (transformational grammar) seems to be inadequate for picture description in that relationships ar not overtly characterised. Clowes'(1968b, 1968c) picture grammar captures the relational aspects of syntactic descriptions and as such provides a model for the descriptio

* As computing science develops, this issue will of course become a very important one, particularly since it will probably underlie any attempt to determine code equivalence.

** There is one aspect of secondary storage economy which is open for investigation given that data structures can be related to grammars. That is not to store the structure at all, but to reconstruct it (using the grammar) each time it is required, and that may be for a small amount of the data. This approach would allow us to store data in a 'minimal' form. It is worth noting that the ability to reparse seems to be a significant dimension of human performance (cf. centre of attention).

of domains. The idea that several domains are involved is implicit in the traditional linguistic components of syntax and semantics. That the problem domain (at least a part of a semantic component) could probably be characterised in a similar manner to the language domain (the traditional syntax component) was proposed by Clowes (1968a). In Section 2, the multi-domain* approach is argued for and the nature of the relationship between domains (in mapping) is presented.

Based on these concepts, an approach to the plane regions problem was formulated in terms of 'two views', corresponding to a view of the problem domain and a view of the language domain, and a mapping relationship between them. It is felt that this approach facilitated the implementation of the system.

Writing adequate grammars (syntactic description) is at present very difficult. However, as the issues become clearer, we can expect a language to emerge which facilitates the writing of grammars. Such a language would form the basis for discussing relative merits of information representation. Our preoccupation with data structures (in this and most other computer graphics papers) is testimony to the need for a common descriptive language. The point is, that to date, data structures (and perhaps some English) have been the main descriptive devices employed and for reasons outlined in this paper, they are just not adequate, (in spite of some good work done by Gray (1967) in attempting to formalise data structures).

One important aspect of the domains theory is not exemplified by the regions problem. The regions domain is characterised by pictorial-like objects which exploit pictorial relationships for their definition. The approach is, however, general enough to handle problem domains which are clearly non pictorial and, of course, the mapping of such domains into picture or any other language. Thus we would expect to characterise electrical circuit theory quite independent from the language in which circuits (which reflect this theory) are presented.

Finally, a remark on a possible qualification to the premise that machines be able to 'see' in order to communicate graphically. It may be that a particular problem requires that the machine recover only certain, quite specific, things in a picture such that we would not identify these things with an intuitive description of what we see. We would expect such systems to be based on ones which do recover a complete description, but where computational 'short-cuts' had been utilised.

* By way of contrast, see Shaw's (1968) picture description language which is essentially formulated within a single domain.

At this stage then, it seems that the only way in which we can come to grips with the whole area, is to exhibit pictures in terms of overt, competence models.

ACKNOWLEDGEMENTS

I am indebted to Dr. M.B. Clowes of the Computing Research Division, C.S.I.R.O., for encouragement and many fruitful discussions over the course of this work, and for his valuable suggestions in the formulation of this paper.

This work has benefitted from discussions held within the VERBIGRAPHICS project; other participants were M.B. Clowes and D.J. Langridge of C.S.I.R.O., C.J. Barter of the University of N.S.W., and R.J. Zatorski of University of Melbourne.

Much of this work was carried out while I was at the Department of Electronic Computation, University of N.S.W.

I am grateful to Dr. G.N. Lance, Head, Division of Computing Research, for allowing me access to C.S.I.R.O. equipment during this time.

5. REFERENCES

ANDERSON, R.H. (1968), "Syntax-directed recognition of hand-printed two-dimensional mathematics"; Ph.D. Thesis, Harvard University, Cambridge, Massachusetts.

BARTER, C.J. (1969), "Data structuring and question answering"; This conference.

BLUM, H. (1964), "A transformation for extracting new descriptors of shape"; In, "Models for the perception of speech and visual form", Ed. Wathen-Dunn, M.I.T. Press, Cambridge, Massachusetts, pp. 362-379.

CHOMSKY, N. (1957), "Syntactic structures"; Mouton & Co. Press.

CHOMSKY, N. (1958), "A transformational approach to syntax"; In "The Structure of Language", Eds. Fodor and Katz, Prentice Hall, 1964, pp. 211-245.

CHOMSKY, N. (1965), "Aspects of the theory of syntax"; M.I.T. Press, Cambridge, Massachusetts.

CLOWES, M.B. (1968a), "Paradigms and syntactic models"; Seminar Paper No. 10, Division of Computing Research, C.S.I.R.O., Canberra.

CLOWES, M.B. (1968b), "Transformational grammars and the organisation of pictures"; Seminar Paper No. 11, Division of Computing Research, C.S.I.R.O., Canberra. Presented to the Nato Summer School on Automatic Interpretation and Classification of Images, Pisa, Italy, 1968.

CLOWES, M.B. (1968c), "Pictorial relationships - a syntactic approach"; In Machine Intelligence 4., Eds. Collins and Michie. In Press.

CLOWES, M.B. (1969), "Picture syntax"; Seminar Paper No. 15, C.S.I.R.O., Canberra. This conference.

COOK, B.G. (1967), "A computer representation of plane regions boundaries"; Australian Computer Journal, 1, 1, pp. 44-50.

GRAY, J.C. (1967), "Compound data structure for computer aided design; a survey"; Proc. ACM Nat. Meeting, pp. 355-365.

GUZMAN, A. (1967), "Decomposition of a visual scene into bodies"; Art Int. Memo No. 139, Project MAC (MAC-M-357), M.I.T., Cambridge, Mass.

LOOMIS, R.G. (1965), "Boundary networks"; Comm. ACM 8, 1, pp. 44-48.

MARTIN, W.A. (1965), "Syntax and display of mathematical expressions"; Memorandum MAC-M-275, Project MAC, M.I.T., Cambridge, Mass.

NARASIMHAN, R. (1964), "Labelling schemata and syntactic descriptions of pictures"; Information and Control 7, pp. 151-179.

NARASIMHAN, R. (1966), "Syntax directed interpretation of classes of pictures"; Comm. ACM 9, 3, pp. 166-173.

PFALTZ, J.L. and ROSENFELD, A. (1967), "Computer representation of planar regions by their skeletons"; Comm. ACM 10, 2, pp. 119-122, p. 125.

ROSENFELD, A. and PFALTZ, J.L. (1966), "Sequential operations in digital picture processing"; J. ACM. 13, 4, pp. 471-494.

ROSS, D.T. (1967), "The AED approach to generalised computer aided design"; Proc. ACM. Nat. Meeting, pp 367-385.

SHAW, A.C. (1968), "The formal description and parsing of pictures"; Tech. Rept. No. CS94, Computer Science Department, Stanford University.

SUTHERLAND, I.E. (1963), "Sketch pad - a man-machine graphical communication system"; M.I.T. Lincoln Lab., Tech. Rept. No. 26.

SUTHERLAND, W.R. (1966), "The Coral language and data structure"; Tech. Rept. No. 405, Lincoln Lab., M.I.T., Cambridge, Massachusetts.

TOMLINSON, R.F. (1968), "A geographic information system for regional planning"; In "Land Evaluation", Ed. Stewart, Macmillan Press, Australia, pp. 200-210.

6. APPENDIX

Implementation

The data structures are of the ring type, and are set up and manipulated through a basic plex processing language (PLEXPACK) which is similar to the CORAL language (Sutherland, 1966). The data block form is shown in Fig. A1. Rings may be set up either with forward pointing elements using a half word (24 bits) or with 3 pointer elements (back, to the ring start, and forward) using a full word. Each of the structure processing routines operate on both single pointer and three pointer rings, allowing comparative economies to be investigated.

Three types of structural blocks are used in the regions/road map structures. They are OBJECT, RELATION and LINK blocks shown in Fig. A2 (cf. {LINE, POINT, PICTURE, etc.}, CONSTRAINT and INSTANCE blocks respectively in SKETCHPAD). The LINK blocks sit on the REL ring of OBJECTS, allowing the one object to enter into many relationships. Fig. A3 shows a structure which would be constructed to describe two curves joined together. For simplicity, the curves have two points only, P1 and P2 for CURVE A, and, P3 and P4 for CURVE B. The two curves are joined since the two endpoints P1 and P3 are coincident.

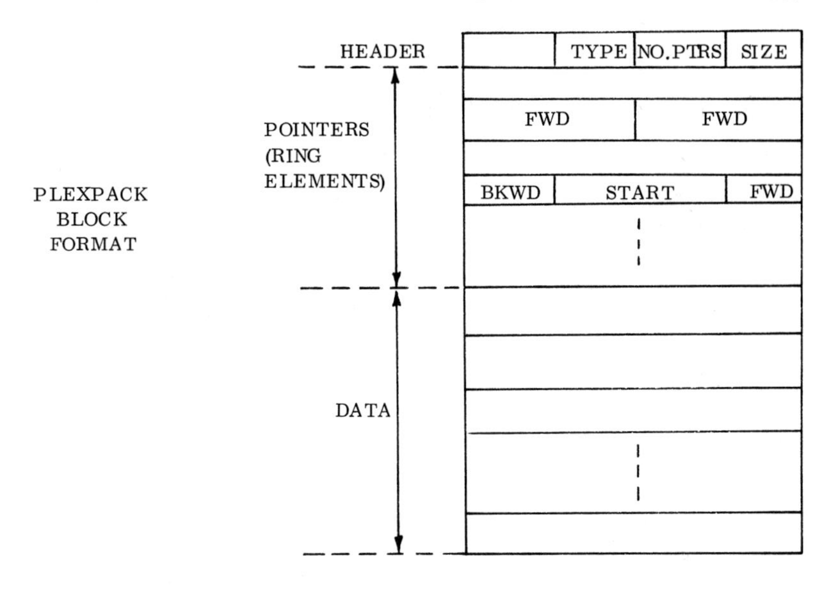

Figure A1 - Data block form.

OBJECT
HEADER
GENERIC (M)
PART OF (M)
RELS (S)
PARTS (S)
HISTORY / DISPLAY
X VALUE
Y VALUE

MAP, CURVE, POINT, CLOSURE

S - START OF RING

M - MEMBER OF RING

RELATION
HEADER
GENERIC (M)
SUBORDINATE TO (M)
OP 1 (S)
OP 2 (S)
SUBORDINATE RELS

ADJ, JOIN, INSIDE, EQUALS, COINCIDENT

LINK
HEADER
LINK (M)
LINK (M)

Figure A2 - OBJECT, RELATION, and LINK blocks.

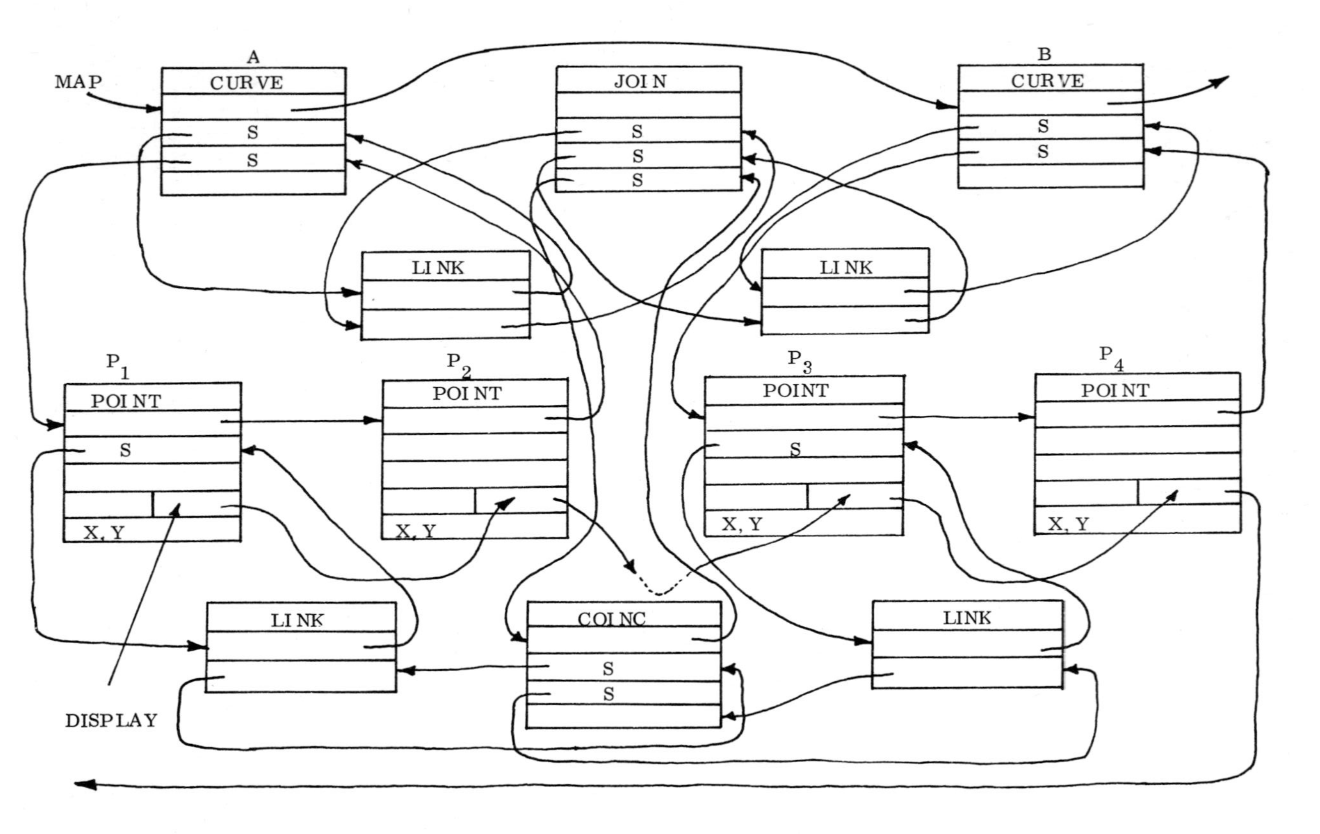

Figure A3 - Structure to describe two curves joined together.

7. DISCUSSION

Kovarik: Do you regard attributes as objects?

Stanton: Yes.

Kovarik: It seems that there's some need for specification when relations are defined over attributes and could not be defined over objects. How do you sort out if a relation doesn't define attributes but just another object?

Stanton: Within this system, relationships can only be defined on objects. What sort of relationships were you thinking of?

Kovarik: Let's say a line being thick or thin, is an attribute of the line and maybe either to the north, or to the south of another line, but this relationship of being to the north or being to the south does not apply to objects such as thickness and thinness.

Stanton: In the rule schemata that we are using, if we wanted to say that north and south didn't apply to say, thickness, then we just wouldn't have that in the description. This descriptive apparatus will allow us to describe all the possible things it can be, when the rules are written out.

Clowes: Do you suggest that the curve you draw in your road map domain is between two sets of points, I mean each position on it is between corresponding pairs of positions. This implies a higher resolution to draw the road map than to plot the regions - isn't that roughly how we think about it; this line lies between the two sets of points?

Stanton: You mean that the positions on that line do not belong to any region?

Clowes: Yes. I mean the relationship to those two by virtue of the line between them in the demarcator.

Narasimhan: What he has given there, 'between' is not defined in the region domain at all.

Stanton: No. I couldn't just do that.

Narasimhan: If you are operating in the region domain, and you want to arrive at something in the curve domain, you would have to operate in terms defining the domains.

Clowes: Yes, but there is no difficulty in having relations which operate in both domains. The only question is where there are some operations which are peculiar to one domain.

Stanton: I haven't considered that. My process has just assumed that we take one set of positions and map these.

Narasimhan: Do your attributes have values?

Stanton: Attributes have values - those values I refer to as being objects. If you want to do arithmetic with the lengths of lines, say, these become objects that can be related by additions and so on.

Narasimhan: You say the length is an object, the values and lengths are objects?

Stanton: Line, is an object which has the attribute length.

Narasimhan: And what is number?

Stanton: Number is the length of the line.

Clowes: Number is the value.

Stanton: I have not used value as such, because I don't know the distinction between just objects and what you call values, so I just leave them all as objects.

Clowes: No, but its clear that if you want to say something of the kind, "this line is longer than that one" then this is a 'greater than' relationship on the values of the two attributes.

Stanton: I have said that relationships can only relate objects.

Clowes: That's right, number is an object - I am saying something about the 'greater than' relationship, now this is essentially arithmetic.

Narasimhan: You started off your lecture by saying that you would like to see neutral descriptions. It seems to me that everything that you have been saying seems to deny that case. In fact in going inside the whole sequence of domains you are implicitly arguing for appropriateness of descriptions, relevance of descriptions and so on. I would agree with that - this is the way one should proceed. I would interpret that by saying that your behaviour seems to be more rational than your representation of the behaviour.

Stanton: I can give an example of what I mean by that. Let us take a certain language that we can describe in a phrase structure grammar, and further that we can give an account in a precedence grammar. If we take the phrase structure grammar in its minimal adequate descriptive form, and if we take a precedence grammar for the same thing, these two are often different, and it seems to me that all we have done is thrown into the descriptive device, say the grammar in the case of precedence grammar, we have thrown into it ideas about how we want to use it, i.e., the parsing process. We have complicated the description in order to make it easier to parse. My personal position is that we may not want to use the description to parse, we may want to use it as a basis for answering questions, we may want to use it as a base for synthesis, and so on. Since these uses often do place different requirements on the description, if the description was to reflect the usage, then we should try to divorce the two, in other words I would say, let us have the description separately, have the process, and work on how it is that, given a particular process in this description, we can get hold of perhaps a more efficient way of running this on a machine. That's what I meant by trying to get neutral descriptions.

Narasimhan: Yes, but that's not what you are doing. There is no such thing as a single description that is universally valid. What you want to have is a variety of descriptions, each of them general to a particular usage, and you want to be able to go from one description to another - which is essentially what you are doing.

Clowes: Supposing you have one particular set of descriptions which you accept as being characterised as Stanton's work, well suited towards particular usage, i.e., parsing road maps to set the regions out. Supposing that you now have the task of going the other way, or answering questions about this work in some Q.A. system: would you want to say that there is nothing to be found of use in that original description? Is there no common element?

Narasimhan: There may be or there may not be.

Clowes: What Stanton is arguing for, and what I'm arguing for, is that there is a common element. Bobrow has said that what we should do is parse it this way, and then generate it that way, and then intersect two such descriptions, and that is a common element. (To Bobrow) Is that right?

Bobrow: I'm saying that if you want to find a common element, that's a way of finding it, but the important thing is that you choose these things for their purpose. Another important test is to know what the mappings are, so that you can tell which things are common and which are things that have been added. So the important thing is how do you go from one domain to the other, and what are good sets of descriptions in each of the domains? As Narasimhan said, that's what you are doing, but that's not what you claim you are doing. I think what you are doing is very good.

Narasimhan: I would say that in the artificial intelligence problem, the core of the problem is to be able to provide varieties of descriptions, and to be able to move, from one to the other, and see what the common parts are, what the disjunctions are, and so on. I don't think it is very meaningful to talk about a description as if it is something that stands all by itself. Descriptions are always relative to what you are going to do with them.

Stanton: If what we are going to do with them encompasses all the variety of usages that we can imagine say, for pattern recognition, or synthesis, or as a base for Q. A. and so on; if you lump all those into the one purpose, how do you direct it then?

Bobrow: You mean more than the union of all the descriptions?

Narasimhan: Why cannot you have union in all descriptions? You seem to be arguing against redundancy inside the computer.

Stanton: I can't imagine what the union of all the descriptions would really look like.

Narasimhan: You have it inside. You have shown a picture with all the data sitting inside.

Bobrow: Like the union of the road map description, and the regions description.

Stanton: I don't understand your point, if you're saying that we can in the one representation, give all the things that we'd ever want to know for all uses.

Narasimhan: No you don't have to. What I am saying is that you can give them different representations, but why do you have to ask for a single representation from which you want to be able to generate all these things in some easy way, and then argue that that single representation is a neutral description?

Stanton: Because I believe that it would be nice to think that: given these descriptions, a common universal analysis procedure could be discovered. If that were true, this would be a very powerful way of constructing this system.

Narasimhan: It seems to me that it is completely misguided. If history teaches anything at all, it teaches it is the wrong way of looking at things - that you don't want to ask for these universal things, even if they exist, as there is no reason to believe that even if they exist they are bound to be extremely efficient.

Macleod: How do you relate the neutrality of the descriptive apparatus to the neutrality of the particular descriptions? Is what you are arguing for a neutral descriptive apparatus, rather than neutral descriptions?

Stanton: What I am arguing for is a representation which is independent, which, when you look at it, captures what is, and not how it was got there, or what you want to do with it now, and so on, in terms of these objects, relationships and attributes.

Narasimhan: That sounds terribly Aristotelian.

Overheu: This sounds a little like the universal computer orientated language problem which I recall John McCarthy considered to be a step up a blind alley. On the other hand, I find it very difficult to believe that one has to take the union of all descriptions and pack it into this thing. If you are going to talk about artificial intelligence, there must be some median point between these two.

Bobrow: I think a critical point is that if you have, for example, the predicate calculus which is thought by very many people to be fairly universal, then theorem proving techniques are considered the appropriate way to solve all problems; it turns out that the major problem with using it is that in some sense the theorem proving techniques don't know how to take advantage of knowledge that is somehow built in. That is, if you can take advantage of specific knowledge that you know about, you can do things a lot more easily. If you say, now, if we can get a universal process, by going to universal Turing machines, it takes a lot of time, and we do a lot of interpretation. If we really want artificial intelligence, getting intelligence somehow means doing it in a smart way, rather than just fumbling around and doing a complete search. I think that is the reason why one does not like neutral descriptions, because neutral descriptions are ones that don't help. I'm saying that if you just use the standard predicate calculus techniques, you don't take advantage of all the things you know about - the subjects that you are expressing, and it is the capability of taking advantage of those subjects, that we are looking for when we want other descriptions in addition to the predicate calculus techniques, and that is what people are now trying to do in theorem proving in trying to build in techniques to escape from the predicate calculus when it looks like its reasonable.

Overheu: Is our speaker interested then in his neutral descriptions as a basic formalism, or does he want it for a single process or at the moment?

Stanton: Certainly the latter. I would expect this to be satisfactory to be used for a particular problem, which is not to say that there may not be better ways because of that specific application. What I would like to do is, given a neutral description, show how a specific application could be done more efficiently, or better, because of that application - because of the usage of the description in a particular application.

PROBLEMS IN ON-LINE CHARACTER RECOGNITION

J. F. O'CALLAGHAN*

Division of Computing Research
C.S.I.R.O., Canberra

1. INTRODUCTION

Many computer programs for recognizing patterns such as hand-drawn characters, bubble-chamber tracks and biological cells have been reported in the literature (4, 7, 8, 12, 13). However, most of the procedures place some restrictions on the inputs in order to achieve a high recognition performance in a sufficiently short time. Examples of these restrictions are:- only one character can be processed at a time; the amount of "line-like" noise in a bubble-chamber photograph must be small; cells are not allowed to overlap. No program has been developed which can recognize a set of patterns under the variety of conditions that a human being can recognize them. That a framework for such a "general program" has not emerged reflects the difficulty of the various recognition tasks.

It is the theme of the current research to develop recognition procedures which might form part of a "general program" to recognize characters. Previous efforts have involved the formation of an approach to character recognition and the development of a program for the on-line recognition of user-specified classes of line drawings (11). This basic program is currently being extended and improved, to operate on a wider range of inputs. It is anticipated that in the future, more complex inputs, such as those having connected and overlapping characters, and those with a matrix-cell representation, will be considered using some of the procedures which have been developed.

This paper outlines the approach taken to character recognition and the basic program, now operating on the CDC DD250 (VISTA) display at C.S.I.R.O. Also discussed are some of the extensions made to the above program - namely the ability

* Previously with the Department of Engineering Physics, Research School of Physical Sciences, The Australian National University, Canberra, A.C.T.

to process multi-stroke figures, a tolerance for local variations in characters, and the incorporation of a "don't know" classification. While the main purpose of the research is to investigate such extensions, consideration has been given to reducing the computation time of the programs so that they may be employed in a practical system, e.g. editing alphanumerics on graphics (maps) or writing mathematical functions on VISTA.

2. THE APPROACH TO CHARACTER RECOGNITION

Many procedures proposed for character recognition incorporate an input representation which consists of a set of unrelated features - typically groups of matrix cells (14) or line segments from an analogue follower (6). These features provide a weak representation, for classes have a wide variety of distortions, e.g. sloppy hand-drawn characters. Furthermore, for complex inputs such as line drawings of houses, a recognition procedure requires information about parts of the input and their inter-associations - that is, a structural description (5) or model of the input.

Research in man-machine communication systems has emphasized the importance of procedures for analyzing and describing classes of objects. In this context, the classification of figures is only part of the overall function of the system (10). Consideration must be given to how the machine can present information about the input picture in a form readily interpretable by the user.

The above points have influenced the type of descriptive model employed to represent the characters in the current approach. The model consists of an hierarchical set of parts, with the following levels:-

Level 1 : Line segments
Level 2 : Curves
Level 3 : Figures

Thus a figure consists of an ordered set of curves, which in turn consist of connected line segments.

The model for the character classes is a set of "average characters" or "patterns" (AVS), each of which has the above hierarchical structure. At least one AV is stored for each class, there being more than one for classes which have subsets (e.g. *2*, 2). There are five curve types - 2 sharp corners (one for each direction of travel), 2 curves and a straight line. It is suggested that an articulation of characters into these curve types is an efficient and useful "coding" (1) for the characters, and is also a representation, "meaningful" to a human observer.

Recognition as considered in this approach, requires a two-stage process:-

(1) In the first stage, objects are found in the input which are similar to the patterns stored in the AV. On the curve level, an attempt is made to group line segments into the same curve types as the AV possesses. On the figure level, a description which is similar to each particular AV is formed - the result is a set of descriptions for the given input. Thus, for example, an input ")" may be described by the AV for a "2" and a "7" as:-

(a) the first part of a "2";

(b) a "7" with a not-so-sharp corner, respectively.

(2) In the 2nd stage, values for "similarity" measures, which define a similarity between an input description and its generating AV, are calculated. The measures depend on factors such as the magnitude of the slope differences between segments in a curve and the orientation and beginning - end vector of a curve. The values for each curve are summed to give a total between 0 and 1. The totals for each AV are compared and the class with the maximum value receives the decision concerning the input's class.

The overall procedure is inductive - the decision of the class membership of the input is a predictive inference (3) from a sample of evidence (the AVS). It has been instructive to consider the processing as hypothesis testing, in which the hypothesis is a statement that "the class represented by the AV is the class to which the input belongs". Rules are embodied in the program to generate the descriptions - it is possible for an input to be quite dissimilar from the generating AV, (e.g. "2", "8"), in which case the hypothesis is purged. The credibility (or confirmation, in this study) of an hypothesis (15) is given by the similarity value. This value may also be thought of as a "subjective" inductive probability (3), which is used in a comparative manner to determine class membership. An advantage of the hypothesis testing is that descriptions can be conveniently compared, by having the curve types as a basic unit of correspondence. If an objective or unique description were formed (using the same descriptors) then there would exist a problem in providing rules for comparing different curve types. The method adopted does appear more feasible and has been employed by Marill et al. (9).

Because of the correspondence between an input description and the AV, learning, in the sense of model construction, can be easily incorporated. Hence the program can construct a model of the classes from given samples. Because characters of different shape may possess the same name, the user is required to provide the correct classification of the input during learning. If the program's decision is correct, the input can be averaged with the AV, to "enhance" the patterns; otherwise, if not correct, a new subset of the correct class or another class (if previously unknown) can be added to the model.

3. THE BASIC PROGRAM, RECØG

3.1. Outline of the Method

The above overall approach has suggested a recognition procedure, which is incorporated in the program, designated RECØG. The method may be divided into the following steps (shown in Figure 1):-

(a) (X, Y) coordinate positions, equidistant along the drawn line (currently 1/6"), are obtained from the VISTA routines; the orientation and slope differences of the defined segments are computed.

(b) For each AV, the slope differences in the input are successively compared (in sign and magnitude) and matched to those in the AV. A particular input value may be assigned either to the current curve, the preceding curve (if possible) or the succeeding curve. In certain cases when a match cannot be found, the hypothesis is purged.

This procedure is a sequential synthesis - a grouping of successive line segments into the expected curves. There is no back tracking if inconsistencies occur and there is no attempt to search forward in the input to look at future variations.

(c) Similarity values based on the following factors, are calculated for the input and AV segments for each curve:-

<u>Curvature</u> : ratio of number of matched segments to the number expected, modified by a size factor (based on the ratio of the total number of input segments to the number of AV segments).

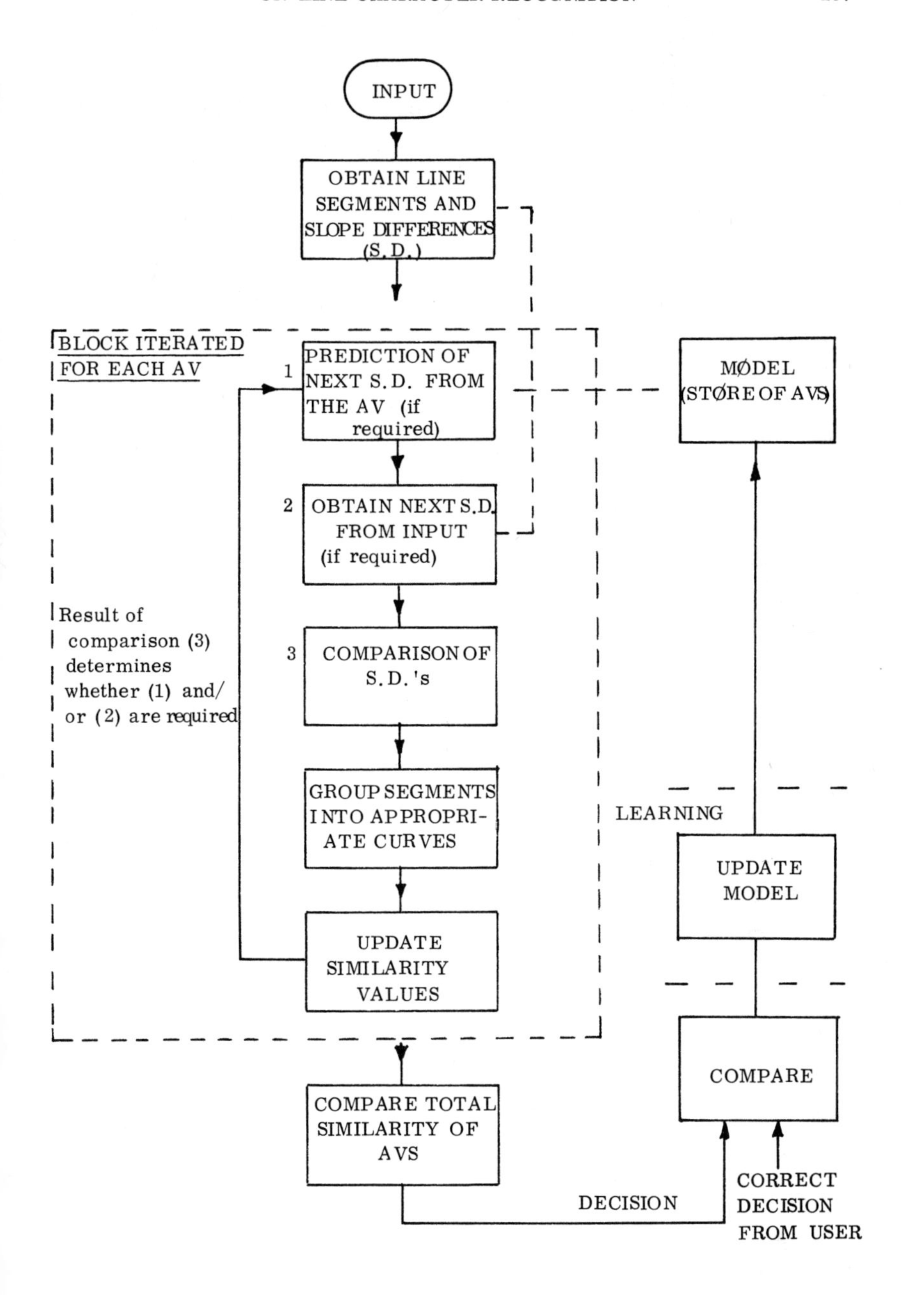

Figure 1 - Procedure incorporated in RECØG.

Orientation : the difference between the slopes of the first line segments.

Beginning-end Vector : the difference in magnitude and direction of the vectors - the expectation is modified by the size factor.

A total value is obtained by averaging the values for each curve.

(d) The decision of class membership is given to the class with the maximum total similarity.

3.2. An Example of the Processing

Figure 2 shows typical AVS for the characters "5" and "8". The information stored for each AV includes:-

(1) the number of curves and their types (see Section 2);
(2) the orientation of the first line segment in each curve;
(3) the slope differences between line segments in each curve;
(4) the vector from the beginning to the end of each curve;
(5) the name of the AV (e.g. A, B ...).

An example of this information for the curve EF in the "5" AV is presented in Figure 2.

Given an input such as that in Figure 3, the AVS for the classes "0", "1", "2", "3", "4", "6", "7" and "9" would be purged and the input would be separately described in terms of the curve types for the remaining "5" and "8" AVS (as pictorially shown in Figure 3). Typical similarity values between the input and each AV are presented in Figure 3. Note that because

(a) of the rounded part at the beginning of the character, poor confirmation for the sharp corners is credited to the "5" hypothesis;

(b) the end of the character is extended more like an "8", the beginning-end vector measurement for the "5" is zero.

The total values indicate that the "8" receives the decision of input class membership.

3.3. Comments on the Performance

RECØG is currently operating on the VISTA display for the numerals 0-9. Although it is incapable of learning, another program LERNREC can construct a user-specified character set. (Multi-stroke figures are allowed in LERNREC, see Section 4.1). During learning, the user receives an articulation of his inputted char-

"5" AV

CURVES			
	AB	-	straight line
	BC	-	(+)ve corner
	CD	-	straight line
	DE	-	(+)ve corner
	EF	-	(-)ve curve

Parameters for CURVE EF:-

Type:	No. 3
Orientation:	-30°
Slope Differences:	-20°
(between segments)	-30°
	-40°
	-40°
	-30°
	-30°
	-20°
	-10°

B.E. Vector:

Magnitude : 5 units

Direction : -135°

"8" AV

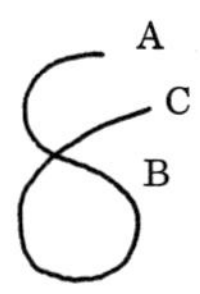

CURVES			
	AB	-	(+)ve curve
	BC	-	(-)ve curve

Figure 2 - Examples of AVS.

INPUT

Purged Classes '0', '1', '2', '3', '4', '6', '7', '9'.

Match with the "5" AV

(1) Description

B'
A'
C'
D'
F'
E'

Curve Match

A'B' - AB
B'C' - BC
C'D' - CD
D'E' - DE
E'F' - EF

(2) Similarity Values

Orientation	B-E Vector	Curvature
1.0	0.9	0.9
1.0	0.7	0.8
1.0	0.9	0.9
1.0	0.7	0.8
0.9	0.0	0.5

Total = 0.75

Match with the "8" AV

(1) Description

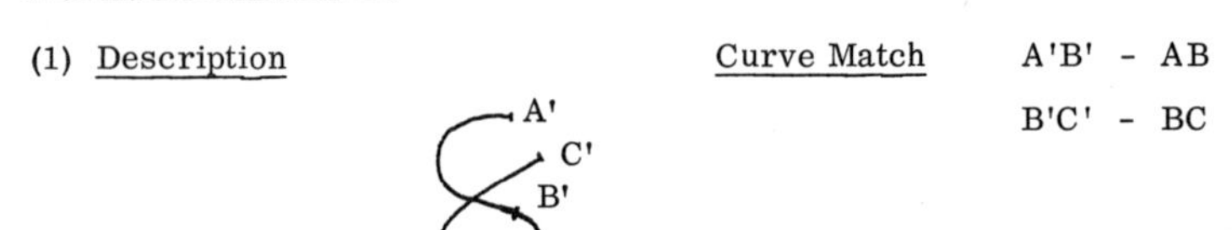

Curve Match

A'B' - AB
B'C' - BC

(2) Similarity Values

Orientation	B-E Vector	Curvature
1.0	1.0	0.9
1.0	0.9	0.9

Total = 0.95

Figure 3 - Processing results for an input to RECØG.

acter into certain curve types, on the screen. If he is satisfied with the articulation, the figure becomes an AV; if not, he is able to erase the figure and draw another.

Important features of RECØG's ability to recognize are:

(1) Characters may have sharp corners where curves are expected, and vice versa; in general, figures which are consistent with the AV in slope difference sign, are recognized.

(2) The input may be in any orientation and have a size variation (typically between $\frac{1}{2}$ and 2).

(3) The time taken to recognize an input is typically less than 0.5 second, depending considerably on the number of predicting AVS, and on the number of input segments.

Some of the limitations of RECØG are:-

(a) It does not allow multi-stroke figures (e.g. X, T) (a "4" must be drawn as "4").

(b) Errors occur when local variations appear in the input, due to the synthesis procedure (see above, Section 3).

e.g. 8, 3

(c) A decision is given to every input. If all the hypothesis are purged, the decision is given to the hypothesis which matched the most number of segments. In effect, the program does not have a "don't know" category.

(d) Two problems exist with the determination of size, because

(i) of the nature of the sequential comparison - large characters can have premature curve change points, i.e.

(ii) the calculated value is an approximation. Consequently for some shapes (i.e. "⊐" compared to "D") an incorrect value for the beginning-end vector may be expected.

(e) The characters must be drawn in a consistent direction of travel. However, this is not a major objection to a learning program, because a particular user is usually consistent in the method of character tracing and the direction of travel can be stored by the program during the learning phase.

(f) Some extraneous segments are allowed on the ends of the lines, but segments which change the sign of the slope differences are not allowed (see (i) above); such characters are considered to be writing, and require (higher-level hypotheses for) segmentation.

Because of RECØG's ability, it did appear advantageous to extend the program to overcome some of its limitations. Specifically, the first three of the above (a), (b) and (c) (and as a result d(i)) have been overcome by programs which are modifications of RECØG. The following section discusses these extensions.

4. EXTENSIONS TO RECØG

4.1. Recognition of Multi-Stroke Characters

The extension of the approach to multi-stroke characters has been made by including another level in the descriptive hierarchy - namely "strokes". Thus a figure consists of a number of strokes each of which consists of a set of curves.

The problem of describing such figures becomes a trivial exercise with the VISTA routines currently being employed. An interrupt is given to the program when the user attempts to move the tracking cross on VISTA to draw another stroke. Consequently, the inputted slope differences are obtained and analyzed into the separately drawn lines. A check is made on each hypothesis after the slope differences have been calculated, to determine whether the number of expected strokes is the same as occurs in the input; if not, the hypothesis is purged.

The second stage of the recognition procedure requires some measure which compares the spatial relationship among the strokes, in the input and the AV. This factor must provide for example, the basis for deciding whether two lines (e.g. /\) are "near" as in an "A", or are "apart" as in an "H". The measurement could be defined on the minimum distance between every pair of strokes in the given figure. However this would require a considerable amount of computing to

(a) compute this distance;

(b) perform (a) for every pair of strokes;

and furthermore additional information would be required to specify the location on the lines at which this minimum distance is expected.

The method employed is a modification of Bernstein's proposal (2). The (X, Y) coordinates of the centre of the minimum (X, Y) coordinate rectangle surrounding the given line is determined, for each stroke and for the whole figure. The measure is based on the differences, in the input and the AV, between the vectors from the centre of the figure to the centre of the corresponding strokes. It will be noticed that these vectors specify the spatial relationships between the strokes, implicitly. In Figure 4, the rectangles and their centres (Z's) are shown for a figure "A". The vectors for the strokes are given in terms of units of length and direction. Given the orientation and lengths of the sloping lines of the "A" the "nearness" of the strokes is contained in the information of the vectors joining the centres of the rectangles.

MENIREC is the program currently on VISTA which can recognize the capital letters A - Z, the numerals 0-9 and the sumbols *, +, -, /, (,). It is possible for the user to draw the strokes in any order, but as in RECØG, the direction of travel along the track must be consistent. MENIREC is not a learning program - its "learning" counterpart is LERNREC, discussed previously. The above characters are recognized in only one orientation ($\pm$ 20° overall), because the computations on the strokes are completed before the orientation of the curves is determined. The time to process a character varies mainly on the number of strokes in the input and number of competing AVS - typically the time is about 1 second. For LERNREC, which requires a consistent direction of travel the time is reduced considerably.

4.2. Ignoring Local Variations in Characters

Often characters are drawn which have local variations in the sign of the slope differences, but in which the variations are ignored as being unimportant by a typical human observer recognizing a given character. As explained previously, the synthesis procedure of RECØG (see Section 3.1(b)) considers that every sign-change in the slope differences reflects the presence of a new curve type, and hence local changes are not ignored. For example, discovery of the second corner, B, (after the first, A) in the two figures below, would cause the hypotheses for the "8" and the "3" in each case respectively, to be purged:-

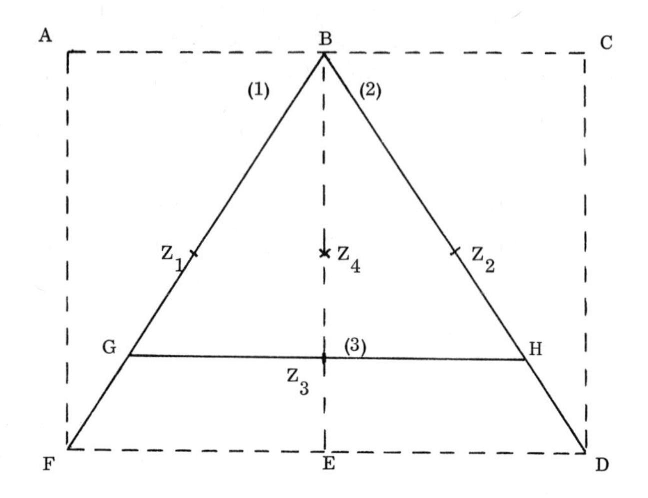

Stroke	Rectangle	Centre	Vector	Vector Length	Direction
(1)	ABEF	Z_1	$Z_4\ Z_1$	$2\frac{1}{2}$	180°
(2)	BCDE	Z_2	$Z_4\ Z_2$	$2\frac{1}{2}$	0°
(3)	GHHG	Z_3	$Z_4\ Z_3$	2	-90°
Figure	ACDF	Z_4			

Figure 4 - Stroke specifications for multi-stroke figure "A".

It was felt that to include forward searching or backtracking in the synthesis procedure (of RECØG) would be difficult. A more feasible approach has been to allow the description generation to proceed by an analysis. That is, instead of grouping the line segments, the procedure articulates the line segment string into the curve types expected for each AV. The method consists of the following steps:-

(a) the slopes of the line segments are quantized into 60° partitions ($\pm$ 30° fluctuation in slope differences are considered to be "local");

(b) a "local" length is defined in terms of a number of line segments viz.

Local length = no. of line segments/no. of curves expected/3 i.e. a local change must be confined to less than or equal to 1/3 of the average length of the expected curves;

(c) the quantized sequence is sequentially searched for the curve changes, reflected in the changes in magnitude and sign of the slope differences. Unexpected changes are examined in a "local" area around the changes to determine if the sum of the changes in the local region is acceptable (i.e. zero or of the expected sign); if not, the hypothesis is purged.

The result of the procedure is the definition for each AV, of the expected curve break points in the input.

Other advantages of the analysis procedure (not available in the synthesis procedure) have become apparent. Purging of hypotheses can be made (at the same position obtained by the synthesis procedure) with less processing, and hence at an earlier stage. Furthermore, it is easy to purge hypotheses which are found to have too much curvature (e.g. a sharp corner with greater than 180° total curvature). The procedure is size independent - consequently the problem with defining curve breakpoints in large figures (d(i) Section 3.3) is overcome.

ANALYZE is the program currently operating on VISTA effecting the analysis, and producing the articulations formed for each AV. It is difficult to assess the performance of such a program because observers often differ as to whether a given variation is local or not. It would appear that ANALYZE is rather conservative in its acceptance of local variations. The program can consistently ignore "odd" variations, e.g.

and can to some extent articulate characters having many variations; e.g.

Because of the advantages with the analysis, it is expected that if the procedure formed part of a recognition program, the time to process an input would be shorter than with the synthesis procedure.

4.3. "Don't Know" Classes

RECØG always gives a decision of class membership to every input - it does not have the facility for eliminating characters into a "don't know" category. This attitude is satisfactory for a user who draws figures which are intended to be similar to the machine-stored classes. However, the procedure cannot be said to recognize the members of each class exclusively.

An attempt has been made to extend the processing rules in RECØG so that inputs which are not "similar" to any known category, can be eliminated. It is intended to write the program so that it can output the reasons for eliminating a particular character.

Four kinds of rules have been considered:-

(1) exclusion on the basis of not enough or too much curvature,

e.g. - a "6" with too much curvature;

(2) exclusion on the basis that the line length is too short or too long,

e.g. - a "9" with the length of the "corner curve" being too long;

(3) requirement that intersections in the AVS be present in the input (and vice versa),

e.g. - an "8" without the intersection.

The cross-over point of the second (crossing) line defines the breakpoint for a new curve (this can be easily detected with VISTA routines). Because the input must have the same (number of) curves as the AV, the existence of the intersection is required in the input.

(4) exclusion on the basis that the distance (and direction) between strokes is too large,

e.g. - a "T" with the strokes separated too much.

A program DØNTREC embodying the rules is currently under test. (Results of this program will be given at the Conference.)

5. CONCLUSION

An outline of the approach to the problem of recognizing hand-drawn characters on-line using the VISTA display, has been presented (Section 2). It is suggested that the articulation of characters into the curve types (and strokes) employed in the programs presents a meaningful description of the characters. Furthermore, the curve level is a convenient compromise between the lower level of segments and the higher level of figure, on which to base the similarity comparison between given descriptions. The performance of RECØG (also MENIREC and LERNREC) indicates that quantitization of the information relative to the similarity is effective. It is to be noted that this information (which provides the basis for the decision) could be outputted in a form readily understandable by a user (e.g. in a subset of English statements).

In order to recognize multi-stroke characters, properties of the spatial relationships between the strokes must be determined. An efficient procedure for (implicitly) specifying spatial features of the strokes involves computations on the (X, Y) coordinate rectangle surrounding each stroke (Section 4.1). In contrast to this implicit representation of relations, a suggestion for the explicit representation of "intersection" (e.g. as in the "8") has to be made in connection with checking the existence of an intersection (Section 4.3). It is perhaps a matter of conjecture whether an explicit representation of such properties is required in a "general program".

A "local area" in a character string has been specified as a function of the size of the string and the number of curves expected in the character. By considering total slope differences in a local area, it has been possible in an analysis procedure (ANALYZE) to ignore small minor variations in a character, and to define the "correct" curve breakpoints (Section 4.2). The analysis appears to be more efficient than the synthesis procedure (incorporated in RECØG) because rules for purging hypotheses (on account of the slope differences as well as the curvature and length of string) are more easily incorporated in the analysis procedure.

The primary aim of the research has been to investigate various problems in character recognition. The problems discussed, while being only few of many, have nevertheless given rise to an approach which might be useful for specifying procedures to be incorporated in a "general program". A by-product of the research is the potential use of the method for a program to be used in a practical recognition system. A version of MENIREC is currently being tested for its performance over a variety of users.

6. REFERENCES

1. ATTNEAVE, F. (1954), "Some information aspects of visual perception"; Psych. Rev., Vol. 61, No. 3, pp. 183-193.

2. BERNSTEIN, M. (1966), "An on-line system for utilizing hand-printed input"; SDC Tech. Memo., 3052/000/00, (November 1966).

3. CARNAP, R. (1962), "Logical foundations of probability"; Chicago : Uni. of Chicago Press.

4. CHENG, G., LEDLEY, R.S. POLLOCK, D. and ROSENFELD, A. (1968), "Pictorial pattern recognition"; Washington : Thompson Book Co.

5. CLOWES, M.B. (1968), "Transformational grammars and the organization of pictures"; paper presented to the PISA Conference on Automatic Interpretation and Classification of Images, Pisa, Italy, (August 1968)

6. GREANIAS, E., MEAGHER, P., NORMAN, R. and ESSINGER, P. (1963), "The recognition of handwritten numerals by contour analysis"; IBM Journal of Research and Development, Vol. 7, pp. 14-21, (January 1963).

7. GRONER, G. (1966), "Real-time recognition of hand-printed text"; Proc. FJCC, Vol. 29, pp. 591-601.

8. LEDLEY, R. (1964), "High speed automatic analysis of biomedical pictures"; Science, Vol. 146, pp. 216-223.

9. MARILL, T. et al. (1963), "Cyclops-I : a second generation recognition system"; Proc. Fall Joint Computer Conference, Vol. 24, pp. 27-34.

10. NARASIMHAN, R. (1968), "On the description, generation and recognition of classes of pictures"; Lecture Notes for PISA Conference on Automatic Interpretation and Classification of Images, Pisa, Italy, (August 1968).

11. O'CALLAGHAN, J. (1968), "Pattern recognition using some principles of the organism - environment interaction"; Ph.D. Thesis, A.N.U., Canberra, (September 1968).

12. SPINRAD, R. (1965), "Machine recognition of hand-printing"; Information and Control, Vol. 8, No. 2, pp. 124-142.

13. TEITELMAN, W. (1964), "Real time recognition of hand-drawn characters"; Proc. FJCC, Vol. 26, pp. 559-575.

14. UHR, L. and VOSSLER, C. (1961), "A pattern recognition system that generates evaluates and adjusts its own operators"; Proc. WJCC, Vol. 19, pp. 555-570.

15. WATANABE, S. (1960), "Information-theoretical aspects of inductive and deductive inference"; IBM Journal of Research and Development, Vol. 4, pp. 208-231, (February 1960).

7. DISCUSSION

Bobrow: In your system, does each individual have to train on his own handwriting?

O'Callaghan: What I've got at the moment is one program which has a fixed way of writing the characters, and provided you are not too eccentric, you can get your characters recognised. There is another program which I have mentioned which incorporates a learning aspect and in which you can sit down to the console and specify characters.

Bobrow: If you start writing a sequence of symbols, for example, 12 ....., do you have to wait until it recognises the "one" before you can start the "two"?

O'Callaghan: This program is designed to recognise only one at a time, but there are techniques for being able to segment such characters, for example, the time between completing one character and starting the next.

Smith: It seems that the overall approach is based on the assumption of a fairly restricted character set; just how large a character set, efficient in recognition and in time can you have?

O'Callaghan: As size of character set increases, recognition time increases exponentially. The more AV'S you have, the AV'S become closer together, and take longer to process.

Smith: How do you solve the classical problem of O and zero, or are they the same character?

O'Callaghan: As in FORTRAN.

Bobrow: No troubles with S and 5?

O'Callaghan: In RECØG there seems to be good separation between these two classes. RECØG went through a series of tests on U and V, and 5 and S, also through a series of 2's and Z's, 6's and zeros, and again there was no problem - the separation that RECØG did was similar to what a human observer achieves.

Moore: What is the average time to recognise a character?

O'Callaghan: Something less than a second. This could be reduced by making the user draw the strokes in a specified order; there is quite a lot of complication to match up the order in which they draw the strokes in the input and the order in which they are stored in the AV.

Holywell: You have generated a very nice system, which you state will recognise with accuracy. To what extent do you need care in drawing characters to achieve good performance?

O'Callaghan: To take my experiences when I've used VISTA, I think I could draw the characters as fast as I could write them on a piece of paper. The accuracy I would use to write for say a girl to punch-up a FORTRAN program (provided the system was fast enough to respond to me), I think I could write the characters and get them recognised at the same rate as write them in the coding sheets.

Holywell: You do your recognition essentially by using a number of parameters; have you done any work to determine whether these are an optimum set? Do each contribute the same amount of information in your decision making?

O'Callaghan: The functions we use have been changed considerably over a period of time - the whole programs were developed over some 9 months. On the question of information from each measure, the orientation is really separate because that contributes its own information when the characters are rotated. Given the same orientation, we can talk about the curvature and the beginning-end vector. What I have tried to do is weight them; I picked various characters, like 6 and 0, for instance, and tried to match the characters so that variations in one are counter balanced by another. There is some discrepancy, because when a 4 is rotated, this is very dependent on orientation - when it gets to something like 60°, it starts getting recognised as a 5 or a 9 and then back to a 4 again. If you rotate a 5 or a 9 you never get a decision of a 4 in the samples I have used. The measure values which I have suggested are typical and have to cope with a wide range of variations in figures - these measures seem to work.

Benyon: Why can't you distinguish between direction of travel by having 2 AV'S for each character?

O'Callaghan: You can do it this way.

Smith: Do you always give equal weight to the various similarity values? Might not you get better discrimination if you gave more weight perhaps to the beginning-end vector, than to the orientation or the curvature?

O'Callaghan: Yes, there is an optimum of weighting these values but it is a broad hump, not a peak, and you don't have to be spot on it to achieve good discrimination.

Bobrow: How would you characterize the principal difference between your approach and that of others?

O'Callaghan: Probably the closest is Bernstein's. The paper I had on his work was from 1966 and his system wasn't operational. He was suggesting curve types, loops and straight lines, and was also determining these curves on the basis of total curvature and not slope differences. He had an attribute of a curve as total curvature. The other thing that is different is the second phase - I used numerical measures as a basis of decision. I think Bernstein incorporated just a set of rules which partition characters. He has subsequently used a beginning-end vector different from mine - whereas the information of total curvature would be important in Bernstein's program, mine would be incorporated in this curvature measure as a numerical value.

Bobrow: The basic differences are in the primitives that you have chosen for the characterization?

O'Callaghan: That's one basic difference, another is in the attributes of these primitives. There is something attractive about using just three numerical values to discriminate.

Pratt: Using the present method, you seem to have applied the technique of trying to find the general answer, and then from the general description try and deduce the properties. Couldn't you start at the other end and say if you want to purge character one, for example, and you say, 'with one, we expect the following properties, and we expect them in the following order or way', then instead of looking at what we have analysed we just look at the pattern as it stands and say, 'is there any conceivable way in which I can match this pattern with the character?'. That simplifies the analysis problem, in that you don't have to have an initial analysis, instead of trying to make it general and do anything for you.

O'Callaghan: You are suggesting a set of tests which will rule out various characters - such purging does go on in the descriptive phase.

Pratt: What I mean is that if you just start with your pattern, instead of trying to analyse it, supposing it has little squiggles in it, then all sorts of things are going to be purged accidentally, and you may have to go back to it; for example, if you have squiggles in the '2' - your '2' is going to go away for a while - what provision do you make for bringing it back when you find the squiggle was irrelevant?

O'Callaghan: As I mentioned, there is none in RECØG. The program ANALYSE does it. When I was talking about local area, I was defining a local area in which I could allow slope difference values between segments in this local area, and this is really the test which can ignore such local variations; you go right through the character and pick out the major variations, and on the basis of this again if you find variations which are not expected, you purge an hypothesis. The point with this scheme is that you can purge these hypotheses sooner than you can with RECØG by tracing round.

Pratt: So what you have done is, you haven't purged anything, you have analysed the whole lot before you start recognising and you can hold onto a character the whole way through.

O'Callaghan: No, as soon as you get one that is unacceptable, you throw it away. Its not referred to again. I've defined local area, and I've said that if the slope difference is not what you expect, not of the expected sign, then it is a change. If this change doesn't fit in with what you actually expect, then that hypothesis must be purged.

ON THE ROLE OF LEARNING IN PICTURE PROCESSING

S. KANEFF

Department of Engineering Physics
The Australian National University
Canberra

1. INTRODUCTION

Developments in the philosophy and technology of picture processing have been leading, to an increasing extent, to the representation of pictures as large matrices of numbers representing gray shades of 'points'. However, pictures are not simply arbitrary matrices, as Rosenfeld (1968) has pointed out, but are pictures of something - they represent, for example, real objects or symbols: it is this representational aspect which gives rise to many of the basic picture processing problems. To handle the situation, (and irrespective of the nature of the representation chosen for convenient processing), the need to devise adequate picture descriptions has been stressed previously (Kirsch 1964, Narasimhan 1966, Clowes 1967), and one of the central problems has been recognized to be the extraction of a description of the objects(s) portrayed in a picture: other related problems involve image enhancement, abstraction or simplification, and so on.

Solution of these problems seems to presume knowledge, acquired or able to be acquired in some way, of the objects represented pictorially, whether these objects have previously been encountered or are novel. To handle those aspects of pictures which are normally inferred by human beings from prior knowledge (not directly portrayed in the pictures themselves), it seems necessary, as indicated by Rosenfeld (1968), that models of classes of pictures and/or objects should be available. Factors relevant to the acquisition of such knowledge and class models by a picture processing system, form the main concern of this paper.

2. ACQUISITION OF INFORMATION

It will be convenient to consider differences in approaches to solving picture processing problems which hinge on just how information is acquired - whether all information is imparted to the system directly and explicitly by a programmer, or whether the system is provided with learning capabilities, for example. At one extreme, one might consider the syntax directed approach (expounded by many, including Narasimhan 1962; Kirsch 1964; Evans 1868 a, b: Clowes et al., this conference)which requires in essence that the system be given complete information to cover all eventualities - the models of picture classes and objects are then formulated by regarding these classes and objects as sets of grammatically well-formed sentences in a picture language, the grammatical rules constituting the models. Operation might be illustrated as in Figure 1, where an input picture is parsed with respect to a syntactic specification in an appropriate meta-language or formalism, the analysis being completely under the control of the prespecified grammar rules. The output from the parser is a syntactic description on which appropriate operations may then be performed to achieve the desired output.

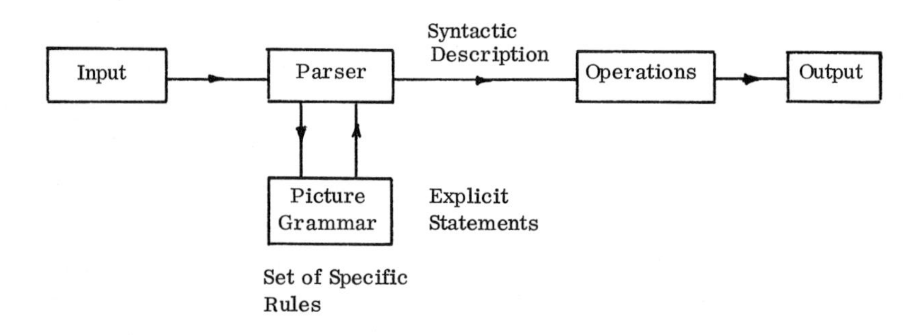

Figure 1. Syntax-Directed System

In contrast to the above, one form of learning approach might be as characterized in Figure 2, in which a description of the input is generated through the operation of a set of general rules and implicit statements; as in the previous case, the description may be operated on to secure the desired output. The vital difference is to be found in the feedback present, whereby an external (supervised learning) or internal (unsupervised learning) teacher is able to compare the output with the desired goals and the real-world situation, make decisions, abstract information and cause useful rules and implicit statements to be stored and selected, thereby allowing close adaptation to the problem and a continual updating of capabilities in the light of new experience.

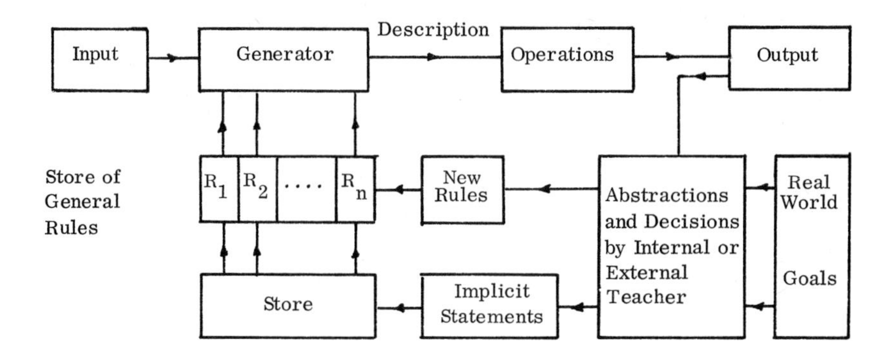

Figure 2. Illustrative Learning System.

If an all-embracing picture grammar could be devised, the necessary information for required picture processing assembled, and the whole imparted to a machine, such an 'Informed Machine'* operating in the mode of, say Figure 1, might be thought able to obviate the need for incorporating learning. However, as has been found in the corresponding case of languages and mechanical language translation, such formalizations remain elusive; even if they were available, it remains to be demonstrated that effective operation would be practicable in a real-world environment: the situation in picture processing appears to be inherently similar.

2.1 An Argument for Learning

Given a basic structural and functional capability, any organism (or machine) needs an adequate set of concepts (or models), explicit and/or implicit, to allow it to operate and survive. We generally base our assessment of the relative levels of organisms on their corresponding capabilities in acquiring new concepts by learning, as judged by our observations of their actions and behaviour. Study of evolution has demonstrated that those concepts (or stimulus-response schemata or models of behaviour) which are rigidly pre-specified, can lead an organism into serious

*The term 'Informed Machine' is sometimes employed (for example by Clowes) in preference to 'Intelligent Machine', of which it might be considered forerunner (or equivalent, depending on view point).

difficulties in a changing environment, while flexibility in adaptation or learning can be vitally advantageous - one of our indicators for intelligence is the ability to select an appropriate response from a set (which may be ever changing) of alternatives.

Our operational concepts appear to be acquired through a consideration of instances or examples. In the language of Narasimhan (1969b), we operate Paradigmatically, at least initially; subsequently, through training, we may attempt to 'talk about' and formalize what we know, and use this to enhance our capabilities through determining and applying specific explicit aspects of concepts in appropriate situations (Syntagmatic operation). Inherent in the Paradigmatic mode is the fact that, because each individual's operational concepts have been acquired through a unique and complex set of experiences, even if all human information-processing capabilities and structures were identical, expression of these concepts seems practicable only in general terms, as their precise meaning will differ for each individual: to this fact must be added the matter of an ever-changing environment. Indicative of this state of affairs is the immense difficulty in formalizing concepts by verbal expression, and the consequent 'circular' nature of word definitions in dictionaries. Indeed, endeavours to achieve formalization of our concepts have often been characterized as futile, as illustrated by Wittgenstein's comment:

> "We are unable clearly to circumscribe the concepts we use, not because we don't know their real definition, but because there is no real 'definition' to them. To suppose that there must be would be like supposing that whenever children play with a ball they play a game according to strict rules."

Evidently, our models or concepts are experience and environment dependent, and subject to continual revision and up-dating - a feature conducive to viability in a real-world environment in which the new, the unrecognized and the unforeseen, are always croping up.

Thus, although a great deal of effort has been directed towards the devising of grammars and formalization of problems and their solution, a case can readily be made that the inherently 'fuzzy' nature of our operational concepts, works against successful verbal or picture processing based purely on syntax-directed approaches, except in completely defined, that is closed, environments - which have, to date, hardly touched on real-world problems in picture processing. (However, operation in the syntagmatic mode may enhance operation of systems having more-general capabilities). These considerations alone, point to the desirability of seeking alternative methods of approach; other important factors also suggest the need for breaking out of the restrictions imposed by closed systems.

2.1.1 Requirements for Man-Machine Communication

It is becoming recognized that for effective man-machine communication (as in man-man communication), both should be able, to an appropriate degree, to view their respective concepts or models in the same way. Thus Stanton (1969) comments, "Man-machine graphical communication is possible only if the machine can see the same things in a picture that man does". Because of the previously-discussed properties of man's concepts which involve redundancy, 'fuzziness', and allow subjective interpretation, it is suggested that these features should also figure in the machine's models, thereby giving some hope for mutual adjustment and readjustment between man and machine, to arrive interactively at a common view-point: this applies equally to the acquisition of new concepts as it does to the refinements of those already existing.

For the purpose, adequate communication capabilities must be available.

2.1.2 Expansion of Behaviour-Repertoire

A serious restriction imposed by closed systems involves the inability to augment their capabilities, whereas given such a capacity through interaction with a human callaborator, the possibility arises of building up behaviour repertoire and thus coping with the ever-increasing demands for greater performance, so tending to alleviate a characteristic situation in which capability invariably lags behind requirements, because the particular mode of operation not provided has not been originally envisaged.

2.1.3 Purposive Operation in a Changing Real-World Environment

Certain extra-linguistic goals are involved in the activities of picture processing. These goals, which provide cues to reduce ambiguity by placing the situations being processed in adequate context, themselves carry the previously-mentioned features in relation to our operating concepts - that is we are unable clearly to circumscribe them, the more so because normally there are many paths open to a possible over-all goal (nor is the concept of optimal path necessarily relevant because as circumstances differ - for example, different constraints, different performance criteria and so - different paths may become optimal).

The function of pattern recognition, allows not only appropriate processing procedures to be invoked, but also produces a tremendous coding, structural and memory economy, by allowing categorization. In the process, the problem of similarity (whereby objects which are similar, but not identical, may be classified together) arises, and itself is a concept whose formalization has remained elusive and outside the competence of formal systems.

At present there seems no prospect that any of the above factors can be handled satisfactorily, in a real-world environment, by formal closed systems. On the other hand, we can handle the complex situations successfully through the operation of our learning behaviour. Accordingly, it is suggested that serious study of learning systems is warranted for possible application in picture processing machines, which might thereby learn to acquire and work with real-world concepts.

2.2 Relevance of Problem Solving Objectives

In the quest for methods of obtaining picture descriptions, it could be profitable to reflect on whether particular approaches are soundly based - do they view the problem in the most fruitful context; are the objectives in fact irrelevant? Is the problem being solved in the right way? The history of scientific discovery is rich in instances in which rules or theories devised by man have been mistaken for laws of nature, and even where the closed nature of the man-made laws has been recognized, it has often been found ultimately that the particular method of viewing the problem has been quite inappropriate. It is recognized therefore, that a searching appraisal before deep involvement in particular lines of approach, is of considerable value.

In the realm of language, Narasimhan (1969b) has argued that there is no such thing as natural language, only natural language behaviour. He further argues that whereas the behavioural machinery itself can be formalized, the output from this machinery cannot; if this argument is accepted, it must be considered futile to attempt formalizing natural language behaviour. This is consistent with the informal view-point that real-world situations do not behave necessarily in accordance with grammars.

To throw further light on these aspects, consider the construction of a honeycomb. One can marvel at the apparent intelligence of bees in producing such regular geometric shapes, and can speculate on the complexity of the program needed in the bee's brain in order to execute such a structure. How can one devise an appropriate grammar to describe the situation and program a bee to construct the hexagonal lattice of the honeycomb? How would one represent the apparent co-operation needed between adjacent and distant bees and their means of communication used, to ensure that the overall structure has the requisite regularity? Such an approach turns out to be largely irrelevant in relation to how the honeycomb is actually constructed, and to what information and capabilities are employed in the process - it is an inappropriate and unprofitable way of looking at the problem. Beer (1967) suggests that what really happens is that each bee is able to build a cell (approximately cylindrical) around himself by moving round and exuding wax. Because of the great density of bees in the hive, the cylinders, while being formed, are jostling together under gravitational pull - each one will move down as far as it will go, before settling and forming a closely packed structure in which all spaces are filled because of the forces and density of bees involved. Any given bee is thus surrounded by six other bees (all being of roughly the same size); hence his cylinder is touched by six others approximately equally spaced around him. The wax is still very malleable and capillary and gravitational forces cause the regions of contact to squeeze together and to become plane, due to the mechanical pressures involved - thereby ensuring that the six contact areas define a straight-sided, hexagonal figure.

Whether or not Beer's explanation is correct in fine detail, is not vitally important. What it suggests, however, is that the organization of the honeycomb which is apparent to the beholder, has arisen through the operation of a set of behavioural capabilities on the part of the bee (the production and working of wax, for example) and the operation of a set of constraints (including the dense population of bees, the restricted space available, the action of gravity, capillary forces, and the physical properties of wax) - the result of the system interacting inevitably leads to the formation of the honeycomb. The organization of the honeycomb itself, is not identifiable as a specific goal as far as the bee is concerned; if the observer tries to understand or describe the honeycomb in terms of this organization alone, his thinking may well be misdirected, unless he probes more deeply.

The view-point that real-world situations come about as a result of the operation of sets of constraints and behavioural or action capabilities, is central in the theory of Self-Organizing Systems (Beer, 1967). The apparent (to a casual observer) goals of such systems are not directly specified (or even specifiable) ab initio - given the constraints and action capabilities, the system inevitably moves to a particular stable point which an observer subjectively interprets as the goal. The end point (or apparent goal) changes only if one or more of the constraints and action capabilities change. Regarding the operation of natural self-organizing systems as the result of a set of constraints and action capabilities is an extremely useful concept, facilitating design of man-made systems, and is an important aid to elucidating fundamental aspects of systems.

It is interesting to note that the above approach to understanding aspects of natural systems is hardly new, (but seems to have been ignored until recent times) - one of Leonardo de Vinci's main tenets throughout his career was that the detail of nature is significant because of the fact that this determines the way the action flows from the structure - that is, action is determined by structure. There is little doubt that Leonardo de Vinci had a visual understanding of Science - in the language of this Conference, through a paradigmatic approach; his tenet is consistent with the above view-points.

The above considerations illustrate the illusory nature of searching for understanding and trying to build machines having performance similar to that of natural systems by attempting merely to formalize their end result (or product) which is in fact the culmination of the interaction of sets of (often complex) constraints and action capabilities. To approach a language or a picture and try to define structure may prove irrelevant. It seems preferable to direct our ingenuity to teasing out and elucidating the constraints and action capabilities; then, having discovered and understood the fundamental factors involved and their interrelations, it might be profitable to attempt formalizing the 'machinery of action' which produces the particular situation in question.

We need to discover the necessary constraints and action capabilities necessary to build systems which, as a result of their operation, produce learning and picture processing. This seems a preferable objective to attempting to formalize only the results of real-world systems having the capabilities to learn and to process pictures, because in the process we are unable to foresee all possibilities involved, and the magnitude of the task of handling even the known output of the systems may be too great. In an analogous manner to the acquisition of language by children without a knowledge of the grammar, it is suggested that a system with learning capabilities might be used to learn procedures for the processing of pictures and provide required descriptions, without a knowledge of picture grammar.

2.3 Information Acquisition and Learning

We use the term 'Information' in the broadest sense to include factual, relational, structural and procedural information, and learning is considered to be relevant in all cases.

The experience and environment-dependence of our concepts has already been touched on, as has the desirability of incorporating these features in picture processing machines, together with adequate communication capabilities and capability for expansion of the behaviour repertoire to cope with a changing environment. It is central to our argument that successful operation within this environment has been achieved by organisms as a result of learning capabilities. Avoidance of viewing problems in the wrong way has been stressed and a strategy based on discovering fundamental constraints and action capabilities has been suggested. Introducing learning capability holds promise of success as it enables open systems to be contemplated.

We conclude that acquisition of information in general is successfully accomplished in a real-world environment by organisms which learn, consequently our picture processing machines would become viable through the incorporation of appropriate learning capabilities. Although stress has so far been placed on the probable fruitlessness of purely syntax directed approaches - we learn concepts before formalizing them - it is necessary, however, to reiterate that grammars and formalizations are not rejected, but play a part in enhancing performance capability - this is certainly the experience of scientific discovery.

3. LEARNING MACHINES

We have so far avoided here a definition of learning, in conformity with the remarks of Section 2.1, as no adequate formal definition is available. Nilsson (1965) has informally described a learning machine as: "A learning machine, broadly defined, is any device whose actions are influenced by past experience".

The notion of learning involves the presence of a model which can be updated continually in the light of new experiences - this is in harmony with our introductory remarks on the necessity of employing models to handle those aspects of pictures which are normally inferred from prior knowledge.

3.1 Some Classes of Learning Machines

Depending on the a priori information included on the characteristics of a machine, its state and external influences, so we may classify machines as

Deterministic - in which all necessary information is supplied rigorously,

Stochastic - a priori information is supplied on the probabilistic characteristics of the inputs and environment, or

Adaptive - characteristics of inputs and environment are not provided and must be determined by the machine.

Adaptive machines are the only serious contenders to meet our requirements: learning in this case may be supervised (by a human teacher) or unsupervised (where the machine acts alone) - Tsypkin (1968). At the present stage of development, it seems unrealistic to contemplate machines operating with completely unsupervised learning; co-operative man-machine systems might, however, provide the desired performance.

A number of principles and techniques has been developed over the past two decades towards the realization of learning systems; in order of increasing requisite complexity, techniques for incorporating some form of learning in machines include: Memory Updating, Weight Adjustment, Parameter Adjustment, and Structure Adjustment. Discussion of the corresponding details of such techniques is outside the scope of this paper, and reference is made to the literature (for example, Nilsson, 1965). It is worth noting, however, that two lines of development may hold ultimate promise of significant contribution towards effective learning systems:

Structure Adaptation - in which the structure of the machine networks is arranged to adapt to the pattern classes involved, requiring a matching of the structure to the input data.
(Uhr and Vossler, 1961, have rudiments of structure

adaptation. Chow and Liu* discuss the problem more directly).

Pattern Feedback - in which the network is trained to produce a model output pattern which is then fed back to the input - recognition of an input takes place if a stable model output pattern is generated. This principle differs from conventional feedback in that the frame of reference of the output is made identical to that of the input (Aleksander and Mamdani**)

Notwithstanding the considerable amount of effort expended in developing learning features, (largely in relation to pattern recognition tasks), the 'art' is still at only a very early and inadequate stage of development.

As far as picture processing is concerned, learning features have not yet figured prominently. Typically, elements of learning have been employed in Character Recognition (O'Callaghan, 1969), Fingerprint Identification (Hankley and Tou, in Cheng et al., 1968), Game Playing (Koffman, 1968), and discussed in relation to Game Playing (Clowes, 1969), and Photo-Interpretation (Wilcox, in Cheng et al., 1968). Bobrow (1969), has outlined approaches (particularly in the Teachable Language Comprehender) which involve aspects of learning in Natural Language Interaction Systems, illustrative of a situation depending on man-machine co-operation to teach the system requirements.

3.2 Requirements of a Learning System

Although progress in devising learning machines has so far been slight, it is still possible to infer the kinds of capabilities which, if implemented, might lead to adequate learning behaviour, by studying human performance. Reasons have already been suggested in Section 2 for retaining a significant degree of similarity between learning machine models or concepts, and human concepts.

It is proposed that the task may be viewed essentially as the discovery of constraints and action capabilities which, when incorporated in a learning machine, cause it inevitably to learn. Central to the requirements is an appropriate model (or group of models) which can be modified, (including structural modification) in the

*IEEE Vol. SSC-2, No.2, December, 1966, pp 73-80.
**Electronics Letters, Vol.4, No.20, October 4, 1968, pp 425-6.

light of experience, to reflect the operational concepts within its competence. The addition of a sufficiently rich communication capability between the machine and its environment seems vital to functioning.

The learning system, as a result of the operation of appropriate structures and procedures (all of which might be learned), may be considered to have a behavioural repertoire: as Narasimhan (1969a) suggests, "One primary aspect of learning is the capacity to generalize available behavioural repertoire to new situations by perceiving structural analogies between these new situations and the already known situations". To carry this further, by interaction with the environment, the competence or behaviour repertoire may be enhanced continually through the incorporation of new information and the re-structuring of that already present - this aspect of learning machines is clearly of vital importance to operation in a real-world changing environment.

The role of learning is involved with the learning of concepts or rules which, apart from the resulting economy in coding, structure and memory, facilitate reorganization of information for future use. Pierce (1968) indicates the nature of some of these concepts in relation to language learning, "The grammar a language teacher needs is rather like tips for playing golf or tennis, or rules of spelling. What is needed is reminders that can be remembered and used by human beings: that is, useful rules to supplement and guide the acquisition of implicit knowledge of language".

The organization of concepts into appropriate conceptual frames of reference, is itself not only an economy measure, but a vital ingredient in augmenting behavioural repertoire. (The relationship of representation, which has become a recurring theme in 'behind the scenes' discussion during this Conference, is here relevant). When, say, an object or concept has been recognized as fitting within a particular frame of reference, all those aspects of the behavioural repertoire relevant to that frame of reference may be invoked for processing. At the highest levels of operation, discovery of new relationships or frames of reference implies creativity.

In relation to picture processing, as Macleod (1969) has suggested, the ability to isolate objects and their sub-parts and to recover organizational detail, implies the capability of forming descriptions of individual objects - with the addition of capabilities of abstracting from object descriptions to class characterizations, a machine would be able to learn new classes. Further extension is then possible.

Operation of such a learning system may be envisaged as occurring through communication in conjunction with a co-operative human collaborator. By exposing the system to various situations, and issuing appropriate information, the teacher can cause the system to build up its models (and behaviour repertoire) to an appropriate extent. Eventually, the system might have sufficient capability to enable it to study situations, comparing the information so obtained with its own concepts or models, and on the basis of this comparison, generate hypotheses which are tested by deliberate experimentation. Confirmed hypotheses could lead to new relationships, concepts and procedures being added to its behaviour repertoire.

4. CONCLUSIONS

The view-point has been expressed that purely syntax-directed approaches for picture processing systems seem unlikely to solve real-world problems, largely because of the inherently 'fuzzy' nature of our operating concepts. It is suggested therefore, that rather than attempt to write picture grammars, it appears more profitable to study the processing 'machinery' itself, inspecting those systems which are successful (that is, biological processing systems) and trying to find the fundamental constraints and action capabilities which produce successful performance.

It is considered that a particularly profitable line of attack is to face real-world picture processing problems squarely, resisting the urge to simplify, at least in order to ascertain the key aspects of the problem situations, then attempt to devise learning systems which, in co-operation with a human collaborator, may eventually learn appropriate procedures and concepts to achieve a useful behavioural repertoire, and for this repertoire to be enhanced through interaction with the environment. Later still, such a system might be able to enhance its own capabilities unaided. Learning would play a central role in such a machine - the learning of concepts, procedures, and in general, action capabilities. What it is that must be done is becoming clearer; how to do it seems an immense (but stimulating) task.

5. REFERENCES

BEER, S. (1967), "Decision and Control", Chapter 14, Wiley: London and New York.

BOBROW, D.G. (1969), "Natural Language Interaction Systems"; This Conference.

CHENG, G.C., LEDLEY, R.S., POLLOCK, D.K. and ROSENFELD, A., Editors, (1968), "Pictorial Pattern Recognition"; Thompson Book Co.: Washington D.C.

CLOWES, M.B. (1967), "Perception, Picture Processing and Computers"; In "Machine Intelligence 1", pp 181-198, Eds. Collins and Michie, Oliver and Boyd: Edinburgh.

CLOWES, M.B., (1969), "On the Description of Board Games"; This Conference.

EVANS, T.G. (1968a), "Descriptive Pattern Analysis Techniques"; Lecture Notes, NATO Summer School on Automatic Classification and Interpretation of Images, Pisa: Italy.

EVANS, T.G. (1968b), "A Grammar-Controlled Pattern Analyzer"; Reprints IFIP Congress 1968, pp H 152 - H 157.

KIRSCH, R.A. (1964), "Computer Interpretation of English Text and Picture Patterns"; IEEE Transactions on Electric Computers, EC-13, No. 4, pp 363-376.

KOFFMAN, E.B. (1968), "Learning Games Through Pattern Recognition"; IEEE Transactions in Systems Science and Cybernetics, Vol. SSC-4, No. 1, March, pp 12-16.

MACLEOD, I.D.G. (1969), "On Finding Structure in Pictures"; This Conference.

NARASIMHAN, R. (1962), "A Linguistic Approach to Pattern Recognition"; Tech. Report 121, Digital Computer Laboratory, University of Illinois.

NARASIMHAN, R. (1966), "Syntax-Directed Interpretation of Classes of Pictures"; Communications of ACM, Vol. 9, No. 3, pp 166-173.

NARASIMHAN, R. (1969a), "Picture Languages"; This Conference.

NARASIMHAN, R. (1969b), "Computer Simulation of Natural Language Behaviour"; This Conference.

NILSSON, N.J. (1965), "Learning Machines"; McGraw-Hill: New York.

O'CALLAGHAN, J.F. (1969), "On-Line Character-Recognition"; This Conference.

PIERCE, J.R. (1968), "Men, Machines, and Languages"; IEEE Spectrum, July, pp 44-49.

ROSENFELD, A. (1968), "Picture Processing by Computer"; Computer Science Center, University of Maryland, Technical Report 68-71, Nonr-5144 (00), June.

STANTON, R.B. (1969), "Plane Regions: A Study in Graphical Communication"; This Conference.

TSYPKIN, Y.Z. (1968), "Self-Learning - What is it?"; IEEE Transactions on Automatic Control, Vol. AC-13, No.6, December, pp 608-612.

UHR, L. and VOSSLER, C. (1961), "A Pattern-Recognition Program that Generates, Evaluates, and Adjusts its own Operators"; Proc. WJCC, Vol. 19, pp 555-569.

6. DISCUSSION

Pratt: Do you think there is any essential difference between learning and finding the values of a set of variables?

Kaneff: It depends on how you do it. If you have already written down the equations which imply the values of the set of variables you define the structure involved. By finding the values of the variables, you might be able to say this structure learns to set itself up to a certain value. In this case it is not very different from the adjustable parameter systems. However, if the process is achieved as a result of a completely specified program of operation, say by designated operations on a calculating machine, or the equivalent, I would not call this learning. I have suggested that learning is involved if the behaviour of the system itself is influenced by past experience. If the behaviour repertoire or performance capabilities of the aforementioned machine is not influenced by past experience, then no learning is considered to have taken place - finding the value of a set of variables would therefore normally not be considered learning, unless you have in mind a more unusual situation.

Pratt: Well, say some of the variables are history dependent.

Kaneff: It is not the value or behaviour of the variables in this case which determines whether or not it is a learning system. It is a learning system if its own behaviour is influenced by past experience - this was the informal notion of learning which I offered.

Pratt: Surely learning the value of a set of variables implies a framework into which you fit them.

Kaneff: Yes, that is so, but learning the value of a set of variables is different from saying that the structure is a learning structure. The question arises as to who or what does the learning - that is, what behaviour is changed as a result of the experience in finding the values. If the programmer finds out the values which his learning behaviour has given him, the machine could possibly be merely a desk calculator whose behaviour repertoire is unchanging. On the other hand if the system itself, as a result of its operation, now has its behaviour repertoire changed, then I would call it a learning machine.

Macleod: In Figure 2, with an outside teacher, just what sort of information do you see coming back into the system, is it going to be simply yes or no?

Kaneff: This depends on the complexity which has been incorporated.

Macleod: I am thinking here, if you do say simply yes or no, this would seem to be a rather poor sort of teaching. I'd question how any child would fare in this environment where it was just told yes or no and not told why something isn't a particular pattern or why it is. It would seem that if you are going to teach this system in any real sense, it has got to have some notion of the structure of the input pattern so that you can then tell it in what sense this particular input is different from the present class. Then it becomes a question of, if you are going to do this, just how much of this learning do you incorporate in the outside teacher, and let him change the program, and how much do you incorporate in the machine. It would seem to me that the second diagram is very similar to that of the first, during the development of the grammar, except that in this case the outside teacher, if you like to think of him in this way, is the person who is developing the grammar.

Kaneff: Within the restricted system you are postulating, I think this is very true. In answer to the first part of your question, whether you can say 'yes' or 'no' or something else, depends on the reasons for the whole structure. If it is able to accept only 'yes' or 'no' and nothing else, then even if you gave it a sentence in fluent Spanish, it could still give only the same output - this doesn't help you at all. The whole point of my paper is that you must have your models, your rules and your structures, so rich that you can take advantage of what the outside teacher is able to say - there must be adequate communication capabilities. There is another aspect which favours the human being: a simple 'yes' or 'no' answer to him may speak volumes, because of the richness of his concepts and associations, but in a very restricted machine which doesn't have anything like this richness, the saying of 'yes' or 'no' may simply lead to changing a weight.

Macleod: At least the child can come back and say is it not an 'A' because it has an extra part to it.

Kaneff: It is important to stress that the learning systems discussed are not intended to be simple-minded. I suggested that they would need considerable communication capabilities able to handle complex messages for interactive operation with co-operative human collaborators; their models or concepts would be quite complex and extensive, and would be able to be modified and enhanced through the interactive process. One cannot expect such machines to become available overnight, but in spite of the considerable difficulties involved, the viewpoint was expressed that the attempt to produce these systems should be made - we might as well face up to the whole problem right from the start in order to discover and cope with the interrelated factors which would not be elucidated in a more restricted attack.

Clowes: You said in your talk that TLC, the Teachable Language Comprehender, was a learning system.

Kaneff: You can train this system.

Clowes: Could you and Dr. Bobrow clarify the sense in which it is a learning system?

Kaneff: I think Dr. Bobrow should do this, he described the system.

Bobrow: I would characterize this as a teachable system, not as a learning system.

Kaneff: Can you have a teachable system without it being a learning system?

Bobrow: I always have a feeling that a learning system is something which you give it some reinforcement of some sort, and each draws its own lesson or conclusions from the reinforcement that you are giving it. In a teaching system what you are doing is trying very hard to make explicit the structures that you are wanting it to know - the difference between punishing a child and trying to teach him something.

Kaneff: You are at liberty to choose that definition, but if you take any effective practical teaching system, it can teach because it learns something about whoever it is teaching, or whatever it is teaching. I should add that there are now teaching systems incorporated in hardware systems for automatic control, where you have a mechanised system which learns, and a mechanised system which teaches; looking into that problem, you find that they both learn together, the teacher learns about the student, if you like, and the student learns something from the teacher, and the interaction means the group learns together and the behaviour of each is modified. You still obviously need an outside teacher even if his only function is to devise these machines that provide the information.

Simmons: I would like to make the comment that I have often had to reject the learning approach because the economics of trying to implement this system was just not feasible whereas the syntactic approach is cheaper and quicker.

Kaneff: This may well be so, but it is because we have not yet been clever enough to devise learning systems - that may not be the only reason. It may well be that your environment was so well defined, so restricted that a closed system was adequate and there was no need to use learning. It is where you have open ended situations that learning has its attractions. This is what I was trying to convey.

ON FINDING STRUCTURE IN PICTURES

IAIN D. G. MACLEOD

Department of Engineering Physics
Australian National University
Canberra

1. INTRODUCTION

Convincing arguments have been propounded for the lack of generality of any descriptive scheme which does not reflect the structure of the objects described (Narasimhan 1968, Evans 1968). If we accept that an encoding of structure is a necessary component of a general picture description scheme, we then have the problem of discovering structure in the picture to be described.

We define a structural description (SD) of a given object, e.g. a picture, as a description which incorporates at the highest level an encoding of the object's organization in terms of its parts and the relationships between them. An SD is hierarchical in so far as each of the parts may, when viewed at the appropriate level of complexity, be considered as an object in its own right, defined in terms of its parts and the relationships between them. At the lowest level of description, the parts are regarded as primitive or atomic. Thus, each part above the level of a primitive is considered as having its own SD.

In attempting to describe structure in a picture, it is often the case that the structure of the situation underlying the picture is that which is of ultimate interest, rather than pictorial structure, i.e. gray-scale or density structure. We call the situation underlying the picture, the "object-world". The actual three-dimensional (3D) scene represented in a 2D photograph is the object-world of prime concern in the following discussion. There is clearly a correspondence, although not necessarily one-for-one, between pictorial and object-world structure. SD's may thus be considered with respect to both types of structure.

The aim of this paper is to determine components of a picture processing mechanism capable of deriving SD's for reasonably complex input material. For simplicity, but with little loss of generality, we consider gray-scale rather than full colour input.

Pictures depicting naturally occurring objects such as snow crystals, pollen grains, and trees, are typical of those pictures for which an encoding of organization would seem to be a prerequisite to effective description and interpretation. Natural objects often possess an organization which arises out of constraints (not usually known to the observer) present during their formation. Because of differing environmental and other factors, instances of the same class are rarely identical and tend to be characterized by the relationships between their parts as well as by the parts themselves.

Apart from the inclusion of an encoding of organization, other requirements for a satisfactory descriptive scheme are that:-

(i) it includes information regarding the attributes or metric properties (e.g. size, orientation) of objects and parts;

(ii) it can produce descriptions which are adequate for the intended use. Taking picture transmission as an example, a test of the adequacy of the description is that it is possible to form output pictures which the recipient considers to be satisfactory approximations to the input pictures. This does not mean that the output picture has to be a good approximation to the input picture when compared on the basis of gray-scale values at corresponding points on a superimposed matrix. For some uses, a very concise description, such as 'it's a "2"', may be adequate;

(iii) it is as economical as practicable (consistent with the requirements of (ii) above and (iv) below). This implies that the recurrence of similar features (i.e. parts, attributes, or relationships) should be detected and used where possible to reduce the coding cost, and that irrelevant features should be omitted. We may replace a portion of an SD by a pointer (usually a name) to a structure which has been previously found or defined, together with a specification of any significant differences. Economy is achieved when the portion so replaced is similar to the replacing structure, in the sense that the pointer, together with the specification of significant differences, is less complex than the replaced portion. Similarity, so conceived, is very much a function of what differences are considered significant. Where several alternative structures are possible replacements, ambiguity exists and may require consideration of a broader context for resolution. To the extent that they are useful in such replacements, the previously found or defined structures are likely to be representative of classes of objects rather than individuals. The recurrences found may be within or between pictures in the pictorial or object-world, or of structures imparted to the processing mechanism by the designer. Clearly, the greatest economies may be effected with highly structured inputs. Economy in coding requires that the input should be articulated into parts appropriate to the class of pictures, and not dependent on the "parts" of the initial coding, e.g. the cells in a scanned matrix.

(iv) it includes an explicit encoding of information which is likely to be required frequently in any given application. The desirable extent of such explicit encoding is a function of the difficulty in obtaining the required information from an implicit encoding and of the relative importance of economical description.

SD's may be considered with regard to description of classes as well as individual objects. A class SD would embody only those parts, ranges of attribute values, and relationships which characterize the given class. It is the inclusion of these skeleton structures which makes class SD's useful components in SD's of individual objects. If required, "flesh" can be added to the "bones" by the inclusion of additional components in the object description. The usefulness of class SD's is a function of their ability to describe structure which is generally (but not necessarily always) present in class members. The non-occurrence in a given member of structure implied by the class SD would have to be indicated by an appropriate correction. What structure should be included in a class SD depends on the economy of description of the total experience and it is a matter of balancing additions against corrections. On this basis, an object may be treated as a member of a given class if the class SD requires only minor corrections and additions to satisfactorily describe the object. Usually it is object-world and not pictorial structure which is characteristic of objects, and therefore class SD's would normally be in terms of object-world attributes and relationships. Classes with distinct sub-classes, e.g. '4' and '4', will require the inclusion of several alternative SD's in the class SD.

As inferred above, the form of a suitable descriptive scheme depends quite heavily on the ultimate use. Applications with different biases are picture transmission, picture storage, and picture interpretation.

2. SOME COMMENTS REGARDING GENERATION OF STRUCTURAL DESCRIPTIONS

To generate an SD for a given input picture, we need to articulate the pictorial or underlying input into parts, assign various attributes to these parts, and discover significant relationships between them. Evans has outlined a system for the generation of SD's and has shown that it is capable of handling fairly simple input pictures (Evans, 1968). The components of the system are shown in Fig. 1.

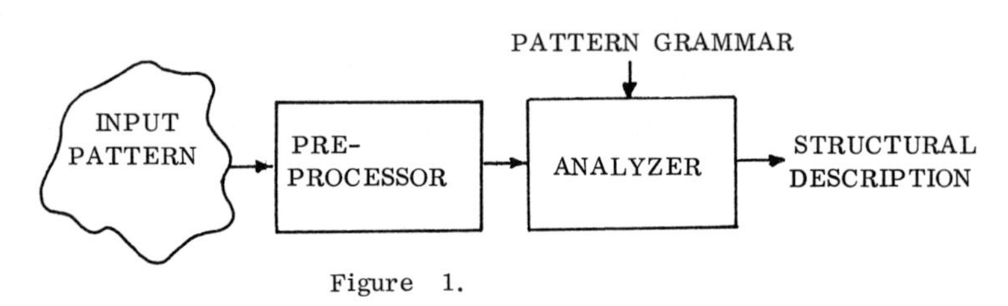

Figure 1.

The preprocessor transforms the input into a list of the lowest-level constituents used by the analysis program, together with any desired properties (attributes and relationships) of each constituent. The pattern grammar defines the manner in which low-level grammatical constructs (perhaps the lowest-level constituents) may be grouped into higher-level constructs (ultimately the patterns of interest such as triangles or squares), and the information (regarding attributes and relationships) to be attached to these latter constructs. The analyzer produces "a list of all the patterns defined by the grammar", which can be built up out of the list of lowest-level constituents passed to it by the preprocessor, together with an encoding of the manner in which the patterns were constructed, i.e. an SD. Examples of inputs handled successfully by Evans are shown in Fig. 2.

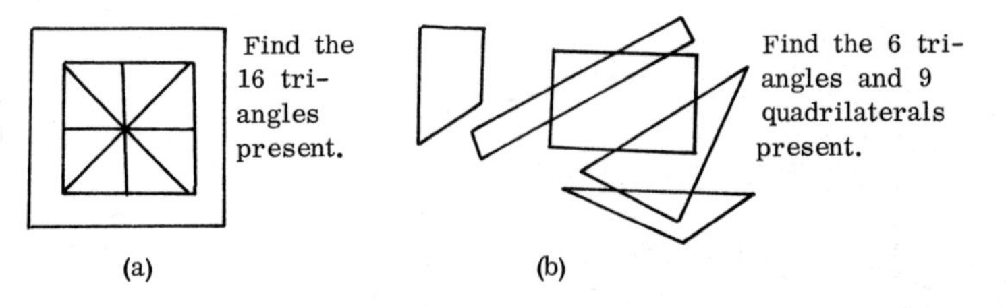

Figure 2.

Evans suggests that "the generation and manipulation of appropriately-chosen descriptions is a highly promising avenue of attack on the machine processing of such complex patterns as aerial photographs and physics photographs (Evans, 1968), and quotes Minsky's discussion of such an approach (Minsky, 1961).

When we contrast general photographic material to the symbolic, abstracted, or relatively simple material which has been the subject of most work with a descriptive flavour (Evans, 1968; Ledley et al., 1965; Grimsdale et al., 1959), we find that the greater complexity, and the presence of several types of so-called "noise", limits the applicability of the techniques employed to date.

One of the most basic problems with photographs is that there may be one or more underlying structures in the density structure. Taking a photograph of a 3D scene as an example, the density structure results from variations in the pigmentation and aspect of the subject matter and variations in the intensity and direction of the light falling onto the subject matter and being reflected towards the camera. In this case, the perception of the underlying structures is interdependent and, in general, we have to perceive all structures to arrive at any one (structure). Fig. 3 illustrates the interdependence of assumptions made about the direction of lighting (often

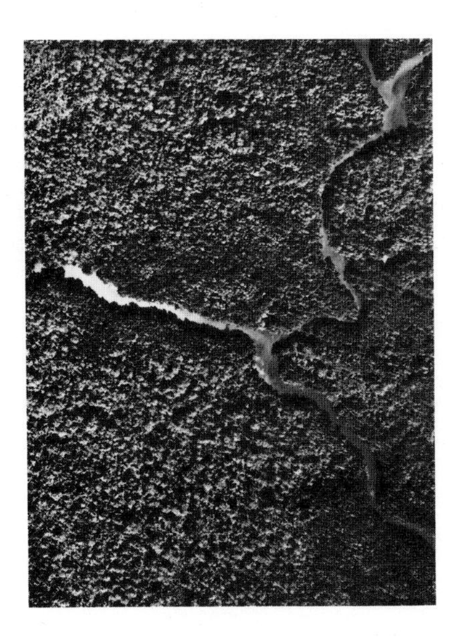

When viewed in the given orientation, this picture may be interpreted as a raised ribbon on a pock-marked background.

When inverted, a somewhat different interpretation becomes apparent!

An erroneous assumption that the scene depicted is illuminated from above, causes errors in perception of relative depth.

Figure 3. Pseudoscopic scene.

assumed to come from the top of a picture), and perception of the aspect and relative depth of surfaces. It may not be possible to perceive the aspect and pigmentation structuring in regions of poor lighting. Gibson (1966) lists some cues which could possibly be used to isolate these structures. His discussion of the visual world versus the visual field, i.e. how normally we perceive surfaces, edges, slants, etc., (whereas to perceive light or darkness requires a conscious effort), would also seem to be pertinent here in so far as it suggests that this type of processing is an automatic interpretation of basic 3D structure from pictorial structure (Gibson, 1951). The problems related to interdependent structures are mitigated if the class of inputs to be analyzed is chosen such that one of the structures predominates, or the structures have sufficiently different characteristics. Ledley et al. (1965), for example, chose a pigmentation-structured input, whereas Roberts (1965) chose an aspect-structured input. In pictures of 3D subject matter, we do not see all the parts of objects in each instance, because some parts are hidden, either by other parts of the same object, or by other interposed objects.

With automatic parsing techniques such as those employed by compilers or in Evans' analysis mechanism, there is a rapid growth in search time as the input becomes more complex, and as the complexity and number of defined types increases. The growth of search time is so rapid that the question is almost one of possibility of analysis rather than one of efficiency. One possible technique for reducing the search time is to segment the input into (hopefully) meaningful small pieces and to analyze these, or at least to restrict the search area by examining only a local context. We note that in parsing with respect to pre-defined types, segmentation of the input may result from the analysis (with inputs such as handwriting) - satisfactory segmentation may not be possible otherwise, but this does not limit the magnitude of the initial search problem. Another technique is to reduce the number of constituents in the input to the parser, by employing constituents which are more complex yet still meaningful, e.g. the "recognizer" phase in some compilers passing integers and variables to the analysis routine, rather than letters and digits.

Many abstract concepts, such as triangularity, are concerned with the shape of regions delineated and not with the difference in properties which suggested the delineation. We would thus require of a general picture processor the ability to describe all the figures in Fig. 4 as triangles. With other objects, such as forests, it is the internal structure rather than the overall shape which is characteristic of them.

It is this characterization by internal structure that enables us to identify many objects even though their boundaries are partially or totally occluded.

Figure 4. Figures which could be described basically as triangles.

The low-level pictorial structure attributed is relatively independent of other than local context and of the ultimate objects that we are trying to find. This applies more to complex pictures than graphics, where the distinction between low and high-level structure is not as clear. It would not seem possible, except in simple or artificial situations, to specify the variety of different types of structuring which may arise in objects and their corresponding pictures. We are thus forced to look beyond the structures themselves to see if we can incorporate general principles which will enable the picture processor to articulate the (perhaps already partially structured) input into parts, and group these parts into reasonable constructs, following this with segmentation of regions on the basis of structural or metrical properties. A further possible advantage of the incorporation of such "innate" rules of organization is that low-level structures abstracted may be fairly specific to the objects being searched for. These structures might then be used to suggest objects in terms of which the input could possibly be construed.

The extent to which the input may be satisfactorily structured by the application of general rules of organization is limited by ambiguity, as is indicated by our experiences on viewing novel material such as a microscope slide - parts and regions are discernible, but a coherent overall organization may not evolve. The mention of a key-word sometimes helps to achieve such organization - probably because we become aware of what object-world is involved and which descriptions and relationships are applicable, and then try to organize and interpret the pictorial input in these specific terms. It is expected that the two types of organizing rules, general and specific, will both be involved in abstracting structure, with the general rules predominating initially, and the specific rules playing an important role in the final organization.

The arrangement indicated above resembles that proposed by Neisser (1966) as a basis for human visual perception, in which the first phase is of "preattentive" processing which operates on the whole visual field simultaneously, and which separates each figure or object from the others in its entirety; followed by a second phase of pattern analysis, which operates on the objects so obtained. The two types of processing may, however, be too interrelated to separate them to this extent. Neisser describes the preattentive processes as being "global" and "wholistic", and does not seem to consider segmentation of textured or micro-structured regions which would require treatment of fine detail, as part of his first phase. For applications such as picture transmission and storage, abstraction of low-level structure may be all that is desired.

The structuring which results from the application of general rules of organization is not expected to be free of errors (it may in some cases prove necessary to return and restructure the input) and as a result we require that routines for interpreting structures in terms of class SD's should be capable of dealing with similarities rather than only identities, i.e. they should be able to interpret a region of the input as an instance of a class, together with changes, corrections and additional specifications. This requirement also arises when we consider that parts of an object may be missing or distorted because of other interposed parts and objects, or unsatisfactory lighting. Analyzing in terms of a predefined class may be viewed as comparison of the input SD and the class SD, given that the input has been structured in terms relevant to the class. If we attempt to find recurrences within the given input by comparing structures abstracted, here too we are interested in similarity as well as identity. The similarity of two SD's is itself structured and so we could generate an SD of the similarity, perhaps including such attributes as relative size and orientation.

Thus, when we examine the applicability to general photographic input of the analysis mechanism employed by Evans for simple line drawings, we see that as well as the component concerned with organization of the input according to relationships specific to known classes, we may require at least five additional (but possibly related) processes. These processes are:- restriction of context and segmentation of the input into regions for further analysis; organization (according to general principles) proceeding in the absence of specification of relevant relationships; description in terms of similarity as well as identity; re-examination of abstracted structure when inconsistencies arise; and derivation from the density structure of object-world structures such as those of lighting, pigmentation, and aspect.

We now examine the problems of choosing suitable primitives and general rules of organization. These problems arise with respect to both pictorial and object-world structure, but the comments below are addressed primarily to the pictorial case because of the specialized nature of suitable descriptors for a given object-world. Simon and Kotovsky (1963) were able to select suitable primitives and organizing rules for a program which abstracts structure from letter sequences and use it to predict continuations. The primitives are simply the letters of the alphabet, and the relationships used as the basis of organization are the "same" letter or relationship, and the "next" letter in a forward or reversed alphabet. For pictorial input, it is not as straightforward to choose primitives which are applicable to a wide range of pictures, but a useful principle is that the primitives should be based on regions of <u>perceptibly</u> constant attributes or properties. The simplest attribute of a photograph is density (at a given location) and thus 2D regions of perceptibly constant density could be suitable primitive descriptors, together with regions of perceptibly constant gradient of density and more complex properties. Such regions are delineated by boundary segments which are located at the junction of perceptibly different regions. These boundary segments may be regarded as lines possessing attributes such as angle and curvature, which could form the basis of suitable primitives for describing these segments. Boundaries are in this case considered as objects, but they may for other purposes be considered as attributes of bounded regions or relationships between adjacent regions.

The pictorial primitives chosen should be appropriate to the type of input material if this is known in advance, and should be as complex as is consistent with their generality. We must distinguish between pictorial primitives and the smallest units employed in processing the picture. As an example, consider the encoding of a smoothly curved boundary AB, as shown in Fig. 5. We might employ an edge-detector to indicate the presence of relationships (between matrix points in a small region) specifying an edge of a given orientation. We could then group a number of such responses, on the base of proximity and similarity of direction, into a boundary curve such as CD. It seems natural to consider the separate responses of the edge-detector, or perhaps even the cells in the input matrix, as primitives. Pictorially though, the curve CD cannot reasonably be seen as consisting of a number of component parts. If boundary curves such as this (with attributes of length, curvature, etc.)

were suitable starting points for SD's of the expected inputs, then (considering efficiency of processing and economy of description) they should form the primitives rather than any smaller units used in processing, which cannot be seen as parts in their own right.

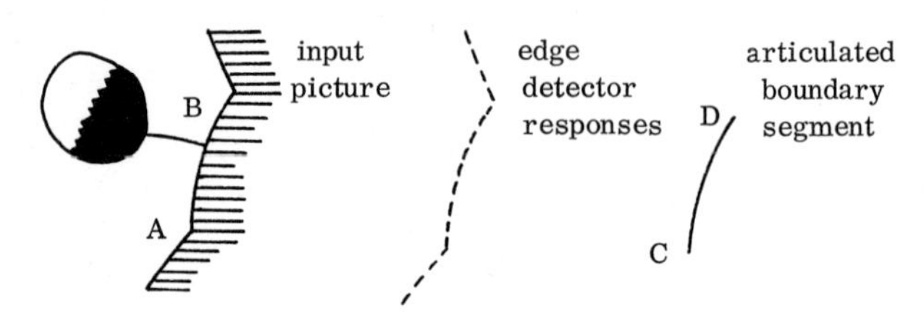

Figure 5. Association of several processing primitives into one pictorial primitive.

The rules of organization are concerned mainly with relationships between parts, and consequently with attribute values upon which relationships depend (e.g. the relationship "parallel" depending on the attribute "relative orientation"). Techniques advocated by Chomsky for the study of structure in language, (viz. ambiguity, anomaly and paraphrase) have been suggested as useful guidelines for the study of pictorial relationships (Clowes, 1968). Another guideline for the determination of significant relationships is the requirement of simplicity (and thus economy) of representation (Attneave, 1954).

A fundamental process in the abstraction of organization is articulation of the raw or partially structured input into regions throughout which some pictorial or object-world property does not change perceptibly. The property involved may be a combination of attributes and relationships but it must be capable of being sensed. Changes of some properties within a region may not be considered significant. There are two relatively independent attributes of regions; their internal properties, and their boundaries or form. We distinguish parts and regions on the basis of the former being concerned more with the structured boundaries and the latter being concerned more with the internal properties. In articulation there appear to be two related processes:-

(i) taking parts which are initially considered as unrelated and seeing if the presence of some relationship between them (and perhaps their context) makes it reasonable to associate them, i.e. synthesizing a higher-level structure;

(ii) taking a region which is initially considered as a connected whole, and seeing if there are perceptible changes in some properties which suggest a delineation of differing sub-regions, i.e. analyzing a region into components. Comparison of the delineated sub-regions may suggest associations additional to, or stronger than, connectivity.

Whether a process is regarded as analysis or synthesis is sometimes simply a function of the way in which we regard the input (Evans, 1968). It is more natural to regard articulation of a region on the basis of a simple attribute such as curvature or density (see Fig. 6(a)) as analysis (because there are no distinct "parts" to be associated), whereas articulation of the two regions shown in Fig. 6(b) could be viewed as synthesis.

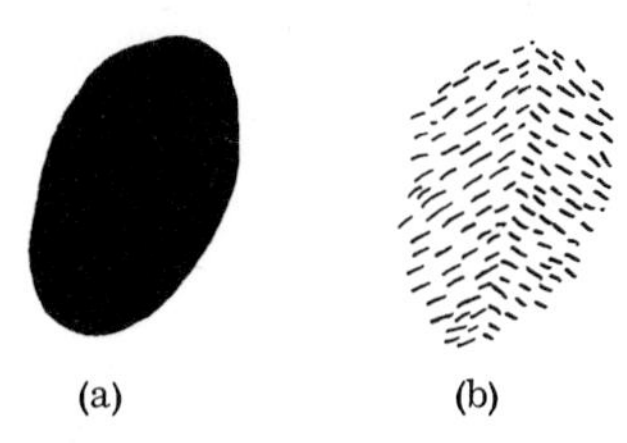

(a) (b)

Figure 6. Articulation by both analysis and synthesis.

The most basic and pervasive of organizing principles is the association of features such as parts, attributes, and relationships, with other similar features in the input - the "law of similarity" (Katz, 1950). Another related principle is the association of parts when the structure so formed is similar to others within our global, as opposed to local, experience - the "law of experience" (Katz, 1950). This law, expressed in a form such as rules of grammar, underlies most of the previous work in articulated processing of pictures. These associations will be examined further below in relation to comparison of SD's. We note that similarity is a relative term, in so far as significant differences depend on the entities being compared, and the local or global context (depending on the type of association and stage of processing). Other factors (not all independent) involved in forming associations are proximity, symmetry, continuation of an interrupted attribute or relationship, and simplicity.

Quite often, there are many possible conflicting associations of one part with others, the problem being to decide which is the best association in terms of overall structuring, including higher levels. Unfortunately, this is not always the locally strongest association. The strength of an association between two parts may be a function of their relationships to other parts, the strength of association of one part to a second being reduced by the presence of a strong conflicting association between the second and a third. When for one part there are conflicting (or alternative) associations with other parts, ambiguity exists. We require the overall structuring in both pictorial and object-worlds to be coherent, and this will in some cases remove ambiguities, perhaps at the expense of restructuring parts of the input. When associations in adjoining regions are inconsistent, and no satisfactory alternatives exist, anomaly results. If, throughout a region containing many parts, the parts have similar associations with each of the surrounding parts, we may interpret the region as one of relatively constant microstructure or even texture. Specification of the microstructure of such a region, together with statistics if there are random components, may be a sufficient (and economical) description for many purposes.

Different associations of various regions of a part (which had formerly been considered as a whole) to other parts, may cause a restructuring of the first part even if it had originally been regarded as primitive, see Fig. 7.

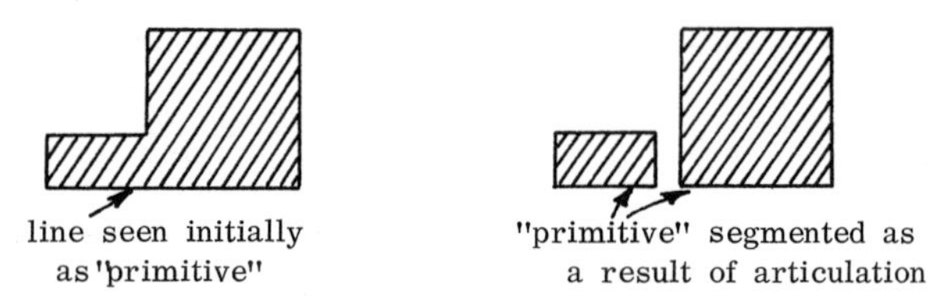

Figure 7. Segmentation of a primitive by higher-level association.

When we attempt to bound or delineate regions on the basis of there being a perceptible difference in the property being sensed, we have to decide what constitutes a perceptible difference, and the level of detail to be examined. Note that the property of interest may be some combination of metric and relational properties. It is possible for several properties to change at one place, in which case the change in the simplest property tends to mask other changes on the same level of structuring (see Fig. 8(a)), but may reinforce changes on different levels (see Fig. 8(b)).

Simple boundaries such as "edges" (changes in density) are usually clearer than more complex boundaries at the same level, such as changes in gradient of density. For this reason, sudden changes in a distinguishing property are generally more

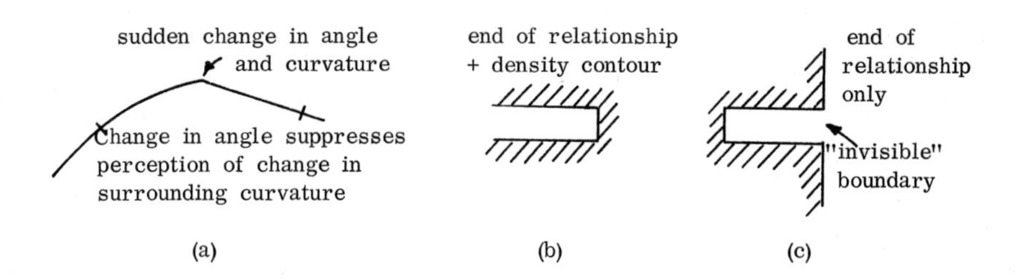

Suppression and reinforcement of one type of boundary by others.

Figure 8.

obvious than gradual changes, because the boundary then depends on a change in a more complex property, i.e. gradient of the original property. When examining regions at a coarse level of detail, gradual changes at a fine level may appear to be sudden. There are many different types of boundaries corresponding to changes in any of the properties being sensed. Thus, we do not require that a region be bounded completely by any one type of boundary. Partly because edge-detection often results in gapped boundaries and by itself does not isolate regions, much early work in picture processing used some form of level-slicing as a technique for delineation of regions. The problem with level-slicing is that it often places boundaries of regions where no perceptible boundaries exist, unless the input material is of a black-white nature. An obvious requirement is that the processor should indicate boundaries between two regions only when they differ perceptibly. Edge-detection is a crude attempt to discover regions by finding their boundaries, whereas level-slicing is an equally crude attempt to discover regions on the basis of their internal properties.

In a picture processor designed for fairly general input material, we should not be concerned if regions are not bounded initially by distinct edges, for some more complex property or relationship is no doubt involved and may enable us to perceive boundaries which are not present as sudden changes in the density. Some boundaries may in fact be "invisible" when the local density values alone are examined. Such boundaries may arise at points where the domain of some relationship ceases - see Fig. 8(c). Thus, the gaps which are liable to arise in edge-detector responses are of no special concern. Also, at higher levels of structuring, some sudden changes in the density will no longer be seen (at that level) as boundaries. When we consider detection of boundaries at a coarse level of detail, as in Fig. 9(a), we must ensure that we have already accounted for the fine detail before we examine what underlies it, otherwise we run the risk of assigning, for example, points 1 and 2 in Fig. 9(b) as boundary points of coarse straight-line regions.

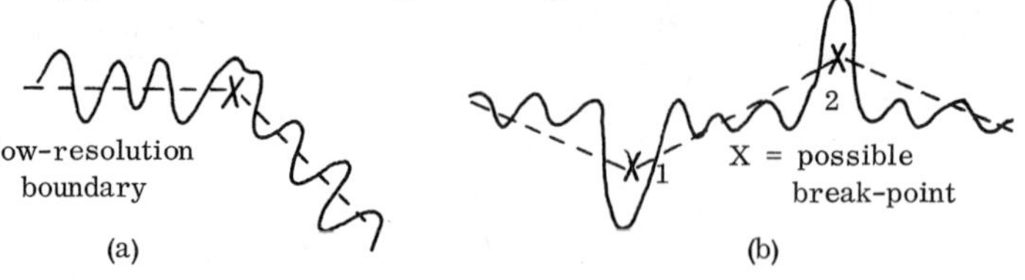

Figure 9. Low-resolution articulation

Closely linked to articulation, whether by association or segmentation, is the idea of "figure on ground". As we get a strong association between parts, a bounded "figural" object tends to appear at the next level of description. The surroundings which are not accounted for, are inclined to appear as one or more types of unbounded ground which the figure is seen as being on. The effect may also appear, or be enhanced, when a relatively constant region (ground) surrounds an interruption which is seen as a figure, see Fig. 10.

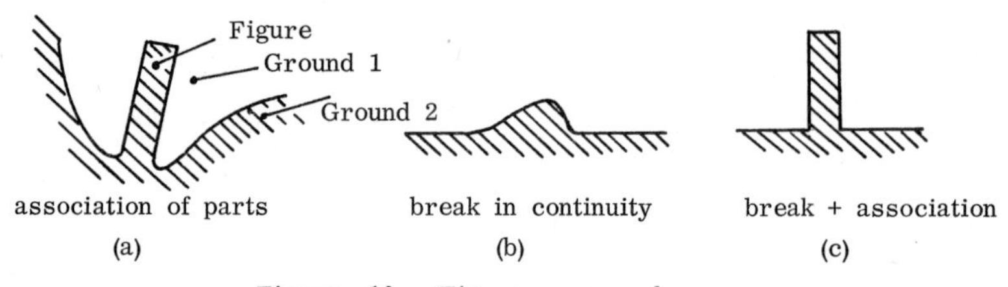

Figure 10. Figure-on-ground.

We may also be interested in the ground after we have accounted for any figures, and there may be several possible stages of figure-ground analysis, in so far as the ground at one level may well be a figure at a higher level (and vice versa). What is left behind as ground after such processing may be of more interest than the figures extracted, an example being the case of seeking trends in noisy data. There is a duality to figure-ground articulation in that it is also possible to think of figures as being what is left behind after the ground is accounted for.

As mentioned above, an important factor in the association of parts is that the structure so formed is similar to that of an already known object, i.e. it is recognizable. In forming this type of association, we have first to attempt to describe the input in terms relevant to the known object, and then to see how effective the description is. This process may be viewed as comparison of an input SD with a class SD. We must first compare structurally to see if we can obtain corresponding parts, the attributes of which may then be examined, perhaps on a statistical basis. An example of a process which is similar in some respects to that advocated above is the recognition of probable accidental errors in the input text by some compilers, when messages such as "Right parenthesis inserted in statement number.... " are issued. Parts may be missing, added, or displaced, and attribute values may differ, such changes being recorded in the SD of similarity.

Some attributes of the structured input, such as absolute size and orientation, may not be included in the class SD because they are not significant for class membership, their omission from the class SD being made possible through relating sizes and orientations, etc. to object-based reference frames. Our class SD's for 3D objects in particular should not be tied down to a reference frame but should include details of the total construction of an object in terms of 3D parts and relationships (not just pictorial (2D) parts and relationships visible from a given aspect). It is not expected that the input SD would incorporate such refinements because of the lack of knowledge during the initial structuring of what constitutes the object and what relationships are relevant to it. It would probably be simpler to relate all parts initially to an external frame of reference, internal relationships thus being implicitly specified. Using a class SD which is not tied down to an external reference frame will facilitate comparisons with an input SD which is so restricted. Detection of the relationships between the object-based reference frames used in the class SD, and the external reference frame used in the input SD, will be necessary for such comparisons. Such relationships may be obtained from the identification of corresponding parts.

As we structure a 3D object from its photographic representation, we are concerned with the 3D disposition of the parts obtained, not with the 2D disposition seen from a given aspect. This requires that our picture processor obtain 3D information from a 2D input. A lot of work has been performed on automatic processing of stereo pairs using cross-correlation techniques, but the work of Roberts (1965) is the only notable example of extraction of depth information from a single static photograph. Interpretation of aspect structure from pictorial structure is obviously a complex process and involves assumptions regarding the pigmentation and lighting unless these structures are derived as well. Also involved are cues such as density, density gradient, interposition, perspective, and texture gradient.

As noted above, with views of 3D objects, some parts will always be missing. It would be desirable, but no doubt difficult, to incorporate within our processing mechanism, sufficient pictorial and object-world knowledge for a decision to be made (given an aspect) as to which parts may or may not be visible, and if not, why not (e.g hidden by other parts of the object).

Another question related to the problem of input complexity is that of how much information from the input should initially be passed to higher stages of processing. Some barely visible details of the input picture may at later stages assume an importance out of proportion to their initial distinctness, because they help confirm or reject a suggested structure. Yet if we were to pass on all such information at one time, we might well "overload" the subsequent stages. Neisser's (1966) arguments for attending to a portion of the input and devoting all the processing power to a small region are relevant here. Thus, we might process the input to obtain very distinct initial structure, and, working from this, return to the input as necessary to obtain additional information as structuring of remaining regions, at higher levels and in different systems, proceeds.

A common problem in picture processing is that of pictorial (or perhaps even object-world) "noise". There are many different manifestations of such noise, e.g. gaps in bubble chamber traces, and it may be helpful to consider noise as extraneous structure, i.e. structure that interferes with the abstraction of "interesting" structure and yet is not of interest itself. It has been customary to attempt to filter out and/or ignore such noise from the outset. The interaction between the extraneous and desired structures is often such that this type of approach will lead to errors in assigning an SD, and a better approach would be to assign noise structure for the purpose of obtaining the structure of interest, and then later to ignore the noise. A simple example of this approach would be to describe a gapped bubble-chamber track as a continuous track with gaps in it, and then later to ignore the gaps and describe it as a simple track, rather than to attempt such an assignment immediately.

3. CURRENT PROGRESS IN IMPLEMENTATION

To clarify and refine the concepts involved, implementations of some components of the proposed picture processor are being attempted. It is not anticipated that useful practical systems will be developed in the near future unless the problems chosen are relatively simple.

Two input classes have been chosen. The first is that of tracks made in photographic emulsion by magnetically deflected low-energy alpha particles resulting from a nuclear interaction. By counting the number of tracks/unit area with the correct direction, length, and density, it is possible to reject extraneous tracks and determine an energy spectrum for the alpha particles. We neglect the thickness of the emulsion and consider it as two dimensional. The structure of the input can be regarded as resulting from pigmentation only. The main problems result from intersecting and "gapped" tracks, and extraneous structure ("noise"). An example of this input class is illustrated in Fig. 11. The second input class is that of micrographs of pollen grains taken with a scanning electron microscope (Echlin, 1968).

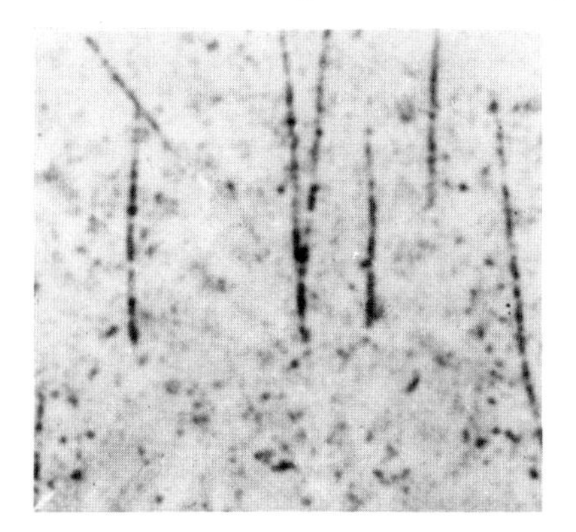

Figure 11. Alpha-particle tracks in photographic emulsion.

A typical micrograph is shown in Fig. 12. A practical system for the automatic classification of pollen would be useful in several fields (Flenley, 1968), but such a system could not be constructed with the current technology and knowledge of picture processing. There are, however, several interesting aspects of this problem, including the various manifestations of structure in pollen grain walls, and the fact that there are several hundred thousand distinct varieties of pollen and it would not seem practicable to compare the structured input against each of the classes in turn.

When the scanning electron microscope is used to scan pollen, raised surfaces tend to be lighter than those behind them. The structure of such surfaces (initially neglecting the three-dimensional nature of the input), is currently being investigated.

Figure 12. Scanning Electron Micrograph of a THRIFT pollen grain. (Copyright Patrick Echlin and Cambridge Scientific Instruments. Reproduced by kind permission).

A digital computer is used to simulate those components of our proposed picture processing mechanism that are being studied. The manner in which primitives or constructs may be put together to make higher-level constructs is sometimes exhibited in a picture-grammar (Ledley et al., 1965). In our experiments, no such explicit formulation has yet been developed. Representation of the structure abstracted from the input picture would seem to be most readily accomplished through a data structure which (conceptually) parallels the abstracted structure. The flexibility required of this data structure indicates a list-processing approach. Weizenbaum's (1963) SLIP has been chosen, not for its symmetry (which exists only within lists and not between them) but for its flexibility, efficiency, and ease of implementation. To facilitate traversal of the list structure and characterization of some relationships, additional pointers are incorporated. With relationships such as "connected to", lists end up pointing indirectly at each other and to prevent the SLIP indexing machinery from looping, the additional pointers are placed on the attribute-value lists of the normal lists.

Photographic input is fed into computer memory via a scanner consisting of a light source, lens assembly, and photosensitive device, all mounted on the pen carriage of a conventional analog X-Y recorder. The scanner operates under control of the Australian National University's IBM 360/50 computer with associated Analogue-Digital Converters and Digital-Analogue Converters. The initial internal representation of the input picture is a 128 x 128 square matrix of density values, corresponding to a 3.2 cm. square section of the input photograph.

The most elementary property of the input is photographic density at a point, and therefore the most elementary boundary results from a sudden change in this property, i.e. an edge. That edges are significant features of pictorial input would seem to be supported by the ease with which humans may interpret differentiated pictures and cartoons, and also by neurophysiological studies.

Aiming initially to construct regions of perceptibly constant density, (out of perhaps several different types of boundaries), we first try to find perceptible changes in the density (i.e. 'edges'). The notion of perceptible change involves the concept of "just-noticeable difference" (JND). To approximate psychological difference by numerical difference, we have to perform psycho-physical scaling of the property being examined. For density we employ a logarithmic scaling based on the classic Weber law for JND, but truncated to allow for the fact that some light is reflected from the greatest density and that light regions are usually in the vicinity of dark. Roberts (1965) employs an exponential (square-root) scaling for density.

We have found that edge-detection based on only nearest neighbours in the scanned matrix is unsatisfactory, a somewhat larger context giving better results. The form of edge-detector finally chosen consists of superimposed displaced exponentials such that the correlation mask is,

$$F(x, y, \theta) = \left[+e^{-\left\{(x \sin\theta + y \cos\theta + d_p)^2/d_p^2\right\}} - e^{-\left\{(x \sin\theta + y \cos\theta - d_p)^2/d_p^2\right\}} \right] \times e^{-(x^2 + y^2)/d_r^2}$$

where θ is the relative direction of the edge and d_p, d_r determine the rate of decay of the exponentials perpendicular to and along the edge respectively. An impression of one such mask is given in Fig. 13.

Various masks corresponding to edges at several orientations, are correlated with the input matrix via Fast Fourier Transform (Rosenfeld, 1968) at a considerable saving of execution time compared to conventional techniques. Edge responses are defined as correlations which exceed a threshold of absolute correlation and decay rapidly in a direction perpendicular to the edge (i.e. are sharply defined). Individual edge responses are associated into boundary strings on the basis of proximity and similarity of direction, being entered as cells in a SLIP list as association proceeds. Bifurcation is avoided by choosing only one continuation at each branch point.

Gray background has weight of zero

white is weight of +1

black is weight of -1

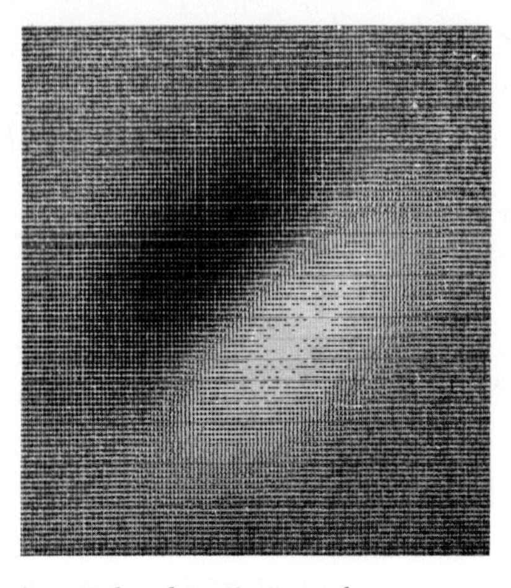

Figure 13. Representation of an "edge-detecting" mask.

The boundary strings so obtained are examined to see if they may be articulated into reasonable parts. The segments finally chosen as reasonable parts were curves whose angle, curvature, or rate of change of curvature with distance does not change perceptibly. The corresponding break points chosen were distinct changes of angle, curvature, and rate of change of curvature. This contrasts with Attneave's (1954) choice of maxima of curvature and Freeman's (1967) choice of points of inflection. The further complexity would seem necessary to deal with the articulation of figures such as ellipses and "hooks", as shown in Fig. 14.

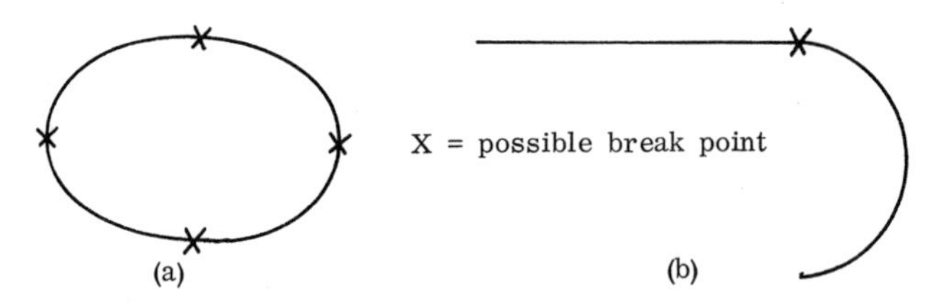

Figure 14. Articulation of an ellipse and a "hook".

Boundary strings are first examined for distinct local maxima of curvature and broken at such points. The resulting segments are fitted (via a linear least squares routine) by a polynomial curve with enough parameters (such as initial angle, initial curvature, etc.) to solve for possible break points. The suggested break points are then examined to see if they are sufficiently distinct to warrant further segmentation. Connectivity relationships between segments, together with the type of break point, are preserved in the SLIP list structure. As the breakpoints that are searched for involve more complex properties, so (as in the case of other boundary points, curves, and surfaces) the context necessary for their detection broadens. For short boundary strings it may not be reasonable to search for more complex break points than sudden changes in angle (curvature maxima).

It would be possible to segment a boundary string on the basis of the magnitude of the edge response at various points along it, but we assume that a sudden change will appear only in the vicinity of a perpendicular edge and may thus lead to eventual segmentation.

With reference to the earlier discussion on primitives, if we treat our articulated boundary segments as primitive, we should then be able to omit reference to the individual edge responses used in obtaining the segments. This is not practicable until boundary segments have been compared with regard to the distance apart of corresponding points, because with long curves and small distances, large discrepancies may occur when the comparison is made on the basis of the fitted boundary curves. These discrepancies result because the fitted boundary curves are only approximations to the actual boundaries.

When pairs of boundary curves are being examined to see if relationships defined on their distances apart apply, "corresponding" points on each curve are required when measuring the distance apart. Investigation has revealed that the idea of corresponding points on curves is related to the presence and attributes of an axis of symmetry. Such an axis need not be straight and may be of the same form as a boundary curve. Fig. 15 depicts one way in which an axis of symmetry may be obtained from a pair of curves.

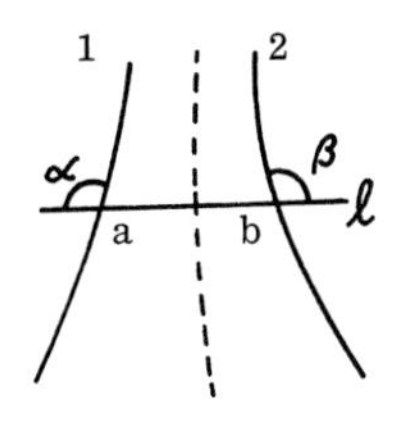

(i) draw line "ℓ" such that $\alpha = \beta$; "a" and "b" are corresponding points.

(ii) point on "ℓ" midway between intersections with curves 1 and 2 locates a point on the axis of symmetry.

Figure 15. Determination of an "axis-of-symmetry".

Using the techniques outlined above, a rather limited depth of structuring (mainly the association of boundaries to give a higher level construct, for example, a short black bar on white surroundings) has been achieved. Work is continuing with regard to deriving more complex structures. A feature which has not been implemented as yet but which it is thought will be of great assistance in forming associations based on proximity, is a matrix which corresponds to the scanned input matrix (but perhaps of great cell size, i.e. reduced resolution) and which "points" to the structures in which the local cells are involved. This will enable "direct addressing" of proximate structures.

4. CONCLUSIONS

The design of a picture processing mechanism capable of assigning a structural description to reasonably complex pictures has been discussed. Several regions of difficulty have been examined:- the contribution to density structure of interdependent aspect, pigmentation, and illumination structures; the complexity of the input; problems arising with 3D objects; the requirement for general rules of organization; the possibility of alternative structures; and the requirement for description in terms of similarity as well as identity.

It is clear that a fairly general picture processing mechanism would be very complex and consist of many interacting components. We may wonder whether it will ever be possible to assemble such a mechanism at all, let alone achieve a practicable one. Advances in our understanding of picture processing, followed by development of suitable technology, are required. As our understanding and technology advance, so will the class of inputs and picture processing problems which can be handled practically, grow.

We have said little about photo-interpretation as such. Photo-interpretation is sometimes viewed as the process of locating known objects in a photograph. We take the somewhat broader view that interpretation involves the assignment of a description in terms of object-world attributes and relationships, on the basis of an SD of the pictorial representation of the object-world. As Kirsch (1964) implies, interpretation is a mapping between structures in different systems. An example of this type of mapping is the derivation of aspect structure from density and gradients of density etc., in a photograph of a 3D scene. A great deal of knowledge about the object-world may be required to effect such mappings.

Incorporation of general rules of organization should facilitate discovery of novel structure as opposed to detection of pre-specified structure. Structure so discovered would not be novel in the sense that the primitives, attributes and relationships involved are new, but rather that the total organization is original. In the proposed system, the abstraction of novel structure might be viewed as pattern discovery, and the detection of other similar structures as pattern recognition. Not everyone would agree that finding some order or structure in the input data is discovering a pattern. Giuliano (1967), for example, requires that meaning be assigned to the order found - as noted above, interpretation (or assignment of meaning) may be viewed as mapping between structures. Hunt (1968) thinks of what he calls pattern recognition (which seems to embrace what we call pattern discovery) as learning a grammar by observation, and then using this grammar to recognize further instances of the class described. A pattern does not reside in the structure of a particular instance but rather in the "common core" of a large number of instances. As noted above with regard to class SD's, this common core need not be present in every class member and will require additions and possibly corrections when applied to any given member. Having found individual structures, discovering a pattern is the process of deriving such a common core. Given an effective mechanism for the generation and comparison of SD's, such derivations might well be possible.

5. REFERENCES

ATTNEAVE, F. (1954), "Some informational aspects of visual perception"; Psych. Rev., Vol. 61, No. 3, pp.183-193.

CLOWES, M.B. (1968), "Pictorial relationships - a syntactic approach"; Seminar Paper No. 12, C.S.I.R.O., Canberra, (August).

ECHLIN, P. (1968), "Pollen"; Sci. American, Vol. 218, 4, (April), pp.80-90.

EVANS, T.G. (1968), "Descriptive pattern analysis techniques"; Lecture notes for NATO Summer School on Automatic Classification and Interpretation of Images, Pisa, Italy, (Aug. 26 - Sept. 7).

FLENLEY, J.R. (1968), "The problem of pollen recognition"; In "Problems of Picture Interpretation", C.S.I.R.O., Canberra, pp.141-143.

FREEMAN, H. (1967), "On the classification of line-drawing data"; In "Models for the Perception of Speech and Visual Form", W. Wathen-Dunn (Ed.), M.I.T. Press, Cambridge, pp.408-412.

GIBSON, J.J. (1951), "The perception of the visual world"; Houghton Mifflin, Boston.

GIBSON, J.J. (1966), "The senses considered as perceptual systems"; Houghton Mifflin, Boston.

GIULIANO, V.E. (1967), "How we find patterns"; Int. Sc. and Tech. No. 62, (Feb.), pp.40-51.

GRIMSDALE, R.L., SUMNER, F.H., TUNIS, C.J., KILBURN, T. (1959), "A system for the automatic recognition of patterns"; Proc. IEE, Vol. 106, Pt. B, (March), pp.210-221.

HUNT, E.B. (1968), "Computer simulation : artificial intelligence studies and their relevance to psychology"; Ann. Rev. Psych., Vol. 19, pp.135-168.

KATZ, D. (1950), "Gestalt psychology"; Ronald Press, New York.

KIRSCH, R.A. (1964), "Computer interpretation of English text and picture patterns"; IEEE Trans. EC-13, 4, (August), pp.363-376.

LEDLEY, R.S. et al. (1965), "Fidac: Film input to digital automatic computer and associated syntax-directed pattern-recognition programming system"; In "Optical and Electro-optical Information Processing", J.T. Tippett et al. (Eds.), M.I.T. Press, Cambridge, Mass., pp.591-613.

MINSKY, M. (1961), "Steps toward artificial intelligence"; Proc. IRE, Vol. 49, 1, (January), pp.8-30.

NARASIMHAN, R. (1968), "On the description, generation and recognition of classes of pictures"; Lecture notes for NATO Summer School on Automatic Classification and Interpretation of Images, Pisa, Italy, (Aug. 26 - Sept.

NEISSER, U. (1966), "Cognitive psychology"; Appleton-Century, Crofts, New York.

ROBERTS, L.G. (1965), "Machine perception of three-dimensional solids"; In "Optical and Electro-optical Information Processing", J.T. Tippett et al. (Eds.), M.I.T. Press, Cambridge, Mass., pp.159-197.

ROSENFELD, A. (1968), "Picture processing by computer"; University of Maryland Computer Science Center, Tech. Rept. 68-71, (June).

SIMON, H.A. and KOTOVSKY, K. (1963), "Human acquisition of concepts for sequential patterns"; Psych. Rev., Vol. 70, No. 6, pp.534-546.

WEIZENBAUN, J. (1963), "Symmetric list processor"; Comm. ACM. Vol. 6, 9, (Sept.), pp.524-544.

6. DISCUSSION

Jacks: How many grey levels do you have in your print out?

Macleod: I have 20 combinations of symbols, but by choosing statistically, can represent pretty well anything. The average over some area will be almost exactly what it is that you want to represent.

Lance: How about the picture scanner output used as input to the computer?

Macleod: I use a fourteen bit analogue - digital convertor, but this is much greater resolution than is warranted by the scanner, or the picture processing, for that matter. There are 30 levels and I would expect this order of accuracy.

Bobrow: Do you normalise the grey scale of the picture so that the blackest spot is zero and the whitest is one?

Macleod: Yes, I have a limited density range, so I always choose so that black is black and white is white, or at least the darkest point in the picture is pure black and the whitest point is pure white.

Cook: I wonder about the advisability of trying to obtain boundary strings from the photographs. I notice that you mention Gibson's paper - Gibson's line on how we are able to recognise things like this is in fact we don't.

Macleod: As I say, this implies that this is a fairly automatic type of process, and that hopefully with some form of mapping rules, we might be able to get somewhere.

Cook: What I am wondering is whether one could do this through a discovery of boundary strings. As you noted in your photographs, there are lots of things that appear as boundary strings which have no intuitive counterpart when we look at them.

Macleod: I think that it is necessary to look, because before we can map from a pictorial structure to an object structure, we have first to try to discover the regions in the picture, the regions of constant density; we have to some extent to obtain pictorial structure, and if we are going to talk about, as Gibson does, densities, gradients of densities, edges and boundaries and so on, then I consider it to be mapping between these and an object world structure, rather than mapping from a picture as such, without consideration of the pictorial structure.

Cook: The question is whether articulation into regions is an appropriate way of doing this. I think Gibson would say it was not - he would examine brightness gradients and would say 'don't draw a line down the middle of the brightness gradient, but take the form of the gradients'.

Macleod: You don't draw a line down the middle because there isn't a reason for drawing such a line - you draw a line down either end, that's where the change in property is. For responses to edges, there must be a sharp peak in correlation.

Cook: I agree, but in recovering edges from a shadowed figure, that would appear to be inappropriate in terms of the actual articulation of the object in space. This is possibly a reason for the difficulty in handling that general recognition problem.

Bobrow: You're saying in the shadowed picture you see shadows as objects?

Cook: I am saying that this is recovering the shadows, in some sense this is recovering the shadows as objects.

Macleod: This is not unreasonable, because pictorially they are, may be they are not in the object world.

Cook: I think Gibson would question whether they are - if you have an object in three dimensions, shadowed on one side, he would say, by a process about which he is extremely obscure, he would say you recover not two parts, a shadow part which is one object, and a bright part which is the other; he would say you would recover a surface.

Macleod: Yes, I think this is just a demonstration of your mapping rules not letting you see the visual field, but letting you see the visual world.

Bobrow: Concerning that, at M. I. T. they have been doing some work of the same kind and an alternative approach they take, having read Gibson also, is not to look for edges per se, that is not have edge detectors but rather to have region detectors. A region is defined in terms of a number of predicates, some of which have, for example, uniform intensity. They use a fairly complex predicate, some Laplacian over the surface, and look at some predicate which they say is true. They define those things which are inside the same region, and those where there is some peculiarity, and use an overlapping predicate in trying to determine what regions are in the picture. This allows you to have complex definitions of what a region is; boundaries are where two regions come together, where something interesting happens between the two regions. This is an alternative approach.

Macleod: Oh, mind you, I am thinking of what I abstract are the regions, not the boundaries. I don't consider I have a region unless I can bound it in some sense or other, and then the region has internal properties which, for example, could be its density, but finding the boundaries is one way of finding the regions.

Bobrow: And finding the regions is one way of finding boundaries?

Macleod: Yes.

COMPUTER SIMULATION OF NATURAL LANGUAGE BEHAVIOUR

R. NARASIMHAN

Computer Group
Tata Institute of Fundamental Research
Colaba, Bombay 5, India.

PART I*

Background

Basically I am interested in models of behaviour. The long term goal is to be able to say something significant about the brain modelling problem. Many of the constraints that I have imposed on myself are constraints which arise out of what I consider to be boundary conditions that have to be met by any viable model in this area. I shall try and motivate some of these during the course of my talk. The others you might consider as stylistics at this stage of the game.

If you are interested in modelling human behaviour as opposed to animal behaviour, say, it seems reasonable to start with an aspect of behaviour which seems to be highly characteristic of human beings. And this is, of course, language behaviour. If you are interested in short term results, there are more utilitarian reasons for modelling language behaviour, because if you arrive at any kind of a viable model at all, hopefully you can implement it and get a computer to interact with you in the natural language mode. And one tends to hope that this will make interacting with computers at least somewhat less painful.

In the last ten years there has been a considerable amount of activity in modelling natural language behaviour and we have already had aspects presented to us at this Conference. Much of this work is, of course, associated with the name of Chomsky and his colleagues. Chomsky starts off with a paradigm which is somewhat as indicated in Fig. 1. He visualizes a language acquisition device with some

* Editorial Note: Part I represents the paper as delivered by Professor Narasimhan at the Conference, to provide a background for a more ready understanding of the written paper, appearing as Part II (circulated as a preprint).

kind of an input, and he assumes that what comes out is language. That is how you acquire language. And so he claims that if you want to say something about the acquisition device, you have to know what it is that comes out. If you are able to articulate what language is, then hopefully you can try to answer the question, "What should there be in the device so that this language can come out?".

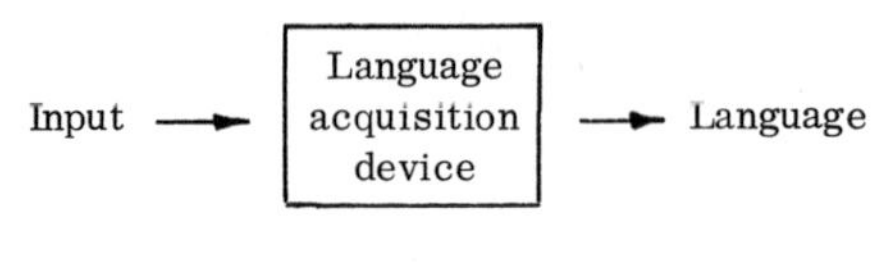

Fig. 1

It seems to me there are several basic fallacies in this approach and I should like to talk about a couple of them. First of all, I do not think this is a correct paradigm at all. One should schematize it not as shown in Fig. 1, but in terms of a language behaviour acquisition device. And what comes out is not language but language behaviour. So one does not ask the question, "What is language?". If you want to, you can ask the question, "What is language behaviour?". But if you do, after some thinking you find that there is very little you can say except: "Language behaviour is the output of the language behaviour acquisition device". So the only thing to do is to plunge in and start to study the language behaviour acquisition device.

It is my view that there is no such thing as natural language; there is only natural language behaviour. And it is also my view that natural language behaviour is no different from any other kind of behaviour. I do not think that the output of the behavioural machinery can be formalized. What can be formalized is only the behavioural machinery itself. So model building must concern itself with the articulation of the behavioural machinery, and then let it function in the kind of environment in which the thing that is being modelled normally functions. One can then evaluate whether the model exhibits the kinds of behaviour that are acceptable analogues, acceptable examples, of the modelled organism.

Basically, then, I have to concern myself with constructing some kind of a metatheory of behaviour in general. To do this, one has to fix appropriate boundary conditions. It seems to me that any behavioural model must satisfy two constraints. The first is, what I would call, phylogenetic continuity. I mean by this the following: If you consider the various organisms - for example, human beings and chimpanzees - you find that there is an enormous amount of similarity in their hardware. And there seems to be a continuity in the evolution of their hardware. It seems reasonable to assume that there should be a similar continuity in the software associated with these hardware. In other words, if a theory tries to account for human behaviour, say, and animal behaviour in totally different ways, I would consider it a priori to be an implausible theory.

This, it seems to me, is what the psychologists have been saying. The Skinnerian approach seems to be interpretable in this light. But the shortcoming of this approach seems to me to reside in the fact that Skinner seems to proceed on the assumption that there is no difference whatsoever between animal and human behaviour. But in fact there must be significant differences between the system organization and functioning in the two cases, since language behaviour is a unique characteristic of human beings. I think if one takes this into account, many of the shortcomings of the behavioural approach can be rectified.

The second constraint is this. I claim that there are basically two modes of behaviour so far as human beings are concerned. I shall call these: the paradigmatic mode of behaviour and the syntagmatic mode of behaviour. The paradigmatic mode is basically associated with behaviour that we refer to as analogical. That is, terms like analogy, inference or reasoning by analogy, naive, informal, intuitive, are all attributes of the paradigmatic mode of behaviour. The syntagmatic mode of behaviour would be characterized by attributes like, logico-deductive, rule-bound, formal, analytic, overtly scientific, and so on. I think this does give some idea of the kinds of distinctions I am trying to make between the two behavioural modes.

The second fallacy in Chomsky's approach, it seems to me, is that he assumes that language behaviour belongs to the syntagmatic mode of behaviour. My contention is that natural language behaviour belongs to the paradigmatic mode. I hope that the theory will ultimately show that the paradigmatic mode of behaviour is really the primitive mode of behaviour. In other words, the phylogenetic continuity that I referred to earlier does arise out of the fact that the primitive part of the metamachinery is set up to function in the paradigmatic mode. Hence, all animal behaviour is paradigmatic in the sense I am discussing. The syntagmatic mode that is so characteristic of human behaviour, I tend to think, arises primarily through our capacity to operate in terms of languages. It is the symbol manipulation aspect which makes it possible to function in the syntagmatic mode.

Since I am saying that the paradigmatic mode is the more primitive mode of behaviour, it must be clear that children when they start out must necessarily function predominantly, I would say almost totally, in the paradigmatic mode. And since language acquisition does take place in these early stages, the plausibility argument is that it must, in fact, be a paradigmatic mode of behaviour. I assert that what schooling does is precisely to teach children how to behave in the syntagmatic mode of behaviour. That is, training is essentially concerned with getting people to behave syntagmatically.

My point is that syntagmatic aspect is precisely what is involved in formal training - in schooling. So when you learn your language in school, this is what you learn. This is where they teach you the grammar of a language, for example. That is, you learn about your language syntagmatically, but all the time you are functioning in the language paradigmatically. What I am asserting here has the merit that it enables you to distinguish between the language behaviour of people who have no formal literacy - that is, people who do not know how to read or write - and that of people who have studied about their natural language in school. So, I think it is completely wrong to talk about the intuitions of native speakers, as Chomsky does all the time, as if intuition is something that is uniform over all native speakers.

If you are willing to go along with me, you will not ask in trying to account for language acquisition in children, how is it that children acquire grammar, because I am arguing that grammar is irrelevant to the paradigmatic mode of operation. This, coupled with the fact that what children acquire is language behaviour and not language, helps to resolve many seeming paradoxes of behaviour. For example, children are able to understand fairly complicated utterances in their home environment long before they are able to articulate such utterances themselves. Transformational grammarians typically use such examples as these to argue that in some obscure way children must have built into themselves the grammar of the language, and so on. It is unnecessary to resort to this kind of explanatory procedure if one recognizes that what the children acquire is language behaviour and that language behaviour has many dimensions. So, when the child's mother says something to the child, there is no reason to assume that it is only the linguistic string that gets decoded and not any of the other behaviour aspects that go along with that linguistic string. If that rich environment is made available, it seems quite natural for a child to be able to respond to complicated utterances before it can generate them by itself.

If one cannot ask "What is language behaviour?", how then does one start to study the behavioural mechanism? My view is that the kinds of questions one should ask are not what it is, but what is it that one does through behaviour. In other words, the questions to ask are: "What is it that people do with behaviour?" i.e. what use do they put it to? And then you ask yourself the question: "What is there in behaviour that enables them to do this and this and this?" i.e. how are they enabled to do this? This kind of analysis I would call the analysis of the pragmatics of behaviour. And it is this analysis that I assert is relevant to characterising the paradigmatic mode of behaviour.

Since one assumes that it is the paradigmatic mode of operation that is inherent primitively in the machinery, the expectation is that if you know how to analyze it properly you would be able to say something about the machinery that enables the realization of this kind of behaviour. Then the question arises that if it is, in fact, true that a child comes into the world with a machinery that primarily can behave only paradigmatically, then, for the theory to be viable, one should be able to demonstrate that, given a machinery which can do only this, one could in fact <u>teach</u> the machine to behave syntagmatically, so it is the <u>teachability</u> which is the characteristic of the syntagmatic mode. Hence the basic constraint for modelling the syntagmatic aspect is to show that it can in fact be taught to a machine that has built into it only the paradigmatic aspect.

It is important to add a word of caution here. One should not visualize the behavioural machinery to consist of two watertight compartments labelled paradigmatic and syntagmatic, respectively. It is the total integration of the two that matters. Once you have trained yourself syntagmatically to perform various actions, in the naive mode of functioning you are in fact behaving paradigmatically. What presumably happens is that your paradigmatic machinery gets enriched in some sense by what you have acquired through the syntagmatic mode.

I think that this distinction which people have failed to see is extremely important in the design of artificial intelligences. People have so far tried to get computers to function intelligently by operating exclusively in the syntagmatic mode; for example all approaches that tend to reduce problem-solving to theorem-proving in an explicitly stated logic. This assumes that all intelligent behaviour can be characterized exclusively in the syntagmatic mode. It seems to me that what has to be done, on the contrary, is to understand the essential aspects of these two modes of behaviour, and to learn how to construct systems that can function in both these modes.

I think that a wide variety of studies of creative behaviour does indeed show that it is the paradigmatic mode that is primarily relevant here. The essential feature of scientific behaviour is that once one has come up with some kind of a construction or hypothesis, or what you will, one immediately goes into the syntagmatic mode to test it out. If one fails to do that, and functions completely in the paradigmatic mode, then all one's reasoning is restricted to reasoning by analogy. This kind of functioning is what one refers to as functioning in a pre-scientific mode. The history of science seems to teach us that it is only after the realization of the fact that scientific behaviour is primarily concerned with the syntagmatic mode that most of the real advances in science have taken place.

It may seem that my current arguments contradict what I was saying in my earlier paper on "Picture Languages" in this Conference, when I was arguing for articulation of pictures, syntax of pictures, and so on. But I do not think that there is any real contradiction. For what I am arguing is not for the one or the other, but a proper integration of both. I do not right now have a well worked out scheme for the paradigmatic approach to pictures. One has tended to emphasize the syntagmatic approach to pictures so far because all the early work in picture processing had taken a completely nondescriptive approach. So I think it was worthwhile emphasizing the syntagmatic approach and bring it up to a certain level. But I think there is a danger one might overdo this. The thing to do would be to pause at this stage and consider whether there is a paradigmatic approach to picture processing, and then see how the two could be combined.

I think this process can be aided by studying in greater detail language behaviour, because the paradigmatic aspect of language behaviour, hopefully, could be more fully worked out. The insights that one picks up could then plausibly be used to study the other modalities of behaviour. In Section 2 (Part II), what I have tried to do is to give some elements of the behavioural machinery which could function in the paradigmatic mode as specialized to language behaviour.

The fact that there are inputs coming in from the external environment through the eyes, ears, nose, skin, etc., and the fact that certain kinds of motor actions are available to manipulate the external world, are characteristics of the system that also exhibits language behaviour. These define various aspects of the world that are available to the system. And this must be mirrored in some sense in language behaviour if it is to be possible to function in the language behaviour modality to accomplish various things in this environment.

An analysis of what these might be is what is contained in Section 2. I have argued there about categories which are basically objects, actions, attributes, relations, states, and other metabehavioural categories such as description, command, and exploration. I have suggested some analogues of these in the language behaviour mode at the end of Section 2.1. The assertion is that these are the kinds of things that one accomplishes through behaviour, through language behaviour in particular.

Next, one wants to analyze what does description amount to in language behaviour. One then says that description is a description of some object, or of some event; that is, a description that this object has this attribute, or that the value of this attribute of that object is such and such, and so on. If, now, one is able to make these kinds of descriptions through language behaviour, then there must be aspects to language behaviour which relates to these various aspects of description one is making. That is, given the language behaviour output, one should be able to say this is the aspect that relates to that object, this is the aspect that relates to the attribute of the object, and it is the totality that says that that object has that attribute, etc.

I have not tried to describe the metatheory of behaviour in this paper because I have discussed it at some length elsewhere (Narasimhan, 1967c). Briefly what the theory says is the following. The behaving organism viewed as a system has various interfaces with the outside world as well as with subsystems making up the organism. In the case of human beings, consider these as the eyes, ears, etc. One of these interfaces is the language or speech interface. Behaviour in each of these modalities is characterized by schemata, what I call action schemata or behaviour schemata and these schemata are nothing but networks of "expressions". So the "expression" is really the behavioural unit that relates, in some sense, to the various aspects of the situation which is being reacted to.

I have tried to analyze in terms of this metalevel characterization, in Section 2.3, some of these expressions as they apply to the language behaviour modality. In the paper earlier cited, I have done a much more detailed analysis. I have also tried to analyze there the output of a child's language behaviour over an entire day in terms of expressions like these, making the plausibility argument that, in fact, these kinds of characterizations work.

I do not want to spend more time here demonstrating this. What I would like to do now is to describe to you a program that my colleague Ramani (1969) has worked out. He has a model which has built into it the machinery for what I would call the paradigmatic mode of operation, that is, it functions by analogy. And it is teachable. So he has a series of sessions with it and teaches it to function in the syntagmatic mode. You can interpret this experiment as teaching a device that is not a digital computer to behave like a digital computer. For a digital computer is the best example of a system that we know of that functions exclusively in the syntagmatic mode.

The point is this. From the examples that I show you, it should be easy for you to satisfy yourselves that through interaction in this manner the system can learn to function in a natural language, i.e. learn to exhibit natural language behaviour. Ramani's particular interest has been to teach it to solve problems, that is, teach it the syntagmatic mode as it applies to problem solving. he has not been specifically concerned with teaching it to function in natural language. But it should be quite clear that through a graded series of exercises it could be taught to function in the natural language mode and we hope to work on it some more.

As shown in Figure 2, the interaction essentially consists in the following: The system is exposed to paradigms. A paradigm of a set of natural language expressions which are labelled by one of three tags: context, stimulus, and response. A given paradigm teaches the system that in this context, this response is related to this stimulus.

Now, what the system does is to schematize it in a specific way and internalize it, so one can expose it to a whole lot of paradigms like these. One can also input informational statements like: "A foot has 12 inches", etc. Then in the test mode of interaction, if one presents it with a stimulus, the system will respond with the appropriate response.

Figure 3 shows a set of paradigms, and some related tests, in terms of which the system can be taught to solve problems in arithmetic, algebra, physics, etc. The system has a built-in desk calculator which can perform the basic arithmetic operations, and various other operations like taking exponents, etc. Action schemata for "find", etc., are built-in. Also built-in are action schemata associated with expressions like "go to" and "stop".

PARADIGMS	TESTS
(1) C: The weight of the ship is 2000 tons S: What is the weight of the ship R: 2000 tons	(1) C: The divisors of 120 are 2,3,4, and 5 S: What are the divisors of 120? R: 2,3,4, and 5
(2) C: The boys are quiet because they are tired S: Why are the boys quiet R: Because they are tired	(2) C: Kids are quiet because their teacher is around S: Why are kids quiet? R: Because their teacher is around
(3) C: Clothes are cleaned by washing them in water S: How are clothes cleaned R: By washing them in water	(3) C: Cakes are baked by heating them in a container S: How are cakes baked? R: By heating them in a container
(4) C: Calcutta is in the east S: Where is Calcutta R: In the east	(4) C: The Astronauts are in Apollo 8 S: Where are the Astronauts? R: In Apollo 8

Fig. 2

PARADIGMS	TESTS
(1) C: The cost of 7 pens is 49 Rupees S: What is the cost of one pen R: 1. Divide 49 by 7 2. That is the cost of one pen in Rupees	(1) C: The cost of 16 tapes is 8010 Rupees S: What is the cost of one tape R: The cost of one tape in Rupees is 500.625
(2) C: A train moves at the rate of 25 miles per hour S: Find how far the train will move in 2.5 hours R: 1. Multiply 25 by 2.5 2. That is the answer in miles	(2) C: A balloon climbs at the rate of 200 meters per minute S: Find how high the balloon will climb in 45 minutes R: The answer in meters is 9000
(3) C: The cost of 16 books is 80 Rupees S: What is the cost of 23 books R: 1. What is the cost of one book? 2. Multiply it by 23 3. That is the cost of 23 books in Rupees	(3) C: The weight of 110 tins is 88000 pounds S: What is the weight of 17 tins? R: The weight of 17 tins in pounds is 13600

Fig. 3

The paradigmatic machinery is a recursive machinery. In other words, if you have exposed it to a sequence of responses in one paradigm, then when it encounters another paradigm in which the stimulus of the previous paradigm occurs as one of the responses, then it will use that as an internal stimulus, so to speak, and call the whole sequence of related responses, basically pushing down the rest of the response sequence in this paradigm. This recursion can be done theoretically to arbitrary depths. The practical limitation is, of course, the memory limitation.

Let me briefly explain how it schematizes paradigms. It analyses a situation in terms of various aspects. A situation is a paradigm. A situation is given to it pre-tagged as the context, stimulus, and response. This is the primitive analysis that is given to it. Then it scans the context, stimulus, and response for the longest expressions that are identical. If there is such an expression that occurs in two of these, at least, then it assumes that that relates to some aspect of the situation. It proceeds thus and schematizes the situation into a matrix and filler form - the matrix consisting of constant expressions and the filler consisting of formal parameters.

So the system merely reacts to a situation by assigning it to a class of situations defined by the matrix with the formal parameters; this particular situation is then defined by assigning these specific actual parameters as values to the formal parameters. Given, now, a new situation it tries to type it by going through all the schemata that are available to it and finding out which one matches. On finding one that matches, it responds by assigning values from the current situation, so reconstructed, to the formal parameters in the internalized response of the matched schema.

Right now it will only respond if the match is complete. Obviously what has to be done next is to get it to respond in plausible ways even if the match is only partial. A much more difficult task is to teach it to acquire behaviour through paradigms which are not tagged as context, stimulus, and response explicitly. One point to note is that one can teach it to ask questions in case of mismatch or doubt. One can teach it to do this paradigmatically. So one does not have to go outside the present system to accomplish this.

PART II

1. INTRODUCTION

1.1. Need for a Theory of the Pragmatics of Natural Language Behaviour

There are, in general, two motivations for studying artificial intelligences: (1) to explicate human intelligence processes by modelling and simulation, and (2) to make computers more human-like in their behaviour so that interacting with them is less painful. Of all aspects of behaviour, the one that is most characteristically human is the ability to acquire and use languages. Also, one can give a variety of plausibility arguments to demonstrate that this competence to acquire and function in a variety of languages is a necessary prerequisite to complex intellectual activities like problem-solving, theory-construction, etc. (Narasimhan, 1966). Hence, it is not surprising that, in the last few years, computer scientists concerned with the design of artificial intelligences have been increasingly preoccupied with the design of computer programs that exhibit natural language behaviour of various sorts. As typical examples of these efforts which have resulted in interactive programs that exhibit nontrivial natural language behaviour, we might mention the following: STUDENT (Bobrow, 1964), SIR (Raphael, 1964), ELIZA and its variants (Weizenbaum, 1966), the DEACON (Craig et al., 1966), and the MITRE query system (MITRE Corporation English Preprocessor Manual, 1965).*

Every program that is designed to exhibit natural language behaviour in some form must be explicitly or implicitly based on a theory of the pragmatics of language behaviour.** The study of the pragmatics of language behaviour is primarily concerned with the study of two interrelated problems: (1) how language behaviour is acquired, and (2) how language behaviour is put to use. It is clear that these two problems are interrelated and cannot be studied to any significant extent in complete isolation, for one acquires a language by functioning in it. Using a language, then, is one necessary aspect of acquiring it. However, since much of the preoccupation

* In this paper, we assume a reasonable familiarity on the part of the reader with the structure and behaviour of these systems.

** See, for example, Bobrow's (1964) paper (p. 592): "We shall describe a theory of semantic generation and analysis of discourse. STUDENT can, then, be considered as a first approximation to a computer implementation of the analytic portion of the theory, with certain restrictions on the interpretation of a discourse to be analyzed".

among computer scientists has so far been in the design of programs that _use_ a natural language rather than in programs that _learn_ a natural language by functioning in it, we shall concern ourselves primarily with this aspect of the pragmatics of natural language. In the last section of this paper, we shall consider tentatively, some aspects of programs that learn natural languages.

The design of a computer program that makes use of a natural language in some well-defined context must have as its basis a theory which concerns itself with the following questions: (1) What is it that human beings do in a natural language? (2) How do they do this? (3) What structural and functional aspects does a language have to enable them to do this? (4) How can a computer program be constructed that can use language similarly and interact with human beings? (5) To accomplish this, is it essential for the program to deal with the structural and functional aspects referred to in (3) above? (6) If yes, should it deal with them in the same way as humans do?

Notice that, in spite of the way these questions are phrased, the concern in them is not with language, _per se_, whatever it might be taken to be, but with _language in use_. An analysis of language in use is an analysis of one aspect of language behaviour. We could just as well have phrased our original set of questions as follows: What is the proper analysis of language behaviour in order to be able to explicate what people accomplish with it, and how they accomplish this? It is, perhaps, much more precise to formulate our problem in this manner as an analysis of language behaviour, rather than as an analysis of language. For, then, there is less temptation to set up a spurious dichotomy between language and language behaviour as, for example, some of the linguists have been doing (Chomsky, 1965).

I have argued elsewhere, at some length, that linguistic theorizings based on the competence-performance dichotomy have essentially no relevance to psycholinguistics (Narasimhan, 1967b). This is equally true as regards the computer sciences. For, in trying to describe a computer program that exhibits natural language behaviour the computer scientist is essentially concerned with saying something in detail about the structure and dynamics of the mechanisms that underlie language behaviour, just as a psycholinguist is. Notice that the competence that we are concerned with studying here is _competence in language behaviour_ and not _competence in language_, as Chomsky postulates (see, for instance, Chapter 1 of Chomsky, 1965). If, on the other hand, one adopts Chomsky's attitude, one would still be left with the problem of explicating how some presumed competence in language issues as the observable competence in language behaviour. In other words, this crucial interface problem between language and language behaviour, would not even have been touched.

1.2. An Analysis of Computer Simulation of Conversational Language Behaviour

Let us now consider in somewhat greater detail, computer simulation of conversational language behaviour. In order to delimit the basic boundary conditions of this problem, consider what the principal phenomenological aspects of conversation in a natural language are. Assume conversation between a man and a machine (i.e. a computer program), and assume, for definiteness, that the communication interface is a typewriter. Let us refer to fragments of conversation as utterances (typewritten "messages" in reality). We shall also at times refer to these utterances as stimuli and responses.

Given an utterance from the man, the machine must assign an interpretation to it. Clearly, this assigned interpretation is a function both of the utterance and the environment assumed to be relevant to it (i.e. context, universe of discourse, etc.). It is by now a commonplace for anyone concerned with simulating natural language behaviour to assume the following as minimum requirements for any viable model: (1) the input utterance must be provided an analysis (i.e. it must be parsed); (2) there must be a representation of the environment in some suitable form inside the machine (the environment, in general, could consist of a local part and a global part); and (3) the utterance must now be assigned an interpretation in terms of its parse and the current "values" of the local and global environment inside the machine.

The really crucial problem in modelling natural language interaction arises at this stage. For, having assigned an interpretation, the machine must generate an appropriate response to move the conversation along to the next step. This response must, clearly, again, be a function both of the parsed utterance and the internal environment (now possibly modified), i.e. it must depend on the assigned interpretation.

Let us now consider some specific cases. If the assigned interpretation reads the utterance as a declarative, then an appropriate change in the internal environment must be effected. The response could be a plain statement that the input utterance has been understood; or the response could be an exclamation of astonishment, or disbelief, or a question of some sort, or whatever.

If the utterance is interpreted as a question, an answer must be generated by operating on the internal environment and ending in a suitable response. Notice that this answer computation could be a simple retrieval from an internally "stored file", or it could be some complicated computation on the internal environment, either local, or global, or both. "Puzzling out an answer", clearly, must involve some such complex computation. In fact, natural language conversation, more often than not, requires naive problem-solving competence. This is a crucial point in modelling, for it sets rather specific boundary conditions for admissible internal representations of environments.

If the interpretation results in a nonassignment of sense to the input utterance an appropriate response must be generated which could be a question for clarification, or tentative readings for verification of what was intended ("Did you mean . . . ", "Are you saying . . . ", etc.), and so on. Notice that such responses which have to be generated for further probing of a situation could be quite complex and sophisticated. That this is true will be seen more readily if one visualizes the machine functioning as a teacher and the human as a student.

What we have discussed above may be referred to as a unit event (or an episodic unit) in a natural language conversation. A unit event consists of three stages:

1. Stimulus analysis
2. Change/search of the internal environment
3. Response generation

The generated response, then, becomes the input stimulus for the next unit event, and so on. Two points must be noted. Firstly, there is no necessary implication in the above that stimuli and responses are single utterances. They could very well be utterance sequences. Secondly, this analysis does not preclude the occurrences of "null" stimuli or responses, i.e. pauses, silences, etc.

Our concern in the rest of this paper will be to consider, first, very briefly, how these three stages have been handled in some of the implemented models referred to earlier, and argue to what extent the theory implicit in these models is viable. Next I shall present a different analysis of natural language behaviour that is designed explicitly to handle the pragmatics of language behaviour. Finally, I shall make some tentative comments about what this analysis of natural language behaviour implies as regards viable structures for, and functioning of, computer simulation models.

1.3. A Critique of Some Implemented Models

Stimulus analysis and response generation: In considering the ways in which implemented models of language behaviour handle a conversational unit event, it is convenient to deal with the stimulus analysis and response generation aspects together. These simulation programs can be broadly classified into two groups: (1) those, like the MITRE and DEACON systems, which use generative grammars based on a system of phrase-structure and transformational rewriting rules for stimulus analysis, and (2) those, like ELIZA, STUDENT, and SIR, which use a format-and-filler type of model for stimulus analysis.

It is relevant to note first that all of them are nonlearning programs, and, hence, are restricted to function only with a limited set of stimulus types. ELIZA, which has the minimum grammatical machinery built into it, is also quite revealingly the one which has the minimum of constraints as regards the stimulus types it can accept. It is also equally significant that all these systems generate their responses through the use of a format-and-filler type of model; that is, they do so when they are not restricted to merely selecting one out of a predetermined list of fixed response tokens.

We have seen earlier that a generated response must be a function, in general, of both the output of the stimulus analysis, and the internal environment. It is, thus, clear that a format-and-filler type of model is a natural one for response generation. The format and filler could both be provided by the stimulus analysis output (as in ELIZA), or the format could be provided by the stimulus analysis output while the filler is provided by a computation on the environment (as in SIR, STUDENT, and the MITRE and DEACON query systems). Notice that this procedure automatically ensures the relevance of the response to the input. Or, to put it somewhat differently, the notion relevance of the response to the stimulus is intrinsically built into this response generation procedure.

Equally clearly, it is seen that generative grammars based on rewriting systems that work strictly from top down, cannot generate responses satisfying boundary conditions of the type we have been discussing. This is because there is no basis for building into these grammars selection mechanisms that could be controlled by external inputs; specifically, in our case, inputs that are generated from stimulus analysis. Thus, there is no mechanism in these models for pairing two utterances as admissible stimulus-response pairs. In other words, there is no basis in these grammatical models for computing relevance.

Notice that relevance cannot be computed in terms of the transformational dependence of two utterances, in Chomsky's sense, as might be thought. In special cases this may work; for instance: "Is this a pen?" "Yes, it is a pen." But, in general, it will not. Consider, for instance, the pairs:

1. "I went to London yesterday."
   "And what happened?"
2. "I am reminded of my mother."
   "Tell me more about your family?"

And so on. In the next section, we shall outline an analysis of language behaviour whose pragmatic basis is capable of pairing utterances such as the ones above as admissible stimulus-response pairs.

If generative models of the type Chomsky has been considering are not useful for response generation, as seen above, the question naturally arises whether they are necessary and/or useful for stimulus analysis. The successful implementation of systems like STUDENT, SIR, and ELIZA, show that they certainly are not necessary. We shall argue later that such models are not likely to be useful either by showing that the category analyses on which these grammars are based have little pragmatic relevance.

Environment computation: The central problem in environment computation is deciding on an appropriate representation for the internal environment. Almost all implementations make use of some form of structured data-base for this purpose. Typically combinations of the following serve as data-bases: list structures, ring structures, tables, dictionaries.

Two important implications of this approach should be noted here. The first is that the environment is essentially considered as a passive store of operands. This is best seen in the SIR implementation where this store basically consists of property lists, i.e. lists of attribute-value pairs. This is not strictly true of STUDENT whose global environment could, in fact, include definitional statements like "A foot has 12 inches", etc., which are essentially rules for interpretation. But by and large it is true that memory, or the internal environment, in these systems is a passive store.

The second implication is that there is an essential abstracting process that intervenes between stimulus analysis and environment construction (augmenting, modification, etc.). The environment is a file of indexed, cross-referenced data, the indexes and cross-references used being abstract labels (noun, verb, transitive, plural, number, etc.) assumed given and fixed. All this is based on some vaguely formulated theory of semantics and not on any analysis of the pragmatics of language behaviour.

One other outcome of this approach is that there is an intrinsic and inevitable dichotomy between the processor and the data-base. A set of procedures constitutes the processor and this processor acts on the data-base, building it, modifying it, searching it, augmenting it. The procedures themselves do not change. Or, if they do (as in adaptive programs), then there is a hierarchy of them and there are always procedures at the highest level which do not change, and they now function as the processor. *

It is clear that this form of representing the internal environment is, at best, a very poor approximation to what is really needed in a truly interactive program. Conversational interaction in a natural language is an open-ended process. The classificatory aspect of the internal environment constantly keeps growing and cannot be tied down to a predetermined set of abstract tags. An object can not only be shaped like a sphere, but it could also be shaped like an orange, like a Florida orange, like a squashed Florida orange, like the squashed Florida orange that we saw in the market yesterday, and so on. The ramifications of the internal environment must, clearly, be as rich as the expressions in the natural language. In other words, if there is an internal data-file, then it must consist of natural language expressions themselves and not merely of some abstracted versions of these. The conclusion to be drawn from this is the following: There have been attempts at formalizations of a natural language in terms of predicate calculus, production systems, and so on, in order to implement interactive programs. All these are misguided efforts for they are all bound to be inadequate by definition. **

* In any case, no-one has so far attempted to build adaptive, interactive programs that function in a natural language.

** It should, perhaps, be emphasized that we are here talking about a natural language system which, by definition, is an open system. Any delimited subset of a natural language (in whatever way the delimitation is effected) is a closed system, and is equivalent to some formal language. Hence, it can be formalized through one of several methods.

As for the processor-data-base dichotomy, it should be clear that this, at best, is again only a poor first approximation to what is needed in a truly conversational program. For such a program should necessarily be a growing program and this growth must include not only an increase in the knowledge of the world but also an increase in the competence to manipulate the situations of the world. That is, a growing program cannot consist only of a fixed processor plus a growing data-base, but also of a growing processor. And it must be possible to grow the processor through conversational interaction. *

This implies that the output of the stimulus analysis must result in additions also to the processor and not merely to the data-base. We have already seen that the data-base must consist of language expressions themselves and not of some abstracted versions of these. The same must, then, apply also to the additions to the processor. In other words, the processor must itself consist also of language expressions themselves. Thus, ultimately, there need be/can be no dichotomy between the internal environment that represents the procedures, and the internal environment that represents the data-base. **

Since a procedure is an action schema, i.e. a complex of response units, it follows from our analysis above that the proper way to represent an internal environment is in the form of interconnected complexes of response units. But we have already seen that these interconnected complexes, in the case of language behaviour, must, in fact, be made up of natural language expressions. Thus, the pragmatic analysis of language behaviour is concerned with exhibiting explicitly these response units of language behaviour. And modelling language behaviour is concerned with exhibiting a viable structure composed of interconnected complexes of these units, and exhibiting a dynamics of functioning of this entity such that interactive behaviour of the type we have been considering results.

We shall outline in the next section, a particular analysis of natural language behaviour that conforms closely to the view discussed above. In the final section, we

* Teaching how should be an open-ended process in much the same way as teaching what is. And this teaching how should, in principle, be done through natural langua[ge]

** Notice that this must not be confused with the uniform way in which instructions an[d] data are stored in the memory of a von Neumann type of computer. In these compu[ters], the processor actually consists of the control unit and the memory forms the data-base. What we are saying here, on the other hand, is that what is needed is a computer system design in which the control unit-memory dichotomy is intrinsica[lly] absent.

shall indicate some implications of this analysis to computer simulation.

2. AN ANALYSIS OF THE PRAGMATICS OF LANGUAGE BEHAVIOUR

2.1. Categories in Natural Languages

The pragmatics of natural language behaviour is concerned with a delimited totality of situations, which may be referred to as the world to which the (particular) language relates. This delimitation is done as follows:

Objects and Actions

Primitively given are sets of Objects and actions. Certain of these Objects can perform certain of these actions. Let us call those Objects which can perform actions agents, and refer to the rest as objects. (Note: We shall write Objects - with a capital O - to refer to the set agents + objects whenever we want to refer to these two categories collectively.) The set of actions an agent can perform is called its action repertoire.

Attributes and Relations

Among the totality of actions, two subclasses are distinguished and referred to, respectively, as computing (evaluating, measuring, finding, determining) values of attributes, and computing values of relations. Attributes are (partially) computable functions defined over the domain of Objects, and assume values from well-defined ranges. The values of attributes are called properties. That is, if A is an attribute applicable to object O, and if the value of A computed with O as its argument is p, then p is called a property of O. We can write this as follows: *

$$A(O) = p.$$

The following example illustrates these ideas. Colour is an attribute applicable to the object this apple. Let colour (this apple) = red. Then, red is a property of this apple.

* Notice that since attribute value computations are actions performed by specific agents, we should strictly indicate this by writing:

$$A_a(O) = p$$

which states that p is the value of attribute, A, as computed by agent, a, on object, O. Notice also that such a statement is valid only if A belongs to the action repertoire of agent, a.

Relations are (partially) computable predicates defined over two or more properties; (see further below for a generalization of this definition). Let R be a relation defined over the values of attribute, A. Let O_1, O_2 be objects to which A is applicable (i.e. over which A is defined). Then,

$$R(A(O_1), A(O_2))$$

is either true or false.

As an example, consider the following. Greater than is a relation applicable to the values of the attribute length. Length is applicable to the objects this stick, that stick. Thus,

greater than(length(this stick), length(that stick))

is either true or false.

Object Types

Since attributes are defined over Objects, Objects can be classified into types (Object types) in the following manner. Objects O_1 and O_2 belong to the same type if and only if all attributes applicable to O_1 are applicable to O_2, and vice-versa. Notice that this type classification defines an equivalence relation and induces a partition on the totality of Objects.

This classification into Object types is a very special kind of partitioning and must not be confused with other kinds of classifications of Objects: e.g. classification according to properties (e.g. all red things), classification according to specific performable actions (e.g. all things that can walk), classification according to specified relationships (e.g. all things greater in length than that object), etc.

States of Objects

The state of an Object is completely specified if all its attribute values are given. Since the totality of attributes applicable to an Object may be a very long list, it is customary to restrict attention to delimited subsets of attributes only, at any given moment. Thus, normally, only partial states are specified. To simplify terminology, we shall use the term 'state' to refer to a partial state, and use the term 'complete state' to refer to the state whenever required.

Since a state is defined by a set of properties of an Object, we can generalize our original definition of a relation by saying that a relation is defined over the states of one or more Objects.

Events and Situations

A consequence of an action is a change of state of one or more Objects. An elementary event is an agent in action, or a change of state of an Object. An event is a structured complex (or a network) of elementary events.

A situation is a structured collection of Objects and events. The totality of situations constitutes a world.

Metabehavioural Categories

The pragmatics of language behaviour in a language that relates to a given world is concerned with describing aspects of situations, and manipulating aspects of situations that make up this world.

Describing a situation consists in specifying what Objects are present (i.e. are aspects of the situation being described), what their states are, what relationships, if any, exist between these states, what events are happening (i.e. are aspects of the situation being described).

Manipulating consists in causing events to happen.

Describing and manipulating are metabehavioural categories. That is, they apply to more than one modality of behaviour. For example, sometimes one can describe by gesture as well as by speech. One other basic metabehavioural category is exploring. As applied to language modality, these categories are identified customarily by a variety of names some of which are listed below.

Metabehavioural Category	Language Modality
Describing	Describing, asserting, evaluating, comparing, etc.
Manipulating	Commanding, controlling, ordering, permitting, requesting, persuading, instructing, coercing, etc.
Exploring	Querying, questioning, verifying, asking, demanding, etc.

lb-Tokens and Utterances

Let us refer to the unit of language behaviour that relates to an aspect of a situation as a language behaviour token (lb-token). The unit of behaviour in the language modality is called an utterance. An utterance is a composition (a linear composition and, hence, a concatenation) of one or more lb-tokens. In terms of the behavioural categories earlier referred to, an utterance is, thus, an assertion, evaluation, a description, command, control, question, etc. The output of conversational language behaviour consists of utterances and utterance fragments; in general, lb-tokens.

2.2. Pragmatics of Natural Language Behaviour

Natural languages primitively relate to the world of naive phenomenology, i.e. to the totality of situations of the physical world as apprehended directly (without artificial instrumental aids) by the sensory mechanisms of animate beings, primarily the human beings. Thus, the utterances of a natural language, describe, primitively, states and relationships as computed by the sensory mechanisms, and command and control actions performable by the motor mechanisms.

Natural language interaction between two human beings, thus, relates to discourse about, and manipulation of, some class of situations of the naive physical world. The interaction consists in making assertions about the states of Objects in a given situation, about the relationship between the states, commanding and controlling actions that cause events relating to the Objects in the situation, and asking questions about the Objects, their states, relationships between the states, and so on.

Since this interaction has to take place through utterances (in the language behaviour modality), the principal problem is to explicate how these utterances relate to the aspects of situations discoursed upon, manipulated, or queried. We said earlier that utterances are composed of lb-tokens, and lb-tokens relate to specific aspects of situations. Thus, the problem of <u>interpretation</u> of descriptive utterances (i.e. the problem of relating them explicitly to the structured aspects of situations), consists in analyzing the utterances into lb-tokens and their interrelationships, and, thus, apprehending the situational aspects and their interrelationships. Similarly, the interpretation problem for command/control, and query utterances consists in performing the appropriate action commanded in the first case, and generating the appropriate description in the second case (which might require performing some attribute or relationship computation). Analogous remarks apply to <u>generating</u> utterances which are descriptions, commands/controls, or queries.

It is clear that for interaction to be possible in this manner, there must be at least one lb-token for each aspect of a situation that can be articulated; and, there must be at least one lb-token which allows one to query that aspect of a situation resulting in the first lb-token as a reply. In general, in most natural languages, there are varieties of tokens that relate to any specific aspect of a situation, and, similarly, a variety of tokens that allow one to query that aspect of a situation.

2.3. lb-Tokens in English: Some Examples

It would be beyond the scope of this paper to attempt an analysis of English which would be comprehensive in any significant sense. However, it is necessary to provide examples to illustrate some of the concepts we have been discussing in the last two subsections. We shall do so now by describing some of the lb-tokens available in English for querying aspects of situations. Later we shall consider briefly some lb-tokens available for location and displacement specification. A much more extensive analysis of English, making use of the ideas developed so far, may be found in Narasimhan (1967c). In what follows, all the lb-tokens are underlined.

Question Tokens (q-tokens)

English makes available a large variety of question tokens (q-tokens); each aspect of a situation can be queried or interrogated by a set of these q-tokens.

Specifically, for example, who whose, whom are q-tokens that elicit information about agent specifications. What, in general, is a question specifier for names. Thus, combinations like do what, doing what, did what, are q-tokens that relate to action names. What state, for example, relates to a state name, what thing to an object name, and so on.

Time tokens relating to time instants could be elicited by q-tokens such as: when, at what time, etc. Time durations could be queried by how long, for what time, since when, in what time, up to what time, etc. Recurrence specifications are elicited by tokens such as: how often, how frequently, how many times, etc.

The following fragment of a dialogue exemplifies the role of q-tokens in the dynamics of language behaviour:

He has been crying.
How long? (or, since when?)
For over an hour (or, since this morning).

Location and Displacement Specifications

In descriptions of events or states, location specifications might arise in different ways as follows:

1. Specifying the place of action or state change.

   This happened in London.
   He grew up in Paris.
   He looked for the book under the table.
   I shall meet you at the station.

2. Action involves placement of object/agent.

   She put the kettle on the floor.
   He put it back inside the box.
   She sat down next to the man.
   He pushed the man out of the wagon.

3. Place of agent/object when in state described.

   The workmen are happy in the factory.
   The fields were green outside Tokyo.
   The sun was bright in Moscow.

The specifications in the above examples may be called "direct" since the intended locations are identified directly through the use of object or place names. In contrast to this, locations could also be specified indirectly. Two major categories of indirect specifications are: (1) through the use of demonstratives, and (2) through reference to another event, or agent, or object.

Some examples of the first kind are the tokens: here, there, this place, that place, in there, down there, under this place, etc.

Some examples of the second kind are the following:

I put the book back where I found it.
I met him where I met you yesterday.

Displacement specifications are closely linked to location specifications since one standard method of specifying a displacement is through the schema:

from (loc. 1) to (loc. 2)

where from, to, are displacement specifiers, and (loc. 1) and (loc. 2) are location tokens, either direct (e.g. from London to Paris) or indirect (e.g. from where you are to where he is).

As in the case of locations, displacements could also be specified indirectly as the following examples illustrate:

Come here.
Move it back up to here.
How far is it from here to there?

Common q-tokens that are used to query locations are: where, at what place, etc. Those used to query displacements are: from where, to where, from which place to which place, etc.

2.4. Artificial Languages and Artificial Worlds

Before concluding this section, it may be worthwhile pointing out that the language categories and the pragmatics of language behaviour that we have been considering so far can be extended in a natural way to language behaviour that relates to situations not part of the naive physical world. Artificial languages, thus, are those constructed to cope with (i.e. describe and manipulate) artificially constructed worlds. In scientific theory construction, for example, these artificially constructed worlds are (considered to be) formalized abstractions of aspects of the phenomenology of the natural world. Such languages, in general, are constructed formally by defining sets of objects with attributes, and action schemata which determine attribute values of given objects, transform one set of objects into others, and so on. *

Artificial languages designed for computer work define worlds whose situations are problem situations. Describing and manipulating these situations are treated as problem-solving activities. In computer languages, in general, manipulations play a much more dominant role than descriptions. **

In story-telling, what presumably one does is to use a natural language to construct descriptions of an artificial world, i.e. a complex of fictive or imaginary situations. We "people" them with objects and agents with fictitiously assigned attributes, states, and action repertoires.

* See Narasimhan (1966), for a more detailed consideration of these matters in terms of a metatheory very closely analogous to the one developed here.

** For a metatheory of computational languages developed along lines similar to the present work, see Narasimhan (1967a).

3. SOME IMPLICATIONS TO COMPUTER SIMULATION OF LANGUAGE BEHAVIOUR

3.1. Statement of the Problem

We can summarize our arguments so far as follows. In Section 1, we analyzed what interaction in a natural language requires (in the case of man-machine interaction) and considered critically the ways in which these requirements have been handled in some of the implemented programs. We came to the conclusion that what is truly needed to cope with this problem is the design of a computer program whose internal structure consists of an interconnected complex of response units, the response units being (in the case of language behaviour) natural language expressions. In Section 2, we analyzed the pragmatics of natural language behaviour and arrived at the notion of lb-tokens which relate to aspects of situations. It is in terms of these lb-tokens, we asserted, that humans using a natural language are enabled to describe aspects of situations of the world, manipulate them, and query them.

These lb-tokens are, thus, the response units in terms of which the interacting program must be structured. And, as we remarked earlier, modelling language behaviour is concerned with exhibiting explicitly a viable structure composed of interconnected complexes of these units, and a viable dynamics of interactive behaviour based on such a structure. The principal thesis of this paper is that it should be possible to design a language behaviour model of this kind. Since no such model has as yet been implemented, clearly, no explicit structure can be exhibited at this time. Many problems of detail remain to be solved before a working model could be exhibited. In the rest of this section, we shall discuss briefly two of these open problems, and make some tentative remarks about plausible approaches to their solution.

3.2. The Problem of Learning

We saw earlier that interaction in a natural language is an open-ended process. Thus, by definition, any program that tries to simulate this behaviour must be a growing program. The central problem in the design of an interacting program, then, is this: how to build into the program the capacity to augment its behavioural repertoire? For, given such a capacity, through continuous interaction with a human partner, the program could build up its competence to an ever growing extent. Notice that this is the best that could, in fact, be done. For, natural language being open-ended, it is not meaningful to talk about a predetermined structure which could cope with all conceivable situations that make up a world. *

Expanding one's behavioural repertoire through interaction is the central aspect of learning. In terms of the model we have outlined, there are two facets to learning. The first is establishing new connectivities (i.e. new "associations") for available lb-tokens in the internal complex. The second is incorporating new lb-tokens together with their appropriate connections in the complex. The first is a form of "generalizing" one's available repertoire to new situations. The second is genuine enlargement of one's repertoire through the addition of new response tokens. One major open problem, of course, is specifying rules that govern both these aspects of learning.

To get some insight into this learning problem we shall now list briefly some necessary properties the response complex should have if language interaction is to be possible. This should give us some idea of what lb-tokens are essential to be incorporated in the complex, and what types of connectivities ("associations") are essential to be made available.

In acquiring language behaviour, then, the following particulars have to be essentially acquired:

1. (Proper) Names of Objects, Object types, actions, attributes, properties, relations, states.
2. Variable names (pro-names) for categories in (1).
3. Associations between these names so that it is "known":
   (1) which Objects are agents and which ones are objects;
   (2) which actions belong to the repertoire of which agents;

* This is the essential difference between naturally given worlds and artificially constructed worlds. One constructs an artificial world by delimiting the <u>types</u> of situations that could occur in it. Hence, the artificial language associated with this world is also a closed structure and can, hence, be exhibited as a formal system. Computer languages are of this kind. So are scientific theories. When scientific theories are confronted with new phenomenological details (i.e. new situations) of the naturally given physical world, they often prove inadequate, and need change or augmentation, sometimes superficially, sometimes quite radically. This, again, indicates why it is futile to try to formalize natural languages as closed systems. <u>The only thing that can be formalized is the language behaviour acquisition machinery; not the output of that machinery</u>.

(3) which attributes are applicable to which Objects;

(4) which properties are values of which attributes;

(5) which properties belong to which states;

(6) which relations are applicable to which attribute sets;

(7) which actions relate to which state changes.

4. lb-tokens which relate to the schema: Attribute(Object).

5. lb-tokens which relate to the schema:

$$\text{Relation}(\text{Attribute}(O_1),\ \text{Attribute}(O_2)).$$

6. lb-tokens that query aspects in (1), (3), (4) and (5).

7. lb-tokens for command.

8. lb-tokens for control with associations as to which actions they apply.

9. Utterances relating to the schemata:

$$O = p, \quad p = \text{property name}$$
$$\text{Att}(O) = p,$$
$$\text{Rel}(\text{Att}(O_1),\ \text{Att}(O_2)) = \text{True/False}.$$

The above is, in some sense, a minimal set. Natural languages contain an extraordinary variety of schemata for specification of situational aspects indirectly. We considered a few of these earlier. Many more such schemata are discussed in Narasimhan (1967c). Part of the thesis of this paper is that, at any given time, the available repertoire for language behaviour of any individual consists of only a finite number of these schemata. Thus, the central problem in solving the language behaviour problem does not consist in being able to exhibit all the schemata exhaustively (since this is not a well-defined set, natural language being open-ended), but in being able to define a metamachinery for behaviour that is capable of augmenting the available repertoire as new schemata are encountered during interaction.

3.3. The Problem of Relevance

We have already seen in Section 1 the central importance of the computation of relevance to interactive behaviour, in general, and to language behaviour, in particular. If learning is concerned, primarily, with <u>augmenting</u> an available repertoire, relevance computation is concerned, primarily, with the dynamics of <u>utilizing</u> an available repertoire. In terms of the behavioural model that we have been considering, there are two aspects to relevance computation. The first is composing an "admissible" utterance out of available lb-tokens. The model assumes that "admissibility" is defined in terms of "learnt" schemata which, in turn, are "exhibited" in the response complex in terms of available directed connectivities. The second aspect concerns making sure that the generated utterance is a relevant one for that stage of interaction.

This second problem can be stated somewhat more descriptively as follows: Assume A and B are interacting in a natural language. If, now, B responds to some utterance A has just produced, A must have a way of deciding whether the response is or is not relevant to his original utterance. For, his next utterance would be determined by the outcome of his decision. Clearly, this aspect of the dynamics of interaction must be incorporated in the model since otherwise it would not be able to interact with humans in a natural manner.

Clearly, there can be no single solution to this relevance computation. For, humans must, in real life, use a variety of heuristics _to make sense out of_ a received response. Depending upon the situation (e.g. problem-solving, diplomacy, psychiatric interview, and so on), this process could involve carrying out quite sophisticated computations. *

One powerful heuristic for relevance computation (at least in the case of naive natural language interaction) is to organize the internal structure (i.e. the complex of response tokens) so that it can generate _monologues_. An interaction could, then, be thought of as an "alternating monologue" between two different individuals. The relevance computation now reduces to deciding whether a received response is one (i.e. belongs to a _type_) which the individual would have himself generated in a monologue at that stage.

One advantage in this heuristic is that it enables one to relate relevance computation and generalization (learning) in a rather natural way. For, now the learning process is reduced to dealing with two cases that might arise: (1) the received response is of an expected type but consists of new tokens; and (2) the received response is of a new type.

We shall not pursue this problem any further here. But it is, perhaps, worth noting that Ramani (1969) is currently working on a language based problem-

* Weizenbaum (1966) has made some perceptive remarks about this problem. Concepts like "credibility", "belief-structure", etc., are closely related to relevance computation as we have been discussing this process.

solver whose design takes explicitly into account several aspects of natural language behaviour discussed in this paper. This program, as it stands now, still makes use of the processor-data-base dichotomy, but the data-base does consist of natural language expressions suitably interlinked. And as we have argued in this paper, the only meaningful way to implement such a data-base is by structuring the program as a growing program. In Ramani's problem-solver, natural language interaction is designed to take place as an open-ended process. The input-output properties of the problem-solver are specified paradigmatically by providing it with sets of input-output pairs, along with their governing contexts. The problem-solver has the built-in capability to analyze the structures of input episodes and relate them to the analyzed structures of past episodes. Based on these associations, it is capable of generalizing its behaviour.

3.4. Concluding Remarks

In this paper, we have been concerned with computer simulation of natural language behaviour. We started by analyzing what the central problems of interactive language behaviour are, and the ways in which these have been approached in some of the implemented programs. This led us to formulate a new analysis of the pragmatics of natural language behaviour, and a structure for an interactive program based on this analysis. Considerations of the dynamics of functioning of a program so structured enabled us to outline two major open problems, and we discussed some tentative approaches to solving these problems. It is our view that simulating natural language behaviour in a non-trivial manner is a necessary pre-requisite to simulating higher level cognitive activities like complex problem-solving and theory-construction.

4. REFERENCES

BOBROW, D.G. (1964), "A question-answering system for high school algebra word problems"; Proc. FJCC, pp. 591-614.

CHOMSKY, N. (1965), "Aspects of the theory of syntax"; The M.I.T. Press, Cambridge, Mass.

CRAIG, J.A. et al. (1966), "DEACON: Direct Access English Control"; Proc. FJCC, pp. 365-380.

English Preprocessor Manual (1965); The MITRE Corporation, Bedford, Mass.

NARASIMHAN, R. (1966), "Intelligence and artificial intelligence"; Tech. Rept. 16, Computer Group, TIFR, Bombay, (Oct.).

NARASIMHAN, R. (1967a), "Programming languages and computers: a unified metatheory"; In F.L. Alt and M. Rubinoff (Eds.): "Advances in Computers", Vol. 8, Academic Press, New York.

NARASIMHAN, R. (1967b), "On the non-relevance of transformational linguistic theory to psycholinguistics"; Tech. Repts. 22 and 22-a, Computer Group, TIFR, Bombay, (Feb. & April).

NARASIMHAN, R. (1967c), "An outline of a metatheory of behaviour with specific application to language behaviour; Part I: Language acquisition by children"; Tech. Rept. 37, Computer Group, TIFR, Bombay, (Oct.).

RAMANI, S. (1969), "Language based problem-solving"; Ph.D. Thesis (to appear).

RAPHAEL, B. (1964), "A computer program that 'understands'"; Proc. FJCC, pp. 577-589.

WEIZENBAUM, J. (1966), "ELIZA - A computer program for the study of natural language communication between man and machine"; Comm. ACM, 9, pp. 36-43.

5. DISCUSSION

Clowes: Would you imagine that the syntagmatic mode of behaviour pre-supposes paradigmatic - is in some sense based upon it?

Narasimhan: Yes. Since I am saying that the paradigmatic mode of behaviour is the primitive mode of behaviour, it must be clear that children, when they start out, necessarily function predominantly, I would say, almost totally in the paradigmatic mode of behaviour, and since the language acquisition does take place in these early years, the plausible argument is that it must in fact be paradigmatic mode of behaviour. I assert that what schooling does is in fact to teach children how to behave in the syntagmatic mode of behaviour, that is, training is essentially concerned with getting people to behave syntagmatically.

Bobrow: Would you think then that when Chomsky talks about competence-performance difference, that competence is this formalism that you call syntagmatic, and this performance which he ignores and says he'll come back to later, is paradigmatic?

Narasimhan: To my way of looking at things, this competence-performance distinction is completely meaningless, so I don't know what in fact he means by competence at all, because in my world all that exists is only performance, and one can talk only in terms of performance; competence, if it has any meaning at all, can mean only potential performance - but I don't know really what he means by competence.

Speight: Would you explain what is the significance of meta-theory and meta category?

Narasimhan: The difference is between meta and object. The functioning of the description of the meta level will concern itself with models, say, for a whole class of organisms. The object level model will be particular organisms. So I would say that when I give a meta description for behaviour, I would assume that this meta description applies both to your behaviour and my behaviour, whereas to characterize your behaviour, I would have to give an object level model, which will then be tied down to your behaviour - I will have to fix parameters; whereas for the meta level, I merely say what are the kinds of parameters that have to go in the model.

Anonymous: To what extent does Ramani's program, in responding to a test, insist upon equivalence with the paradigm. For example, if a test had been: 'A ball is thrown horizontally at 25 metres/sec', whereas the paradigm stated 'thrown vertically upwards', would the wrong answer have come out, or would the fact that on the left we have 'thrown vertically upwards', the fact that there was no paradigm saying thrown horizontally, force it to say, "I have no idea"?

Narasimhan: In this particular case it would just abandon it, it will not work it out, because you see 'vertically' is a crucial statement here.

Anonymous: But how does it know that 'vertically' is the crucial statement?

Narasimhan: I'll explain how it works, then you'll see what expressions are important, and what expressions are not important. All that it does is, it analyses a situation in terms of various aspects. The situation given is a paradigm with three parts - the context, stimulus and response. This is the primitive analysis given to it; what it does is to scan the stimulus, of the context, stimulus, and response, for longest expressions which are identical, because then it assumes that if there is an expression that occurs in two of these, that relates to some aspects of the situation. That is a relevant expression. So it converts the description of the situation into matrix and filler form. What it has to know is how to construct a matrix, how to pull out the fillers; it operates on the assumption that common expressions are plausible values of variables; so it merely says that this situation belongs to a class which can be characterised by a matrix with formal parameters and this particular situation has been arrived at by putting in these particular values to these formal parameters. So it just replaces these actual parameters by formal parameters with identical names. So given a new situation, it tries to type the situation by going through all the schemata that are available to it and seeing which one matches. So essentially what it does is, having been exposed to complete situations and having characterized them in this manner, when you expose it to a partial situation, it tries to fill the situation out in terms of schemata that it matches. At the moment it operates only if the schemata completely match - if there are only partial matches, it will just abandon it. Obviously it now needs extension, by letting it behave in plausible ways if there are partial matches. Given your example, it will abandon it because 'vertical' occurs only in one place, it will know it is part of the matrix - it is not a filler. So it knows that the paradigm applies only to the situation in which 'thrown vertically upward' is an essential aspect.

Kaneff: I take it if its matching only in partial form, there wouldn't be much trouble to incorporate a questioning system which enables further information to be sought and supplied.

Narasimhan: That's what one wants to do. You see, the important thing is that in the paradigmatic mode you can teach it also to ask appropriate questions. You are not going outside the system to be able to do that.

Bobrow: In your examples, Figures 2, 3, how is the response generated in the right hand side? How does it know what that final response is?

Narasimhan: Take example (1) in Figure 3 . The final (Right Hand) response R is the action response corresponding to the response scheme provided in the paradigm. In other words if you are taught in an exercise, C (Left Hand Side) and then you are told R (Left Hand Side) is the cost of one pen in Rupees, in the test mode the equivalent statement to this is taken to be R (Right Hand Side).

Moore: You have characterized paradigmatic behaviour as analogical - naive, informal, intuitive, and syntagmatic as logico-deductive, rule-bound, formal, analytic and so on. I think that paradigmatic statements seem two dimensional and syntagmatic are one dimensional. As example, I consider a flow diagram to be informal and most certainly intuitive; the syntagmatic equivalent of this would be a computer diagram. The flow diagram, by the way, definitely requires a two dimensional representation and the computer program statement is essentially, deductive, formal, analytic and scientific, and is of one dimension. Essentially there exists a one to one transformation from this two dimensional space to the one dimensional space. Man in his processing of information, essentially does it in a parallel manner - in other words, he is capable of processing two dimensional data, graphics, extremely well, whereas a digital computer is a serial device which operates in one dimension. I give that interpretation and put it forward.

Narasimhan: My comment on that is that both these examples you have chosen are just accidental. There is nothing intrinsic to writing programs as linear strings. You can in fact, construct a formal theory for programing languages in which the flow chart is the formal notion, and in fact this has been done. There is nothing intrinsic about a flow chart being naive, and a string representation of that being formal. I don't think so at all.

Moore: In that case you are operating in the paradigmatic mode.

Narasimhan: No. What I am saying is you can give a two dimensional representation, a formal account of computational theory where the input is a formalized two dimensional representation.

Zatorski: Your characterization of the input to the language acquisition device of Figure 1 falsifies Chomsky's view which includes, as part of the input, the innate characteristics, which you have not stated. I have two questions: 'Given that languages are paradigmatic, what is the medium by which we pull ourselves up by our boot strings'? and 'In making discoveries, where does analogy (i.e. the paradigmatic mode) end, and where does logical deduction begin? Or perhaps if I can illustrate this more differently, in the situation involving Newton and the apple or James Watt and the steam kettle, where does analogy end and deduction begin'?

Narasimhan: Regarding your initial remark, merely putting the innate things in, I don't think solves Chomsky's fallacy, because neither he nor I nor anybody else knows what 'innate' is. I can answer the second question more readily than the first because I don't understand the first question. As far as the second question is concerned, I can characterize where analogy ends and when syntagmatic begins. For example, suppose you are trying to prove a theorem, then you dream up ......

Zatorski: But the desire to prove the theorem, where does this come from?

Narasimhan: I am not talking about motivation here. This characterization that I have given, obviously apart from the machinery I have put in, I have said, apart from these interfaces and the behavioural schemata associated with them, I have in my model what I call internal states, state variables associated with the system and these will take care of various things. Everything not there comes under what you might call the effective professional systems and so on - I would consider motivation to be one of these things. If one is trying to build the dynamic behaviour, one would have to go into greater details and then exhibit how these motivational and various other aspects in fact are tied in. One assumes that in some manner there is an interaction of these bits to various schemata. In other words to an outsider, two identical situations produce different language behaviour, and so these things probably will be characterized by behaviour. To continue with the question - if you are trying to prove a theorem, you come up with what you consider a plausible proof, this 'coming up', I tend to think, is done primarily in the paradigmatic mode. Then when you sit down and try to demonstrate the validity of the proof, you have to go into the syntagmatic mode of operation. A very crude way of characterising these two modes of operation is, (for problem solving of any level of complexity, for most people), syntagmatic mode of operation involves computations with paper and pencil.

Clowes: It does seem to me that when you are thinking analogically, what you are doing is simply saying, 'well suppose this is to that as that is to that', that is not asking what sense it is, especially let us not try to characterize that relationship formally, but once you do characterize that relationship formally, then we begin to do it deductively and we have shifted from what you call the paradigmatic analogic mode into the other one.

Narasimhan: Yes, what I am saying is that it is not essential that formal articulation is necessary in order to be able to behave, but unless you formally articulate, you cannot formally move forward, in other words problem solving requires formal manipulations, so if you want to function in a formal world, you have to articulate at the appropriate moments.

Stanton: I notice you give your paradigms in English. Why is this? I would have thought perhaps that if the paradigms were given in some more primitive language, some much simpler language, and the knowledge of English was gained through this simple language, wouldn't that lift the English itself to the syntagmatic state and make Chomsky's investigations valid?

Narasimhan: What kind of language would that be?

Stanton: I am not sure, it may have something to do with lists and objects, much more incremental.

Narasimhan: I do not know how that will work. What I have characterized as objects, attributes and so on, are tags which relate to the interfaces, they are system characterizations. So what I am saying is that the fact that you see the world in terms of objects, attributes, relations of particular kind and so on, in fact must be explained on the basis of your system characterizations. For a different kind of system you would see the world differently.

INFORMATION REDUCTION AS A TECHNIQUE FOR LIMITED SPEECH RECOGNITION

D.G. BOBROW

Bolt Beranek and Newman
Cambridge, Massachusetts

and

D.H. KLATT

Massachusetts Institute of Technology
Cambridge, Massachusetts

1. INTRODUCTION

The two standard approaches to pattern recognition can be characterized as description generation and classification. Those who choose to perform the task as description generation (such as Ledley, Banerji, Evans) first invent a language and grammar for descriptions, and then attempt to define processes which can map the external representation of the patterns into a string in the internal description language. For speech recognition, the usual descriptive language has an alphabet of phonemes, (each of which may have a number of distinctive features (Fant, Halle)) and other phonetic symbols indicating emphasis, etc. The recognition problem in this domain requires mapping a speech wave form into an appropriate string (or perhaps a tree) of phonetic symbols. It requires that individual meaningful segments be isolated, each segment be appropriately identified, and the connections between them made clear, e.g. a sequence of individual phonemes can be identified as comprising a given word. This descriptive language (phonetic) approach has great power, in that it provides a symmetric theoretical base from which either analysis or synthesis can be performed.

However, for the recognition of isolated words we believe it is not necessary to solve the much more difficult problem of phoneme recognition. The latter presupposed a good understanding of the cues that distinguish phonemes in arbitrary phonetic environments. For example, allophones of the phoneme /p/ may be:

20

1. normally aspirated, as in the word "peak" [pʰik]
2. weakly aspirated, as in the word "supper" [sʌpɚ]
3. non-aspirated, as in the word "spin" [spɪn]
4. non-released, as in the word "top" [tʰap˺]

There is no need for a word recognition program to attempt to group this disparate set of physical signals together into one phoneme. However, to the extent that algorithms for deriving a detailed phonetic-feature description of an utterance can be found, they can be of considerable help to a practical word recognition system. Examples of feature approaches that are relevant include the work of Hughes (1961), Hemdal and Hughes (1964), and Gold (1966).

A far simpler task is assumed for the process of classification. Here one assumes a limited set of classes of interest, and assumes that any given input is to be identified as a member of one of these classes. No segmentation of the input is required, nor any description which relates elements of one class to any other. The simplest approach to classification utilizes a set of prototype patterns (or "templates") and a measure of distance between input items and each template. The class for any input is the one whose template is "closest" (has the smallest distance) to this item to be identified. The principal problem with this approach is the need to normalize the input so that it can be reasonably matched against the set of prototype patterns.

In a generalization of the approach, a set of property extractors compute quantities which represent basic features of the input objects. Discrimination between classes is then done by determining in which volume in the multidimensional space defined by the possible property values lies the point for the current input. The critical problem in this approach is to choose a set of properties so that classes lie in well defined regions in this space. A bad choice of properties can make the discriminant function very complex, or non-existent if the areas for each space overlap too much. Much work has been done on the design of discriminant functions. Our approach to speech recognition focuses on the problem of choosing properties which will allow use of a simple discriminant function.

For both simple template matching, and general property discrimination, one can choose to preset the templates or property sets which define each class. Alternatively, one can employ an adaptive classification strategy in which samples of inputs for each class are identified, and the information developed by training the system. By careful preselection one can choose optimum boundaries and exemplars of a class, but this may not be possible where there are wide variations over a controllable input variable. For example, different people have widely different speech patterns; therefore, it may be easier to have a system which is trainable and adaptable to a particular individual, than to define classes which define the form of inputs of particular messages over the class of all individuals. In our system we have adopted this approach to recognition, in line with our general philosophy of information reduction as an approach to pattern classification.

2. INFORMATION REDUCTION AS A STRATEGY FOR PATTERN CLASSIFICATION

We view pattern classification as a problem of defining a process, or set of processes, which reduce the amount of information in the environment. Throughout these reduction processes, information must be discarded which is irrelevant to the current goal of classification. This reduction can be achieved both through limiting assumptions and decisions about the set of possible inputs, and through data reduction processes which are designed to eliminate particular kinds of irrelevant information. Both strategies have been followed in producing the limited speech recognition system described below.

In order to reduce the difficulty of the decision process, we have restricted the information entering the system along a number of dimensions. Rather than use continuous speech in which segmentation is a problem, we work with messages with easily delimited beginning and termination points. The set of messages is limited in number; at any one time the vocabulary to be distinguished can contain up to about 100 items. However, an item need not be a single word, but may be any short phrase. A message list from a NASA mission context, shown in Table 1, was one of three used in testing the system. Note that LISPER recognizes each of these messages as a unit, and does not segment a multiword utterance into individual words for recognition. The system is not designed to work well simultaneously for a number of different speakers or achieve good recognition scores for an unknown speaker. The system is usable by any male speaker, but must be first trained by him. The training period consists of a period of closed loop operation in which the speaker says an input message, the system guesses what he says, and he responds with the correct message. In this training phase, the system will learn the idiosyncratic variations of the speaker's set of input messages. In this closed loop system, it is not unlikely that the speaker will also learn something.

TABLE 1

Message List from NASA Context

one	distance to dock
two	fuel tank content
three	time to sunrise
four	time to sunset
five	orbit apogee
six	orbit perigee
eight	revolution time
nine	closing rate to dock
point	midcourse correction time
plus	micrometeroid density
stop	radiation count
zero	what is attitude
seven	remaining control pulses
minus	alternate splashdown point
pressure	weather at splashdown point
negative	sea and wind at splashdown point
what is yaw	visibility at splashdown point
what is pitch	temperature at splashdown point
end repeat	skin temperature
affirmative	power consumption
inclination	fuel cell capacity
distance to earth	repeat at intervals

By making a number of limiting assumptions, we have reduced the class of inputs and thus the range of information coming into the system. There still remains wide variation and too much irrelevant information in a waveform representing a speech event. Some of the reasons for variation in recordings of the same voice input message are:

1. The fundamental pitch frequency (and thus the entire harmonic structure) of the voice varies for many different reasons, irrelevant to the identity of a speech message. Health, time of day, state of fatigue, and emotional stress are a few of these extraneous factors.

2. The range of intensities encountered will vary because the overall recording level is not fixed. Recording level depends on vocal effort and the distance of a speaker from the microphone.

3. An unknown word is difficult to register with respect to a time origin because word onset time is not a simple feature to detect reliably. For example, initial voiceless fricatives may be missed, prevoiced stops are hard to treat consistently, etc.

4. The total duration of a word is highly variable. An increase in speaking rate is not manifested by a linear compression of the time dimension. Final syllables are often prolonged. Some transitions (for example, the release of a stop consonant) are not as greatly affected by changes in speaking rate as the steady-state portions of vowels. If shortened enough, vowels are likely to be reduced and consonants may not be carefully articulated, with resultant losses in spectral distinctiveness.

5. A speaker attempts to generate an utterance so that it has a particular set of perceptual attributes (or features). We do not know in detail what the acoustic correlates of these attributes are. There is a great deal of variation allowed in the acoustic properties of the signal that will still give rise to the same utterance. There is also sufficient redundancy in the message to permit a speaker to leave out certain attributes entirely. For example, the degree of stress placed on a syllable will determine the extent to which the vowel *may* be reduced (Lindblom, 1963). Consonants in unstressed syllables may contain less frication noise, a weak stop burst release, or incomplete stop closure. Vowels may be nasalized in nasal environments. The substitution of one incomplete gesture for a consonant cluster is also common in unstressed syllables of natural speech. None of these effects would necessarily produce word recognition difficulties if they appeared consistently in the data. Unfortunately, they do not.

6. If a speaker is instructed to speak distinctly and not rapidly, some surprising and unfortunate variability in speaking habits has been experimentally detected. In an attempt to help the system, our speakers released final stops, increased the length of some syllables, and articulated unstressed syllables more carefully than they would normally. Unfortunately, our speakers appear

to have found these speaking habits unnatural, and could not remember from repetition to repetition exactly what they had done to help. For example, final voiced stop releases gave trouble by producing short vowel segments that varied greatly in amplitude, and the words "four" and "core" were sometimes pronounced as if they had two syllables.

3. THE RECOGNITION SYSTEM

Our LISPER (Limited Speech Recognition) System is designed to be an adaptive speech recognizer satisfying the above limiting assumptions. The basic structure of the LISPER system is indicated in Fig. 1. It consists of three principal components: a digital spectral analyzer to provide the raw data; a set of programs for extracting properties of the input message from this digitized input spectrum; and training and recognition algorithms, for storing information about the speech signal and making a decision about the identity of an unknown utterance based on earlier experience. The spectrum analyzer (Stevens, 1968) basically consists of 19 bandpass filters whose outputs are rectified, low-pass filtered, sampled by an analogue-to-digital converter, and then converted into logarithmic units. The 19 filters consist of 15 filters which are spaced uniformly at 180 Hertz up to about 3,000 Hertz, which is the range encompassed by the first 3 resonances of the male voice; and 4 filters which cover the range from 3,000 to 6,500 Hertz in which there is information about the noise component of speech. The 360 Hertz band-width of the low set of filters is sufficiently wide so that one does not see spectral peaks due to individual harmonics of fundamental frequency of the male speaker; a spectral maximum in the filter indicates the presence of one or more formants (resonances of the vocal tract transfer function). The spacing between filters makes resolution of individual formants poor and we do not use formant tracking in our recognition system. The analogue to digital sampling rate is 100 Hertz. Each 10 milliseconds the outputs of all 19 band pass filters are sampled simultaneously, giving an approximation to the logarithm of the short time spectrum of the input message.

The use of this spectrum analyzer reduces the amount of information stored in the computer from a 160,000 bits/second (8 bit samples of the waveform at 20,000 Hertz) to 11,400 bits/second (6 bit logarithms of samples in 19 filters, at 100 Hertz). The information lost is basically that due to variation in pitch frequency of the voice.

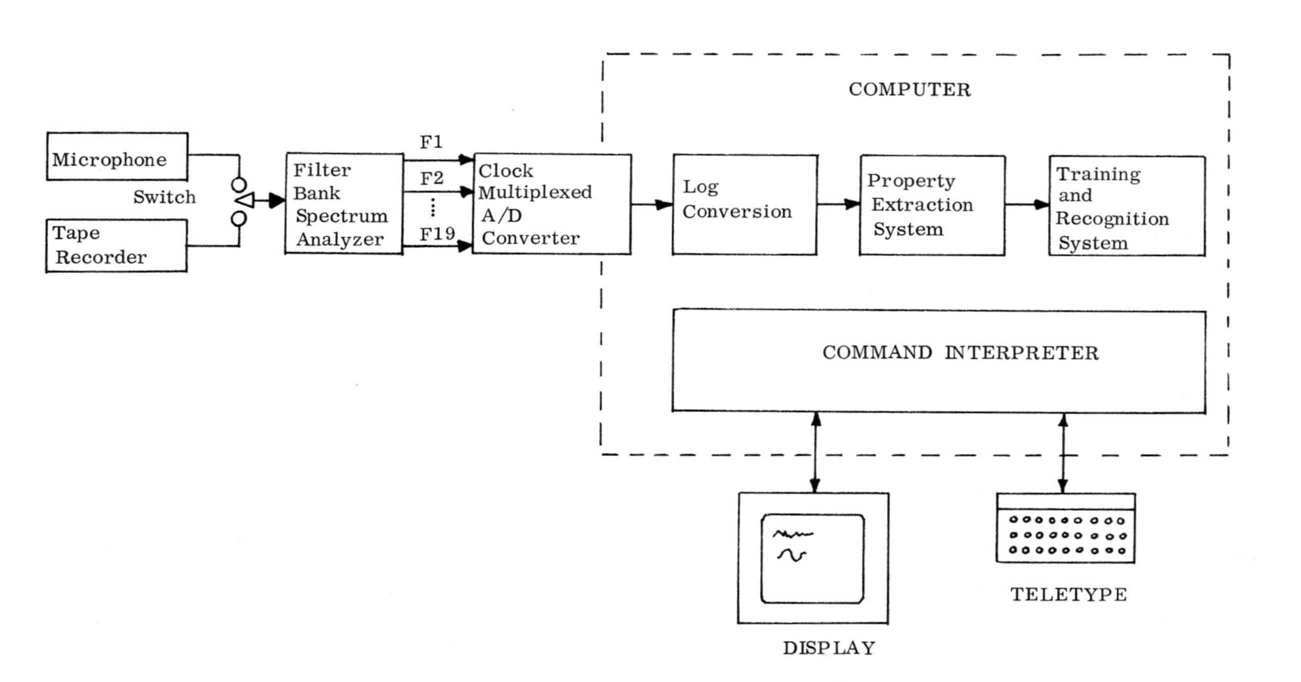

Figure 1. Block Diagram of the LISPER System.

The properties of speech which are used by the recognition program are based on the energy measures derived from the input system. The output of each filter is an elementary function of the speech input signal. We use the notation $F_n(i)$ for the output of filter n at sample interval i; that is F1(i), F2(i), ..., F19(i) are used for the output of filters 1 through 19 at sample interval i. The filter number n ranges from 1 for the low frequency filter to 19 for the high frequency filter; and i=1 for the first sample interval to i=200 for the last time sample of a two second utterance.

More complicated functions of the speech input signal can then be defined in terms of these elementary (base) functions, F1, ... F19, in the LISPER system. For example, the following function has been found to correlate with perceived loudness, independent of vowel quality.

$$\text{Loud}(i) = F1(i) + F2(i) + F3(i) + F4(i) + F12(i) - F7(i)$$

The output of the seventh filter is subtracted from the sum of the first four filters and filter 12 to compensate for the fact that low vowels are inherently more intense when produced with the same vocal effort. We will indicate later how a modified version of this function Loud(i) can be used to correct for differences in recording level between repetitions of the same word.

Loud(i) is useful in reducing the information that must be processed by the decision algorithm, since it helps to normalize the input with respect to a variable which is not important to recognition, namely, the recording level of the input signal. We describe in this section a number of techniques which are used to reduce the information content of the incoming signal in ways that preserve the invariance of message identity over ranges of parameters which are irrelevant to the decision about the identity of the message.

One important way we have found to reduce the information content of a function is to reduce its range of values. Thus, for example, we may define a reduced information property Amp2(i) based only on the output of filter F2.

$$\text{Amp2}(i) = \begin{cases} +1 & \text{if } F(2) > 40 \\ 0 & \text{if } F(2) > 20 \\ -1 & \text{otherwise} \end{cases}$$

Amp2(i) is thus a 3 valued function, where we ascribe no significance to the names of the three values. As described later, a set of non-linguistic threshold properties similar to Amp2 can be used as a basis for recognition. However, we note the following problem. If the signal in F2 were varying around either of the thresholds, 20 or 40, then there would be little significance to the changes of state between -1 to 0, or 0 to 1; they probably would be caused more by noise than a real (significant) variation in the input signal.

To make such properties less sensitive to noise, we introduce a hysteresis region around the thresholds. This helps to insure that a change of state is significant. A revised definition of Amp2(i) is

$$\mathrm{Amp2}(i) = \begin{cases} +1 & \text{if } [F2(i) > 40] \text{ or } [F2(i) > 37 \text{ and } F2(i-1) = 1] \\ 0 & \text{if } [F2(i) > 20] \text{ or } [F2(i) > 17 \text{ and } F2(i-1) \neq -1] \\ -1 & \text{otherwise} \end{cases}$$

Most of the features we use are functions of sampled time having a range limited to a small number of distinguishable states. These functions are very sensitive to slight changes in time scale and origin; however, these variations in the data are irrelevant for recognition. Therefore, time dimension is removed from a feature by transforming the time function into a sequence of transitions of states, as is illustrated in the following example:

i	1 2 3 4 5 6 7 8 9 ...
Voice(i)	0 0 1 1 1 1 1 1 0 ...
Voice =	0 1 0 ...

This transformation reduces the amount of data that must be manipulated by the program. Due to the nature of spoken language, the exact time when features change value will vary from repetition to repetition of the same word, but the essence of the word remains in the sequence of state transitions of an appropriate set of features.

No information about the word onset time, speaking rate, and speaking rhythm can be recovered from the sequence unless these parameters have an effect on the actual states that are reached. To this extent, recognition will be unperturbed by variations in word onset time, speaking rate, and speaking rhythm.

A problem that arises from collapsing the time dimension is an inability to tell whether two features were in specific states at the same time. Time removal assumes that features independently characterize a word. This is obviously false.

A clear example is provided by the words "sue, zoo" which can be distinguished only by knowing whether the strident is simultaneously voiced, or not. It is possible to retain some timing information by the inclusion of features containing states which are entered only upon simultaneous satisfaction of two conditions, or features that count the number of time samples between temporal landmarks in the data and change state if the count exceeds a threshold.

For example, the state "-1" corresponds to voiceless segments in the definition of the spectral quality [r]-like(i). This is used to indicate in which voiced segment an [r]-like phone occurred.

The sequence produced for the word "ratio" should be -1, 1, 0, -1, 0, -1. A more detailed localization of the [r] is probably not necessary and would certainly be more difficult. We found that very gross timing information is satisfactory for recognition. Two features are used to make a preliminary division of a word or message into one or several segments. One feature divides a word into voiced and voiceless segments and the other divides it into syllables. Voice(i) and Syllable(i) are then used to introduce time markers in the definitions of other features.

Time removal achieves the most significant reduction in the information content of the raw data. Information content is also reduced when the intensity and filter number (frequency) dimensions are transformed into changes of state of selected functions. The information retained by these transformations can be increased by (1) an increase in the number of distinguishable states, or (2) an increase in the number of feature detectors.

Information manipulable by the recognition algorithm can also be obtained by mapping the value of a function at a particular time interval into a small range of values. For example, features which classify the vowel quality of the stressed syllable of an utterance into one of 10 categories have been found to be very useful in recognition. The stressed syllable is identified with the aid of a function Loud(i) that computes an approximation to the perceived loudness of a vowel. These features of a stressed syllable are less likely to be affected by the natural variability that characterizes the speech process (Stevens, 1968).

Our recognition algorithm is a program that learns to identify a word by associating the outputs of various property extractors with that word. During learning, the vocabulary of words is presented a number of times, and information is accumulat about the different ways that the speaker may pronounce each word. For example, the result of the training procedure applied to the feature sequence Voice after 5 presentations of the 4-word vocabulary "one, two, subtract, multiply" might be:

Word	Sequence	Number of times sequence occurred
one	010	5
two	010	5
subtract	01010	5
multiply	0101010	3
multiply	01010	2

The two versions of the word "multiply" exemplify a common problem. No matter where we place threshold boundaries, there are some words that are treated inconsistently by the feature detectors. Another way of putting it is to say that there appear to be no absolute boundaries along the dimensions we have chosen. The recognition algorithm takes this fundamental limitation into account and makes a best guess given the imperfect nature of the properties.

The program first reorganizes the training data for each property into a list of these sequences (or other outputs) that have occurred. The list for Voice of our example is shown below:

Sequence	(Word, frequency)
010	one 5, two 5
01010	subtract 5, multiply 2
0101010	multiply 3

If a new utterance is presented to the program for recognition and the feature Voice(i) produces the sequence 01010, then Voice will register one vote for the word "subtract" and one vote for the word "multiply". The unknown word is identified as the vocabulary word eliciting votes from the most features. Typically, the identified word received a vote, signifying a perfect match, from only about 80 percent of the features.

In case of ties, the program makes use of information concerning the number of times a word appears at a node. Thus, if "subtract" and "multiply" tie for first place in the voting from all the features and Voice=01010, then Voice will register 5 votes for "subtract" and 2 votes for "multiply" in the run-off between these two candidates.

The final decision procedure is an attempt to find the message from the set of possible inputs which is most similar to the current input message. Because there is a wide variation in the way people say things, the decision procedure does not insist that the current input must be like one of the prototype input strings in all ways that it was categorized; that is, it need not be suggested by all properties. In this sense, the decision procedure allows a generalization of the original training learning by looking for a best fit without putting a bound on the goodness of this fit.

4. RECOGNITION PROPERTIES

Two different sets of properties were developed and tested for use in the LISPER system. The first is a set of properties and functions which tend to describe the speech signal in more linguistic terms. Transitions of these properties describe features of the input message which can be understood in terms of the ordinary linguistic descriptions of such messages.

The second set of properties used for recognition described below extract features of the spectral shape of a message. Their extreme simplicity makes them seem ideal for hardware implementation, and their high accuracy in recognition testifies to the uniform way that people tend to say a particular message. At least there seems to be a fair uniformity as heard through our spectrum analyzer.

The linguistic properties utilize a number of functions of the input filter bank: loud is a measure of the loudness of the incoming signal and is used to normalize the intensity; ah, ee, oo, and er are functions which achieve maxima for the vowel sounds in "pot", "beet", "boot" and"Bert" respectively; str is used to measure stridency or frication noise; and two derivative approximations dspect and damp give indications of rapid changes in spectrum and amplitude of the signal, respectively.

Three properties are designed to give indications of number and types of segments in the input. Voice is 1 in voiced intervals and Ø otherwise; syllable is Ø in consonantal portions of an utterance, and 1 in vowel segments, although the transitions are not precise in time; stress is identical to syllable except that it takes on the value 2 for the stressed syllable.

Vowel properties are extracted in a number of ways. The properties ahlike, eelike, oolike and erlike are reduced range versions of the functions, ah, ee, oo and er respectively. Each has value -1 in unvoiced intervals (determined by voice) and 1 (or Ø) if the function exceeds (or is less than) a threshold. Hysteresis in the thresholds, that is, reducing the level for change of state after a positive crossing eliminates oscillation of the property if the function is wavering around the threshold. In addition to these direct vowel measures, a number of properties is used to indicate the place of articulation of consonants and position of the tongue during vowels. These properties try to distinguish between vowels classified each side of each dotted line in Fig. 2. Since repetitions of a single word are liable to

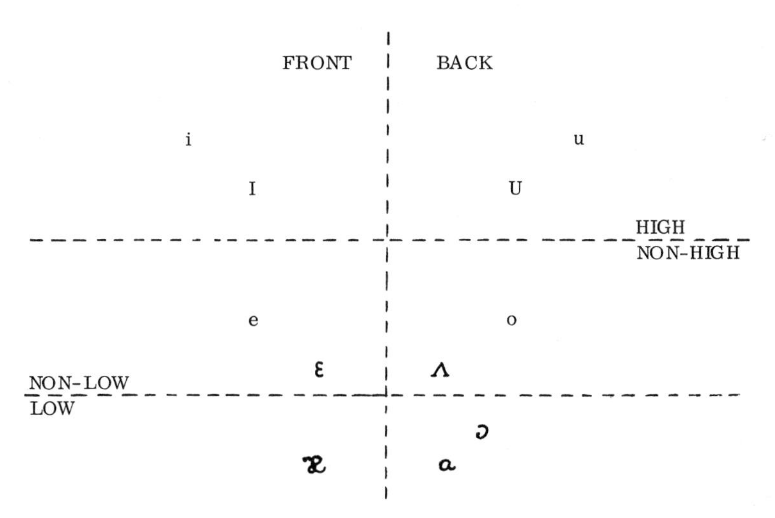

Fig. 2 - Classification of vowels in terms of the phonetic features front, high, and low. These features characterize the position of the body of the tongue and have well-defined acoustic correlates.

be treated inconsistently if only a single boundary is used and the vowel is close to this boundary, we employ two sets of these features with slightly displaced thresholds. The redundancy this adds is of the kind most easily handled by the recognition algorithm.

Consonantal features in the linguistic set include strident, indicating presence of sounds like [s, z sh and zh]; fricative, for high frequency energy characteristic of the initial consonants of fin, vend, thin and then; stopburst for release of a stop such as p, b or d; and nasal which indicates the presence of consonants m and n and sometimes initial l's.

In addition to the above properties which all take on a sequence of values for a given utterance, we found it useful to extract a characterization of the stressed vowel of the unknown message. This was done by identifying the point of maximum stress, and mapping the values of ah, ee and oo into a range of 10 values, to give a representation of the vowel quality at the point, which is usually quite consistent for a single speaker.

The nonlinguistic or abstract spectral properties which were used all were motivated by a desire to crudely characterize the shape over time of the outputs of the 19 filters. Three classes of properties were used. Spectral amplitude properties S_n (for n=1, ... 19) mapped the output of each individual filter into three "interesting" regions: low, middle and high. Amplitude difference properties D_n (for n=1, ..., 10) similarly mapped the differences in output between the first 10 pairs of adjacent filters. A change of state of this property indicates a change in the spectral shape of the input in an interval. Finally, a set of temporal difference properties DT_n (n=2, 5, 8, 10, 13, 16, 19) indicates changes over time (i.e. difference in output of Fn at times i and i-2) in the output of a single filter bank.

The thresholds for these properties were adjusted so that each state could be reached by at least some vocabulary items. The hysteresis regions were adjusted so that very long sequences (longer than about 10) would occur in only very few items.

5. RESULTS

A. Single Speakers

The two sets of properties, both linguistic and spectral, were run on eight speakers on the word list shown in Table 2. On four of these speakers all six rounds of data were used, providing five recognition scores, and on the other four speakers only three rounds were used. The results are summarized in Figures 3 and 4. The scoring algorithm was simplified from the one described earlier. Each word occurring at a node is given a single vote. The frequency of a word at a node is not used to break ties (and hence need not be stored). This simplification was made to save space and time, with the idea that we would not lose much in the way of decision capability. The hope that very few ties would occur proved true on speaker KNS, who provided the data for the design of the original properties, and for early tests of the current more efficient system. Unfortunately, with later speakers, ties

TABLE 2

Expanded Word List

about	four	point
accumulate	from	print
add	generate	push
address	get	quarter
assemble	go	read
at	greater	register
binary	half	replace
bite	if	restore
block	index	revolve
breakpoint	initiate	right
change	input	save
choose	insert	scale
chop	intersect	set
close	jump	seven
comma	left	shift
compare	less	show
compile	list	shrink
complement	load	six
control	location	skip
core	look	space
cycle	make	specify
decimal	memory	square
delete	minus	store
describe	move	subtract
directive	multiply	swap
display	name	tab
divide	night	than
do	nine	think
down	number	three
draw	octal	toggle
dump	of	two
edit	one	unite
end	output	up
equal	overflow	whole
exchange	parenthesis	word
five	plus	yield
		zero

became much more frequent, (about 50% of number wrong) but to maintain data compatibility and the speed and space saving we did not revert to our original scoring algorithm. Ties were assumed broken by a simple coin flip, and the number unambiguously correct was augmented by one half the number of ties to give the scores shown. As can be seen, the average asymptotic recognition rate is about the same (91% and 94%) on these properties for the extended list as our previous results (96%) for the smaller list (Bobrow and Klatt, 1968).

TRIAL

Speaker	2	3	4	5	6
KNS	78	92	90	92	98
DHK	80	81	79	92	90
CD	66	85	77	82	91
DGB	63	76	75	83	86
RST	71	86			
CM	75	83			
DD	75	82			
GJ	65	79			
AVG(4 SPKRS)	72%	83%	81%	87%	91%
AVG(8 SPKRS)	72%	83%			

Fig. 3 - Recognition rate with linguistic properties. Each speaker is treated individually.

TRIAL

Speaker	2	3	4	5	6
KNS	83	94	96	93	96
DHK	74	87	81	98	98
CD	77	90	90	96	94
DGB	52	76	81	82	87
RST	76	92			
CM	75	91			
DD	75	90			
GT	71	82			
AVERAGES(4)	73	87	87	92	94
AVERAGES(8)	74	88			

Fig. 4 - Recognition rate for spectral properties.

B. Performance of the System with Added Noise

Speech shaped random noise, (i.e. white random noise filtered by a band-filter whose frequency response is identical to the long-term RMS speech spectrum) was added to the speech signal to test the degradation of the system under adverse conditions. It was prefiltered in order to (1) obtain approximately equal noise levels in all filter channels and (2) avoid overloading of the filter-networks. This would occur for a white noise input at low S/N ratios, due to the high frequency pre-emphasis in the filter bank. The set up used to add noise to the speech input is shown in Fig. 5.

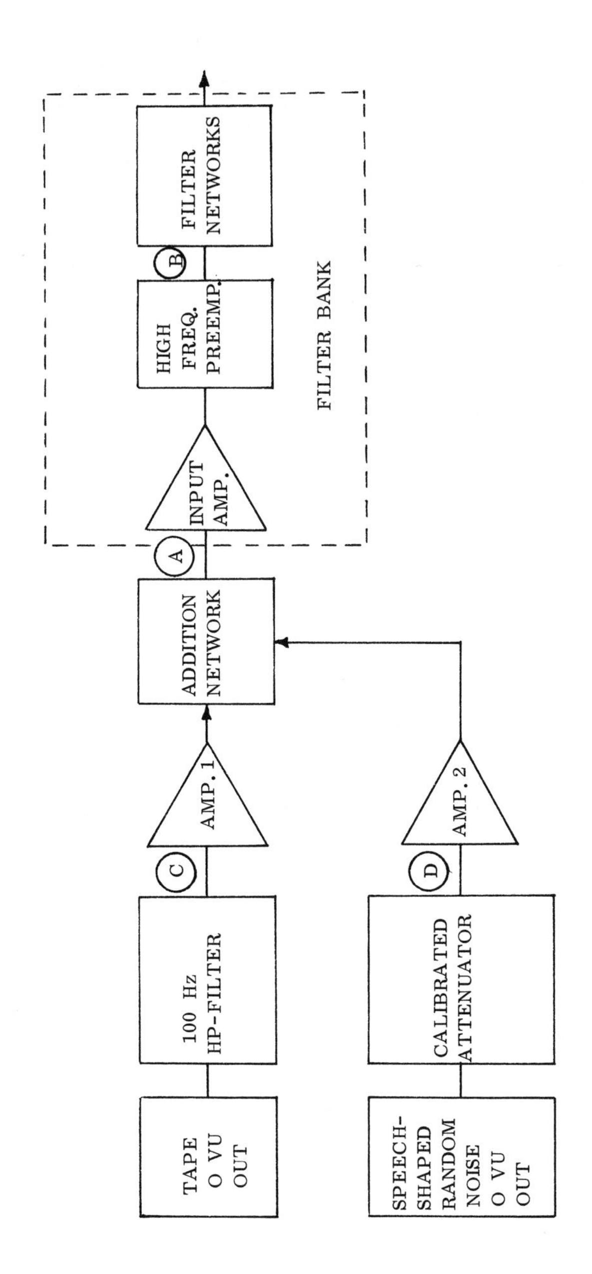

Figure 5. Set-up used to add noise to the speech input.

The system was trained for four rounds with no noise added. The same set of utterances with different levels of noise added were used to test the system. The recognition rates are shown in Fig. 6. Note that DB is a less consistent speaker

Effective S/N Ratio	34	31	25	20	15	10
Number Correct	42	44	43	38	31	23
Percent Correct	78	81	80	70	59	44
Log (%-Correct)	1.89	1.91	1.90	1.85	1.77	1.65

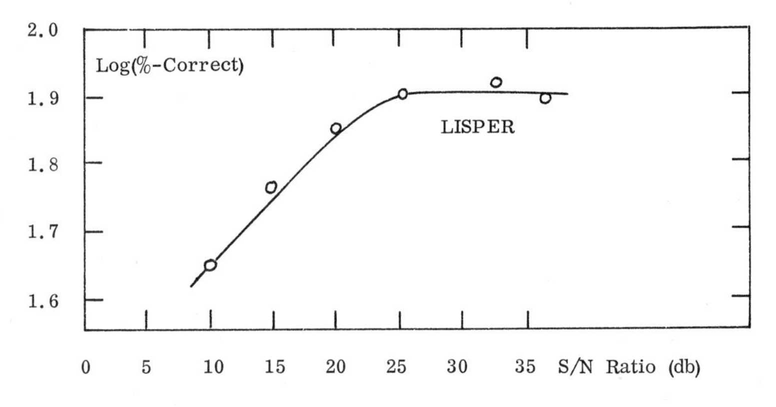

Fig. 6 - Recognition rate with added noise.

than the other two, or the properties need to be tuned for him, because the asymptotic recognition rate is only about 80%. The shape of the rate of recognition curve under noise degradation is identical to that for humans, falling off almost completely in 20 db. However, the break for humans is 10 to 20 db lower, and the asymptotic rate much higher. However, this experiment indicates how the various information reduction techniques help eliminate confusion due to noise.

C. Information Measures of Properties

As recognition scores of the system get better and better, it becomes more difficult to evaluate the individual properties. A measure of information content has been used as an aid in the evaluation of individual features. This measure of the information contained in a feature tree has provided some insight into the desirable and undesirable properties of a feature. The measure will be defined, applied to some examples, and then discussed.

Let: J be an index over the nodes of a tree

I be an index over the words in the vocabulary

n_{IJ} be the number of times word I appears at node J

NV be the total number of words in the vocabulary

NT be the number of rounds of training

R be a measure of the information contained in the tree about the identity of an unknown word

P(I) probability of Ith word (assumed equal to 1/NV)

$$R = H(I) - H_j(I)$$

where $H(I) = -\sum_I p(I) \log [p(I)] = \log NV$

and $H_J(I) = -\sum_I \sum_J p(I, J) \log [p_J(I)] = -\sum_I \sum_J \frac{n_{IJ}}{(NV)(NT)} \log \left[\frac{n_{IJ}}{\sum_I n_{IJ}} \right].$

Example 1:

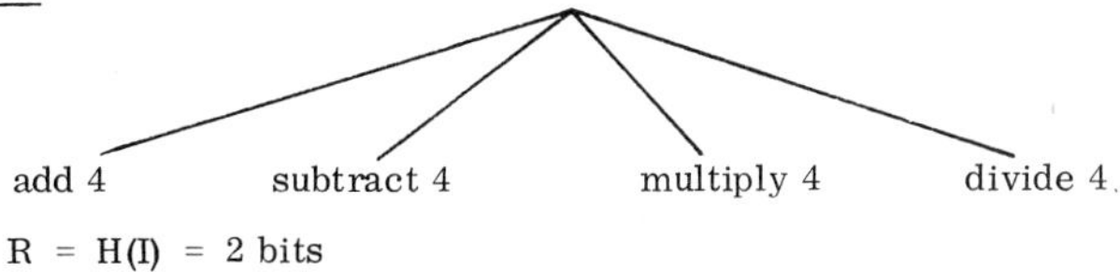

$R = H(I) = 2$ bits

This is an example of the perfect tree, one that assigns a unique node to each vocabulary word.

Example 2:

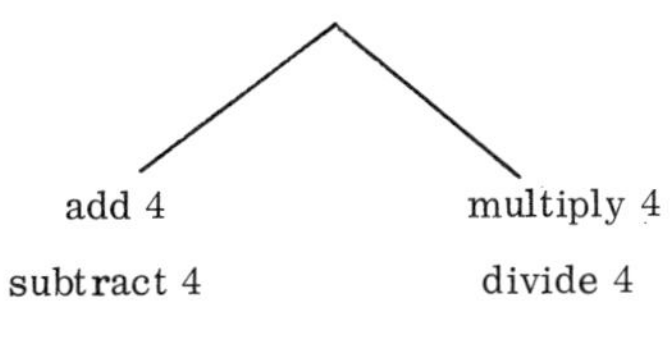

$R = 2 - 1 = 1$ bit

This is an example of a tree that consistently assigns a word to the same node, but does not completely separate the vocabulary words from one another.

Example 3:

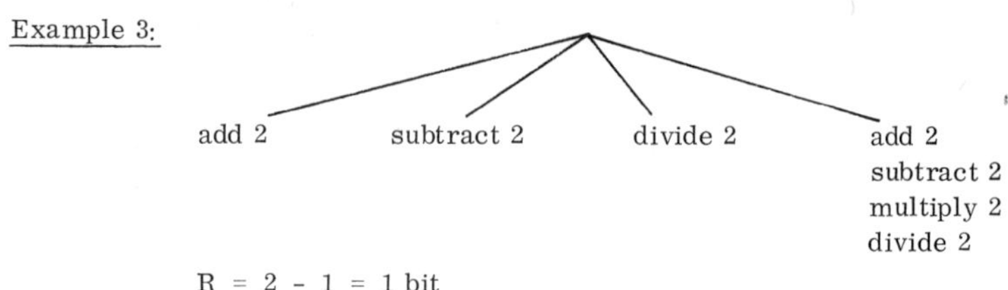

$R = 2 - 1 = 1$ bit

This is an example of a tree with more nodes (greater possible separability), but with an inconsistent assignment of words to nodes.

There is always the possibility that the unknown word will be assigned to a new node of the tree, one that was not present during training. The tree can provide no information; R does not account for this possibility because it assumes incorrectly that the frequencies, n_{IJ}, are accurate estimates of the underlying probabilities.

It may be a biased measure because of this fact. When training grows a large tree, there is, initially, a greater chance of a word producing a novel node than when training grows a smaller tree. The difference between large and small trees in growt of novel nodes after about three rounds of training is not very great. Therefore, we have not attempted to correct for the potential bias in the information measure R. For a 109 word vocabulary, a perfect feature would provide $\log_2(109) = 6.7$ bits of information about the word to be recognized, assuming equal word frequencies. A feature that produced the same sequence of states for any word would provide 0 bits of information.

There are some intuitive measures of the goodness of a feature we have found interesting. The number of nodes N (different input sequences found in training) gives a measure of the range of separation that can be expected of the property. The numbe of messages M, or average number of variations per input message M', gives a meas of the consistency of the property. The spread ratio N/M' of separability to consisten seems intuitively to be a quantity one would want to maximize.

Fig. 7 summarizes the information content of the linguistic and abstract (spe tral) properties. The information measure is computed for trees containing one spea 3 speakers, and 5 speakers. It can be seen that the additional speakers tend to reduc the amount of information about individual word identities in the tree. If speaker 2 is recognized using the combined trees instead of only his own trees, the information measure predicts that his recognition scores will go down. These combined trees were ones used for some cross-speaker identification experiments.

	1 Speaker	3 Speakers	5 Speakers
Linguistic Properties			
Average Information	2.68	2.20	2.13
Average Spread	18.4	16.6	15.1
Total Information	64.32	52.80	51.12
Spectral Properties			
Average Information	2.34	1.77	1.58
Average Spread	14.31	12.6	10.8
Total Information	77.22	56.64	52.14

Fig. 7 - Summary of information and spread measures.

Cross Speaker Recognition

Trees containing training data from a number of speakers were used as a basis for recognition of new speakers. For both the linguistic and spectral properties, two sets of trees were prepared. One set was constructed from six rounds of training from three speakers. The second was constructed from three rounds of training from five speakers. Limitation on the size of the training trees precluded any larger trees. The latter trees were pruned of all branches containing only a single utterance (any one message from any speaker), but this did not affect the recognition rate at all.

The results of these experiments are summarized in Fig. 8. One obvious

		Round	1	2	3
Linguistic Properties	3 Speakers	DGB	66	79	74
		RST	82	83	90
	5 Speakers	DK	74	80	
		CD	83	82	
Spectral Properties	3 Speakers	DGB	65	78	
		RST	75	86	
	5 Speakers	DK	72	82	
		CD	76	85	

Fig. 8 - Recognition rates (percent correct) on pretrained trees. On the first round, the trees contain no information about the speaker to be recognized.

conclusion is that training by more speakers with fewer rounds is better than the converse. Secondly, the results indicate that except for the first round (obviously), the training on other speakers does not particularly help in recognition. That is, the second round recognition rates are comparable to the results obtained by individual speakers on their own trees. This indicates the idiosyncratic nature of the properties extracted for recognition by this system.

We have constructed a theoretical model from which it is possible to predict the applicability of our property detector approach to speaker recognition. The model predicts that, for 5 speakers, and 33 spectral properties, any one of the speakers can speak one word (of the 109 word vocabulary) into the microphone and have 70 to 90% probability of being correctly identified. Of course, if the set of speakers happens to contain 2 speakers who are very similar, the score will be reduced. On the other hand, requiring the speaker to say more than one word will increase the reliability of the identification.

The model is based on the following decision algorithm. If any speaker n ($i = 1, \ldots, 5$) matches the sequence generated by property i ($i = 1, \ldots, 32$), he receives one vote from that property. If more than one speaker matches the input sequence for property i, the vote is cast for one of the speakers determined randomly with the probability of a speaker getting the vote proportional to the frequency of that speaker at that node with that word. The votes from all properties are summed and the speaker receiving the most votes is selected.

The model predicts an expected recognition score on the basis of the data presented in Fig. 9 which indicates the average probability, in percent, that a feature will vote correctly. This average has been computed over all 5 speakers and all 3 properties for each word from the spectral property trees described previously.

If we take, for example, a word like "memory", the expected probability that a feature will vote correctly is 32%. If 33 features vote, the expected score will be $33 \times .32 = 10.56$ votes. The standard deviation of this expected score, $\sigma_{correct} = \sqrt{npq} = \sqrt{33(.32)(.68)} = 2.77$ votes. Assuming that the 4 incorrect speakers are equally dissimilar to the correct speaker, the expected score for an incorrect speaker is 5.5 votes and $\sigma_{incorrect} = 1.75$ votes.

The probability that a _specific_ incorrect speaker will get more votes than the correct speaker is approximately given by integrating the appropriate tails of the convolution of two normal distributions. In this example, the probability is

Pc	Word	Pc	Word	Pc	Word
37	ABOUT	29	FROM	30	PUSH
34	ACCUMULATE	33	GENERATE	34	QUARTER
32	ADD	30	GET	29	READ
36	ADDRESS	32	GO	37	REGISTER
36	ASSEMBLE	33	GREATER	34	REPLACE
31	AT	28	HALF	35	RESTORE
35	BINARY	31	IF	38	REVOLVE
32	BITE	39	INDEX	33	RIGHT
31	BLOCK	31	INITIATE	30	SAVE
39	BREAKPOINT	35	INPUT	34	SCALE
29	CHANGE	34	INSERT	31	SET
31	CHOOSE	33	INTERSECT	32	SEVEN
28	CHOP	26	JUMP	32	SHIFT
32	CLOSE	32	LEFT	25	SHOW
33	COMMA	30	LESS	33	SHRINK
35	COMPARE	33	LIST	32	SIX
34	COMPILE	30	LOAD	34	SKIP
38	COMPLEMENT	33	LOCATION	30	SPACE
32	CONTROL	30	LOOK	36	SPECIFY
31	CORE	31	MAKE	33	SQUARE
30	CYCLE	32	MEMORY	34	STORE
32	DECIMAL	35	MINUS	37	SUBTRACT
27	DELETE	25	MOVE	34	SWAP
35	DESCRIBE	36	MULTIPLY	30	TAB
35	DIRECTIVE	29	NAME	37	THAN
37	DISPLAY	31	NIGHT	29	THINK
37	DIVIDE	36	NINE	31	THREE
27	DO	34	NUMBER	33	TOGGLE
30	DOWN	37	OCTAL	30	TWO
30	DRAW	27	OF	35	UNITE
31	DUMP	31	ONE	30	UP
33	EDIT	34	OUTPUT	26	WHOLE
32	END	31	OVERFLOW	28	WORD
29	EQUAL	31	PARENTHESIS	29	YIELD
36	EXCHANGE	27	PLUS	30	ZERO
30	FIVE	34	POINT		
33	FOUR	29	PRINT		

Fig. 9 - The average probability, P_c, in percent that a property will vote for the correct speaker. Data from 39 properties, each property tree contains 5 speakers.

about 8%. The probability of an error is approximately 4 times this value, giving a probability of correct speaker identification using the word "memory" of about 68%. For the slightly better word "breakpoint" the predicted success is about 88%. Pooling the votes from n words of equal quality decreases the probability of error by $\sqrt{n}$ since we effectively multiply the number of properties being used by n. Four words like "memory" would thus give a probability correct of about 84%, if all the independence assumptions were reasonably accurate (which we know they are not). This redundancy between properties would increase the error rate, as would the fact that on the average all speakers are not equally dissimilar.

6. CONCLUSION

The results of this set of experiments in speech recognition demonstrate that thinking of the problem of classification in terms of irrelevant information reduction can lead to design of properties adequate for the task attempted here. For a single speaker and a vocabulary of 50-100 messages, recognition rates of better than 90% can be achieved even in a relatively noisy environment. By juggling vocabulary items to provide greater phonetic separation, higher rates can be achieved for a particular task. Because the information discarded, in our system, such as time correlation of properties, is important for solving the speech recognition using the descriptive (phonetic) approach, we do not feel this work is directly extendible to that domain. As with hand drawn characters (Teitelman, 1964), this technique will provide an adequate base for a limited speech input system for a computer. Comparable results have been achieved by other investigators (Gold, Reddy).

7. REFERENCES

BANERJI, R.B. (1968), Pat. Recog., Vol. 1, p.63.

BOBROW, D.G. and KLATT, D. (1968), "A limited speech recognition system"; Proceedings of the FJCC, Thompson Press, Baltimore, Md.

BOBROW, D.G., MURPHY, D.L. and TEITELMAN, W. (1969), The BBN-LISP System, BBN Report No. 1677, (April).

CHOMSKY, N. and HALLE, M. (1968), The Sound Pattern of English. New York: Harper & Row.

EVANS, T.G. (1968), Proc. of IFIP.

FANT, G. (1960), Acoustic Theory of Speech Production, Mouton and Co., 'S-Gravenhage:

GOLD, B. (1966), Word Recognition Computer Program. Technical Report 452, Research Laboratory of Electronics, Cambridge : M.I.T.

HEMDAL, J. and HUGHES, G. (1964), "A feature based computer recognition program for the modelling of vowel perception"; In W. Wathen-Dunn (Ed.), Models for the Perception of Speech and Visual Form, pp. 440-453, Cambridge: The M.I.T. Press.

HUGHES, G.W. (1961), The Recognition of Speech by Machine. Technical Report 395, Research Laboratory of Electronics, Cambridge : M.I.T.

JAKOBSON, R., FANT, C.G.M. and HALLE, M. (1963), Preliminaries to Speech Analysis, Cambridge : The M.I.T. Press.

LEDLEY, R.S., JACOBSON, J. and BELSON, M. (1966), "BUGSYS: A programming system for picture processing - not for debugging"; Comm. ACM, (Feb.).

LINDBLOM, B. (1963), "Spectrographic study of vowel reduction"; The Journal of the Acoustical Society of America, 35, pp. 1773-1781.

REDDY, D.R. (1966), An Approach to Computer Speech Recognition by Direct Analysis of the Speech Wave; Proc. FJCC, 1968, Thompson Press, Baltimore.

SEBESTYEN, G.S. (1961), "Recognition of membership in classes"; IRE Transactions on Information Theory, IT-6, pp. 44-50.

STEVENS, K.N. and VON BISMARCK, G. (1967), A Nineteen-Channel Filter Bank Spectrum Analyzer for a Speech Recognition System, NASA Scientific Report No. 2.

STEVENS, K.N. (1968), Study of Acoustic Properties of Speech Sounds. Technical Report, Department of Defense, Bolt Beranek and Newman Inc., Cambridge, Mass.

TEITELMAN, W. (1964), "Real Time Recognition of Hand Printed Characters"; Proc. FJCC, Spartan Press.

8. DISCUSSION

Lance: Am I right in assuming that the time for recognition of a word or a phrase is independent of that word or phrase?

Bobrow: Very much, yes.

Lance: What is the time?

Bobrow: Well, we have done this on two computers - one thing we did on the system (and this made our times much longer than they ought to be), we produced the version in LISP, which made the overhead high, as this higher level language was not designed to do much computation. So when we first did this on the PDP - 1, times were of the order of one minute. Now when we do it on an SDS - 940 and have embedded inside the 940 LISP a good computation language, we are down to the order of 3 or 4 seconds. If we wanted to go into making it fast, we could get it down to within a second.

Lance: How much should the delimitation between utterance gap, that you spoke of earlier, be?

Bobrow: There are two problems there - if you want to do it in real time, then of course that will be the time required to do the recognition; if you are just worried about picking up the message and stopping the tape recorder, we found that half a second was plenty. The gap that we actually used for judging the ends of the word from the tape was about 200 milliseconds.

Lance: One would have thought that you would not have to leave a gap in order to do the computations - the computations would be going on while the sound is being received, is this not so?

Bobrow: If you are talking about 4 or 5 seconds' computation time to recognize a message (of one or two seconds' duration), then you had better stop between messages. We think we could, with not too much difficulty, get it down to being able to recognize it while the automatic hardware is doing the sampling.

Kovarik: I understand you are shifting the time scale; on what basis can you make the distinction between 'e' and 'i'?

Bobrow: Well, those vowels look different in a number of ways, and we get that sort of thing into our system. We are only rejecting the time scale when we want to; there is nothing at all to say that we cannot have a property, and we have had properties which are, for example, that this is the seventh 'br' in a succession of time intervals in which we had 'br'-like properties: that was essentially the case with long 'eeee' and short 'ee'. In fact, one of the properties distinguishes between short stridents like 't' and long ones like 's' on just this basis; that is, it does not turn on until the noise is in there for 5 or 6 time intervals.

Dale: Did you treat each frequency band separately as a sequence in time and then score to the transition matrices?

Bobrow: We treated them separately in time, they came in that way. We used sums and differences of different ones in different bands and we used differences over time, but we just treated them as essentially base functions which we did use.

Dale: What I am wondering is whether you could break each frequency into class and level and treat it as a transition between class and level and just classify the transition matrices.

Bobrow: That was part of what I did in that set of spectral properties - for small sets it works very well. For larger sets, it turns out that the natural acoustic boundaries provided by phonemes, or phoneme classes, cause the properties to be better, and so in general we can do better recognition with fewer properties when we try to do it with the linguistic properties rather than with these special properties.

Clowes: How quiet was the room where you handled speech?

Bobrow: We did a whole series of experiments on how the system handles noise. We did a quiet room where there was 35 db signal to noise ratio (that is, if everybody kept quiet here while I talk, then that would be about right, if a car is not going by): it turns out that you degenerate too rapidly, if the signal to noise ratio gets down to 25 db: performance went off at about the same slope that people's recognition rate goes off with noise, except that people are much better as they can go down to a 5 or 10 db noise ratio before they start losing recognition.

Vazey: You mentioned that the system performed reasonably well with the 114 word set. How well and how long did it take for each to be recognised?

Bobrow: It did not take any longer than for the shorter lists because the only time you are interested in all the properties is in the one scan through the matrix to get the maximum score and that is of course a very fast operation in the computer. Going down the tree essentially does not take very much longer. It made only the same mistakes on the combined list that it made on each of the individual ones before, and there were no new confusions, which we were very surprised but happy to see. So it was much better on the combined list average than it was on say the 54 word list, the other two lists it got 100% correct, or one mistake.

Stanton: In your introduction you mentioned that a certain property extractor operated on 10^5 bits to give the appropriate neutral descriptive information. I fail to understand how you can refer to this description as being neutral since its purposeful nature is clearly exhibited by its position between the decision process and the property extractor. Would you comment?

Bobrow: Well, the neutrality here was with respect to the world, because you do not have to care about what the sensors were very much out there. You could have forgotten, for example, a different hardware sensor, a Fast Fourier Transform Analyser, and obtained these same kinds of properties, and obtained this characterization which was neutral with respect to all the different kinds of things that you might have been doing. But in fact it was very much directed to the actual recognition process. You can have that sort of thing which makes it independent of the kinds of sensors you use, but not of the kinds of tests that you do. So you are right, it was directed towards a purpose but it was neutral in another sense which I think is an important one.

Jacks: You had a recognition rate of 95%. Did you have a rejection rate, or was it strictly substitution errors?

Bobrow: This was a forced choice, so it always said something. If we cut off at some reasonable point, it would have been worse; that is if we had said 'don't know' in some cases, if we had wanted to get rid of all the errors we would have gotten an equal number of drop-outs which would have been correct.

Jacks: What level would that be?

Bobrow: With the 95%, you would have gotten 90% recognized and 10% 'don't know'.

O'Callaghan: How many subjects did you try this on? Would you expect the system to work for Australians?

Bobrow: On the early work we tried it on just 3. We are now in the process of going up to 10, and looking at how much the training on one speaker helps on another; that is, can we in fact drop this cross speaker, this multispeaker restriction? We thought at first that we could, but it turns out that on the first round you do not do too well, but after one round, knowing the particular peculiarities, the system makes use of its general knowledge and will give a very high score much more rapidly, but we still have to give it a little bit of a hint.

Rutter: Did you train it on male and female?

Bobrow: No. It was all men. We tried only some informal experiments with a female programmer. We tried training the system on-line and had a demonstration - she trained it up on a minute vocabulary, 7 or 8 words. On her training it was able to recognize successfully.

Bennett: How does it handle accents?

Bobrow: It does pretty well on across the United States accents - we have tried that - but we have not tried it on any Australians.

Bennett: Can you recognize the individual voices?

Bobrow: We thought about that. We said the following. 'Suppose we look at these properties and two people are on the same tree. Some of these properties will be speaker independent perhaps and some of the properties will not be; we could find the ones that were not. We could tell the system what word we were going to say and record which speaker said it - maybe then we could do some speaker recognition.' We have tried this a little and it seems to work, but I would not swear by it - it is not terribly consistent.

DESIGNER'S CHOICE AT THE CONSOLE

EDWIN L. JACKS

Computer Technology Department
General Motors Corporation
Warren, Michigan

1. INTRODUCTION

In this paper design concepts for an experimental graphics system are described. The objective of the system is to minimize the time required to perform a design task by providing essentially a full graphic environment in which both creative design and design programming may be done. While most of the concepts contained in the paper have been tested in graphic systems in General Motors, no system has been constructed containing all the concepts as described.

The paper emphasizes the man-machine communication aspect of the approach. It is assumed that the reader is familiar with the Associative Programming Language, General Motors Research - (APL)[1], and with the concepts of list processing and ring structures as presented in Knuth's, THE ART OF COMPUTER PROGRAMMING.[2] While the description that follows can be understood without knowledge of the above references, the implementation of the approach is very difficult without good associative programming techniques. These are programming methods which allow relations between things to be readily established and retrieved as programs are executed. The things may be geometric elements such as points, lines, and surfaces needed to define an automobile body surface; they may be a sequence of operations which were used to construct a surface, they may be a set of images to be displayed to represent a surface, or they may be the set of relations which define the relation between the previously mentioned sets. For instance, using associative programming techniques, it is fairly easy to maintain the relation between the operation that was used to define an instance of a line, the resulting coefficients in the definition of the line, the image representation of the line, the use of the line in a geometric model of a surface, and the use of the line in subsequent operations. In the following discussion, the word set, will be used to describe a collection of things which may be either ordered or unordered depending on utilization. Items may be entered or deleted from a set at any time and the members of a set may be all like elements or they may be heterogeneous.

The fundamental tool provided for a design draftsman at a console in a

1: see Dodd, (1966); 2: see Knuth, (1968)

graphic system is a group of descriptive geometry operations which he may apply to the design on the screen. By analogy, this can be viewed as a set of subroutines provided in a program library. By using a light pen on the screen he may select the parameters and variables to be entered into an operator. After completion of the variable selection, he may then ask for immediate operator execution. To provide flexibility in the application of the operator concept, the system concentrates on the control of input to and output from operators. The following topics are covered in the paper:

Operator Concept
Variable Expressions
Screen Space Organization
Variable Identification
Output Control
Sequence Recording
Programming

Before proceeding into the details on each of these topics, a brief example will be given to illustrate our approach to graphics and the meaning of each topic.

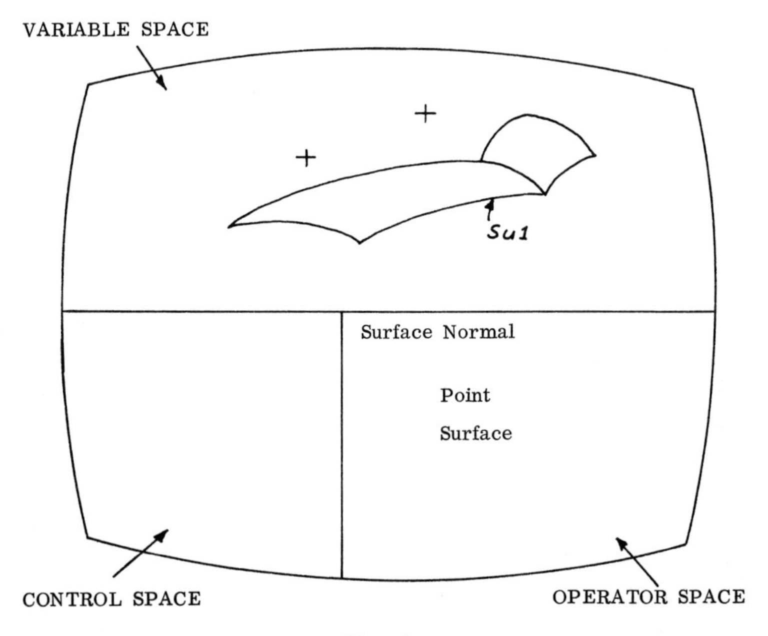

Fig. 1

In Figure 1 a display screen is shown divided into three parts; the variable space, the control space, and the operator space. Any item which is displayed in the variable space may be selected by the user by pointing with a light pen and hence be given to an operator as an operand. The means of control for the input to, and output from an operator will be via the operator space. In general, by a combination of pointing to the operator space and the variable space, variables are selected as operands to the operator. The third area, referred to as the control space, is used for control over both the variable space and the operator space independently of the operator in use. For instance, in the control space will be light pen buttons which will allow the user to suspend his current use of an operator and will permit him to start a new operator. The selection of a new operator is from a list of operators displayed in the operator space. Upon completion of the user's application of the second operator, he may resume applying the original operator at its point of suspension.

The means of variable identification is best described by reference to Figure 1 . In the operator space appear three objects; the operator title - Surface Normal, the word 'point', and the word 'surface'. The occurrence of the word 'point' and the word 'surface' means that a point variable and a surface variable must be given to the operator for it to compute a surface normal. The user's action would be to point at the word 'point' and then at a token (a cross in Figure 1) representing a point in the variable space. He would then repeat the cycle, pointing first at the word 'surface' and then at a token (the characters Su1 in Figure 1) representing a surface in the variable space. After both a point and a surface are selected, a light pen button (Execute) would be displayed in the control space. When the user points at 'Execute', the surface normal would be calculated and displayed in the variable space. Next, an output option operator would be displayed permitting the user to insert the surface normal in any set he desires as well as optionally permitting him to name the surface normal for future direct name reference. On the display screen would appear a vector representing the resulting surface normal.

The above brief description introduces the concept of applying an operator to variables appearing in the variable space, with the result being a new variable appearing in the variable space.

By sequence recording, is meant the system having the capability to remember a sequence of operators, and the system providing a facility for the sequence to result in a new operator. The resulting operator should appear no different to users than the primitive operators in the system. If required, the system should permit sequences to be connected together with conditional statements to provide a programming capability.

2. OPERATOR CONCEPT

The operator concept when applied to graphics, simply means that a function exists which accepts input parameters and input variables, and which produces new values for variables as output. It is assumed that the operator is defined to work on a class of variables, which for instance, may include very complex models of surfaces and three-dimensional objects, or very simple models of two-dimensional drawings. The system controlling the operator, however, is responsible for checking that no input parameter or variable to the operator is of illegal mode for the operator. To do this, the system must have available the following information:

(1) The mode of each variable in the variable space. For instance, a token on the screen may represent either a point, or a line, or a surface, or a name of a point, or a name of a floating point number, etc.

(2) The mode of each parameter and variable requested as input to the operator.

(3) The mode of each variable produced as output from an operator.

(4) The class of models the operator may be applied to.

With the above information, the graphics system can check to see that an operator being requested by a user can be applied to the data being displayed in the variable space.

After that check, as each association (variable name selection in the operator picture, and input variable selection in the variable space) between the operator space and the variable space occurs, the mode of the variable is checked to see that it matches the mode required by the operator. Finally, when the output from the operator is inserted in the display set, the mode of the output is remembered so it may be checked prior to subsequent use as input to other operators.

Variable Sequencing

Explicit in the above description is the concept of the user indicating a variable he wishes to pass to an operator by selecting a token in the operator picture such as the word point, and then selecting from the variable space an incidence of the variable. To simplify this process, the system controls the operator space so that in the majority of variable specifications only light pen selection is needed in the variable space. This is done by having the system intensify the token representing the next variable needed by an operator and placing a number by the token in the operator space. When the user then selects a token representing that type of variable from the variable screen, the system places the same number by the token and intensifies the next generic variable called for by the operator. Figure 2 illustrates the situation - a variable of mode point was called for and the association

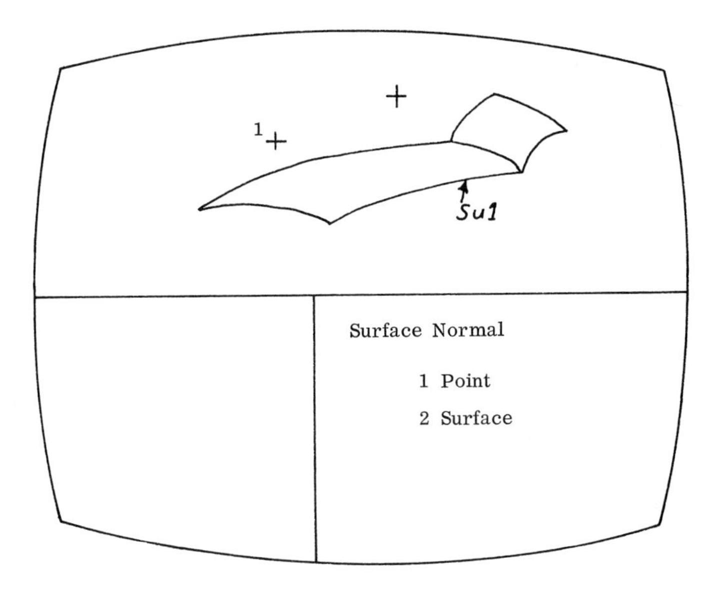

Fig. 2

completed by the user selecting a token for a point in the variable space; immediately the system intensified the word surface, placed a 2 by the word, and is waiting for the man to select a variable of mode surface from the variable space. The combination of variable sequencing with variable mode checking eliminates a very common programming error - the passing of wrong mode variables to subroutines.

The above description assumes that each variable has a representation in the variable space. This, however, may not be true in many situations. For instance, the user may wish to create a geometric point at $x = 0$, $y = 5$, $z = 10$. The concept of variable expression will now be discussed to show how this type of problem can be handled.

3. VARIABLE EXPRESSIONS

At each step of input to an operator, the system knows the mode of the variable being requested. This information can be used to expand the system to handle simple unary resultant expressions of two classes. Expressions of the first class are those characterized by light pen actions. For instance, the operator is calling for a point, and the user selects a line on the screen. This could be interpreted as either an error, or as the user saying, "the point I wish is the point where I am pointing on the line I have just selected", - depending on the application. A second class is characterized by the man keying in an expression such as POINT (.5, 0, .5), or SURFACE (x = 1.15). For easy handling of the above type of expressions, the condition is imposed that each expression must result in only one variable being specified, just as only one variable may be specified with a light pen selection on the screen.

Note that while in the example a cross was used to represent a point, a name of a variable of point mode could appear on the screen, and if the man then selected the name, it should be treated as an expression of point mode.

The object of introducing the expression concept to graphics is three-fold. First, it makes it simple for the user to use both the light pen and the keyboard while in the middle of specifying data to an operator. Second, it permits a uniformity of variable definitions across all operators. That is, independent of which operator is being used; the same set of unary expressions may be used to satisfy a request for a variable of a given mode. And third, it makes it conceptually simple to mix alphanumeric statements in with graphic operations.

Now, while unary expressions can be handled in the above manner, the case where the man has forgotten to calculate several variables prior to calling for an operator on the screen, can be handled better by permitting the man to suspend his use of an operator. This is done by light pen buttons in the control space.

4. SCREEN SPACE ORGANIZATION

In the above description, the use of the variable space for displaying variables and the operator space for specifying the input for operators, has been explained. However, what is the control space used for? It has primarily two functions: (a) to control the use of operators, and (b) to give a level of control over the variable space independent of the operators. At no time is the control space used to produce a new variable. The control space ideally should be application independent.

A typical control over the operator space is SUSPEND. When this light pen button is pushed, the operator being used is removed from the screen, a list of operators that may be used is then placed on the screen, the word RESUME appears in the control area, and the user may proceed to apply operators to his variable space.

Whenever he is ready to go back to the operator he was originally working on, he simply selects RESUME and the original operator appears in the same state as when he suspended using it.

A second function in the operator control is ABORT. For some reason the user may decide that the operator he is trying to use is not the one he needs. On the selection of ABORT, the system stops all actions relative to the current operator, resets the variable space to the state prior to the use of the current operator, and then lets the user select a new operator to use.

Another typical control over the variable space contained in the control space is the selection of field of view. A small diagram indicating an x, y, z, axis is displayed. By selecting the x axis for instance, the image will change so that it represents a view along the axis. Scaling of the image may also be controlled from the control space.

The operations permitted in the control space are hence those needed to aid selecting variables and to control the operator space, variable space and control space. Because of the desire to capture the design sequence, all operations to be captured must be executed from the operator sequence, and all variables must be identified to the operators from the variable space and never from the control space.

5. VARIABLE IDENTIFICATION

The basic process of variable identification has been described as the association of a request for a variable in the operator space with a variable token in the variable space. Aids to this process were described including the numbering of the request and the numbering of the associated variable (as shown in Figure 2). While it has not been explicitly stated, fundamental to the process is the assumption that for every variable required by an operator, there will appear in the operator area a word or token describing the required variable. Furthermore, the controlling system will not permit execution of an operator until all such words have had variables associated with them.

In addition to the variables appearing in the variable space, there is often a need for parameters to appear in the operator space. Now a parameter is really a variable that normally has only several values, such as signifying plus or minus direction in the x, y, or z axis. It has been found for user convenience, that a display in the operator picture of a current value of the parameter is better than having the user associate to a variable to assign a value to the parameter. To signify that a current value is being displayed, an asterisk appears by the parameter in the operator space and adjacent to the current value is shown the alternative values. Figure 5 illustrates the situation where a parameter with the two values "variable" or "constant" is shown. The "variable" has been selected.

Several different techniques may be used. One is simply to display the multiple values with an asterisk by the current value. To change the value, the user selects by light pen one of the choices. A second method is called wheel logic. In this method only the current value is displayed. The alternative values for the parameter are assumed to be ordered on a wheel, and if the user selects the current value with his light pen, the next value on the wheel is displayed and automatically becomes the current value. This method conserves screen space, and is often used where a value is changed infrequently.

Even though parameters are treated as above, it should be remembered that they are variables to the operators and as such, should be fully specifiable as variables. For this purpose, several techniques can be used to trigger the change from parameter to variable mode. One way is to use the asterisks in front of the current value as the trigger. If the user selects the asterisks, it signifies an expression of the indicated parameter mode is now going to be given for the desired parameter. A second approach is to place a light pen button in the control space which, when triggered, means that: (a) a parameter in the operator space is going to be selected, and (b) following that selection, an expression for the parameter will be given.

6. OUTPUT CONTROL

In the simplest mode of graphic console operation, the output from each operation is added to the variable display with the only representation of the output being a token on the screen such as a line, an image of a resistor, or some other graphic representation of the result. The user need not think of the variable by name, but only as an element on the screen. For those applications in which the image on the screen is the final result of the man-machine interaction, the simple no name mode of operation is adequate. However, for more complex applications such as automobile body design, the presence of at least two levels of variable naming appears desirable. The first level is the normal explicit naming of a variable with conventions to aid the user in generating a family of names, such as Line1, Line2, Line3, etc. The second level of naming is indirect. The user may, instead of naming the variable, name one or more sets the variable is to be a member of. This second level of naming, for instance, then permits complete drawings to be displayed by giving the name of the set of images to be displayed, or it permits a collection of points to be given to an operator rather than having each point being selected by the user.

Neither of these two levels of naming may be used until after the operator input data has been specified and the operator has been executed. As part of the execution of the operator, the output from the operator is added to the display in the variable space. In the operator space a naming operator may now appear. The function of the naming operator is to permit the association of names and set names to the new variables. Several ways of doing this can be readily conceived. The following is one relatively simple way. Each output variable from the operator is given a number. This number is displayed by the variable in the variable space and in a table in the operator space (see Figure 3). The numbers start with 1 for each operator, since their only purpose is to serve as a temporary naming convention in a manner similar to the association numbers for data input. The number 1 will be intensified when the name control operator is first displayed. The user may then supply a name mode variable by one of the following actions: (a) pointing at a name on the screen, (b) keying in a name, or (c) pointing at the 'next name' option on the screen and then doing either (a) or (b). If the 'next name' option is selected, the asterisk by the word 'name' as shown in Figure 3 would be changed to the word

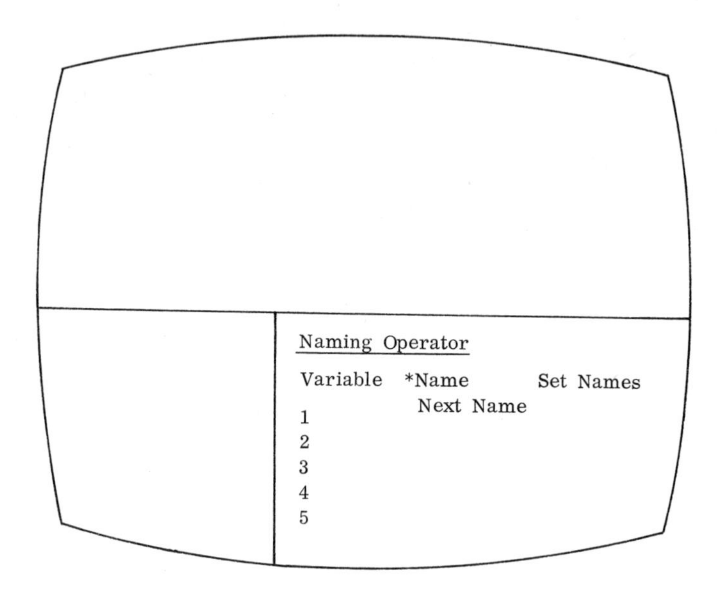

Fig. 3

'next name'. Upon selecting a name from the variable space or keying in a name, the low order digit would be indexed by 1 and that name entered into the name column. If he desires to give no name to the variable, he simply points to the number representing the next variable which he desires to name, and proceeds as above, or he may point to 'set name' and proceed to supply the names of sets which the variable is to be a member of. After supplying all the names he feels necessary, he then selects the EXECUTE button in the control space and proceeds to the next operator of his choice. Note that with the naming operator structured as described, it may be applied at any time by simply permitting the user to associate a token on the screen with the variable numbers in the name operator and then proceeding as described above.

7. SEQUENCE RECORDING

In the preceeding discussion on the operator concept, variable expressions, screen space organization, variable identification, and output control, the focus of the discussion has been on the relation of the preceeding functions to the individual operator. In this section we are concerned with recording a sequence of operators, as they are applied by a designer.

The structure of the graphic system as outlined, implies an interpretive control program which intercedes between the executable operator program and the process of man-machine communication required to associate the input variables with the operator control picture. Under these conditions, sequence recording is quite easy - it consists of remembering all input data to each operator, all output from each operator, and the sequence of operators. Of course, the sequence is not useful unless one can re-use the sequence providing new input data, and if the sequence can in some fashion be named and then referenced by name. Because the variables in the recorded sequence are not limited to the sequence, the sequence basically compares to the group concept in the PL/1 programming language. Essentially, the process of sequence recording produces the code

```
<label> DO;
          .  }
          .  }   Graphic operators as recorded
          .  }
          END;
```

where the label must be specified at the beginning of the sequence recording, and the END statement is a result of ending the recording function. The sequence recording function is started and stopped by selecting a light pen button in the control space. Nested use of the sequence record is permitted with a new label required for each use. An END record button is provided in the control space, but the recording is stopped only when the number of ENDS equals the number of RECORDS (DOs). In the following section the methods of converting a recorded sequence to an operator will be discussed.

8. PROGRAMMING

Before proceeding to define briefly the programming approach, I would like to emphasize the objective of the approach. The assumption has been made that the problem being solved at the graphic console has high pictorial content, such as constructing images representing three-dimensional objects, drawing electronic circuit layouts, constructing piping layouts, developing computer flow diagrams, and specifying control paths for machine tools. Now assuming that the system without sequence recording makes it possible for the user to create pictorial representations of such objects without ever using a classical programming language, the challenge is to allow the user to take constructs and use them repeatedly in the same sense as general purpose subroutines, without forcing the user to change completely to a general purpose programming language to define the constructs. In the method of doing this described herein, it is assumed the user has planned, away from the computer, the decision branches, the iterations, and the input to and output from the operator he is going to define. While this is certainly neither necessary nor desirable, it does ease the explanation of the concept. In the author's opinion, the maximum payoff from graphics will occur when the end user at the console can apply both graphics and programming techniques freely as he works, and the paper is aimed at pointing out one scheme for merging the techniques.

During the process, we will assume the man is alternately working with three separate images; one to represent the application variable space, one image to represent the control picture for the new operator, and one image to represent a flow diagram of the new operator. At the beginning of the defining process it is assumed that the variable space represents a legitimate situation for the new operator to be applied to. This condition is specified so that as the operator is defined, it may be executed step by step, and hence retain the graphic flavor of the application. At the completion of the definition operation, all input to and output from the operator is assumed to be handled as a normal operator in the system.

In the control space, control buttons are provided to provide for switching back and forth from the sequence recording mode and to identify when a variable is to be used in a program loop. In addition, the following operators are provided to permit the new operator to have properties of a classical computer program:

(1) A DO statement operator,
(2) a conditional branch operator,
(3) an operator picture defining operator,
and (4) a save variable space operator.

The function of input definition to the new operator is handled as part of the operator picture defining operator.

DO Statement

In PL/1 programming, the DO statement is assumed implicitly to apply to the group of statements following the DO, and ending with a matched END statement. In the graphics system, the statement applies to the sequence of graphic operators following the DO statement, and ending with a matched END statement. Two forms of the statement are needed.

The first form:

<label> DO;

essentially permits naming the sequence of operators following the DO statement.

The second form:

< label> DO FOR < set name > ;

causes the following graphic operators to be repeatedly executed until the members of the named set are exhausted. Each time through the loop, one member of the set is used. This form of a DO statement is much more restrictive than the normal array type looping statement, but is suggested because of the ease in which the user can work with sets. The appropriate operator space is illustrated in Figure 4.

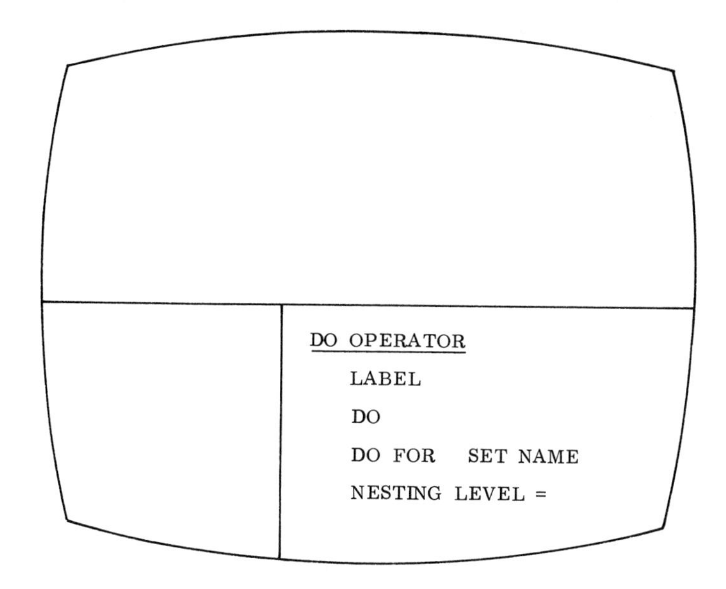

Fig. 4

The implication by the reference to only one set name is that only one variable may be looped on in a recorded sequence. This, however, is not the situation. For each variable that is used for either input to or output from an operator, the graphic system must know its mode whether it be point, line, surface, or set. During the recording of a sequence, a control button is active so that each variable which is to be looped on may be addressed as a member of a set. For example, if the operator being recorded requires a point and that point is to be provided from a set of points during the loop, the user must depress the control button to indicate that the point is to be taken from a loop set. Multiple sets may be used in a loop, but the loop is repeated only until the named set is exhausted.

Conditional Branch

Conditional branches are handled by using the PL/1 form of the IF statement. For single branches, the form is

<label> IF <exp> THEN <statement>

For multiple branches the form is

<label> IF <exp> THEN <statement>
<label> ELSE IF <exp> THEN <statement>
.
.
.

The expressions permitted must be relational expressions based on measures of the variables in each application. For example, in geometric applications, the expression could be used to test if two lines are parallel. The statement following the THEN must be a DO group. This means that after every application of the IF statement, the next operator to be used must be the DO operator.

Operator Picture Defining - OPDEF

The process of operator defining starts by the user selecting the operator picture defining (OPDEF) operator and specifying the name for the operator to be defined. He must next construct in the variable space a data situation which represents the input variables,to which the to-be-defined operator will be applied. He then goes into a sequence define mode, and applies an operator (which I will refer to as the INUSEOP, but which is an arbitrary operator) to the initial situation. At the completion of all input variable associations, and prior to execution of the (INUSEOP) operator, the operator defining operator (OPDEF) is requested by his selecting the OPDEF control button in the control space. Upon doing this, the variable space is completely changed. The operator he has been applying is placed at the top of the screen, the new operator picture (NEWOP) is placed in the middle of the screen, and the OPDEF operator in the operator space is as illustrated in Figure 5. The user must now decide whether any of the tokens in the INUSEOP picture are to be used as tokens in the NEWOP picture, if so he then selects them from the INUSEOP and

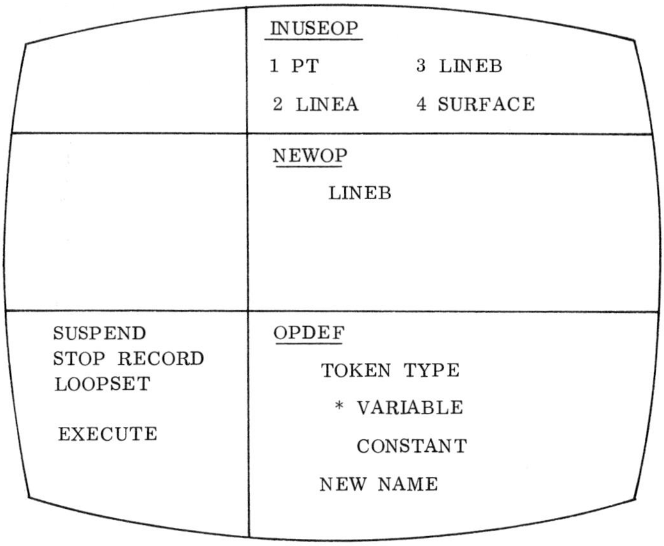

	INUSEOP 1 PT 3 LINEB 2 LINEA 4 SURFACE
	NEWOP LINEB
SUSPEND STOP RECORD LOOPSET EXECUTE	OPDEF TOKEN TYPE * VARIABLE CONSTANT NEW NAME

Fig. 5

places them in the NEWOP picture. If he wishes to change the characters in the token, he selects the new name option and keys in new characters for the name. All tokens moved from the INUSEOP then represent input variables to the NEWOP. Tokens not selected represent constants in the new operator. By pressing 'Execute', the execution of the INUSEOP operator is initiated with the screen changing back to the application variable space. After execution of the INUSEOP, the Naming Operator illustrated in Figure 6 may be applied. The operator is the same one as shown in Figure 3, but with additional functions. For each variable, two additional pieces of information are needed. First, if the variable is to be final output from the NEWOP operator, this must be indicated, and second, if the INUSEOP operator is within the scope of a DO loop, any variables to be placed in a loop set must be indicated. The user indicates these conditions by pointing at the LS (loop set) or OP (output) column in the NAMING operator. Those variables which are to be used as final output are the only ones to be displayed after execution of the new operator. Those variables to be placed in a loop set will be added to the set each time the INUSEOP operator is executed in the loop, the remaining variables will be written over each time through the loop.

The above steps are now repeated for each operator to be used in the new operator.

Save Variable Space Operator

The type of programming permitted is one of programming with immediate execution. That is, the user is not only specifying what operators he wants, but he is executing those operators as he proceeds. Now two situations occur where he may need to back up to a previous situation. First, on an IF ELSE IF situation he needs to create for each branch the proper input situation. By permitting him to save and restore variable space situations as he works along, he can create the conditions he needs for each branch starting from a common situation. The second case is where he applies the wrong operator and destroys the variable space for the operator he is defining. Hence, at any time during the operator defining function he should be able to back up to a previously defined version of his variable space.

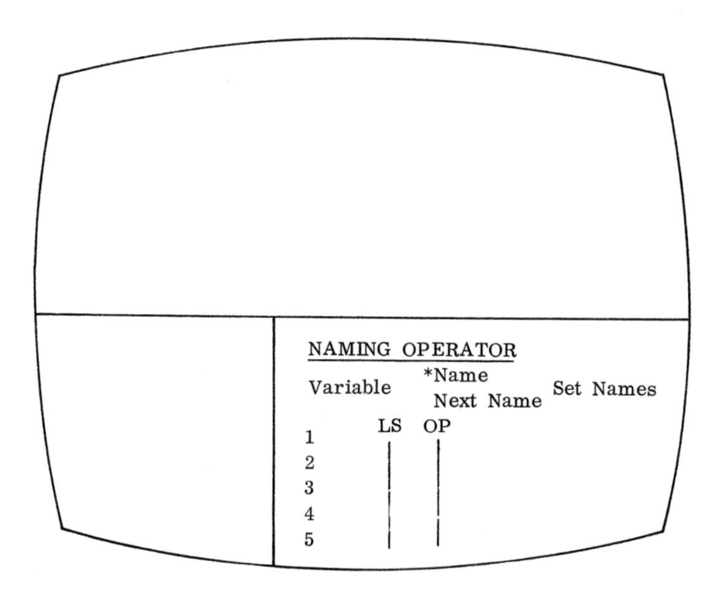

Fig. 6

New Operator Flow Diagram

The language structure provided the user is very easy to flow diagram. In Figure 7 the diagram representations for the DO, the DO FOR, the IF and the ELSE IF are shown. For a DO statement, the label is shown and the nesting level

1. <label> DO

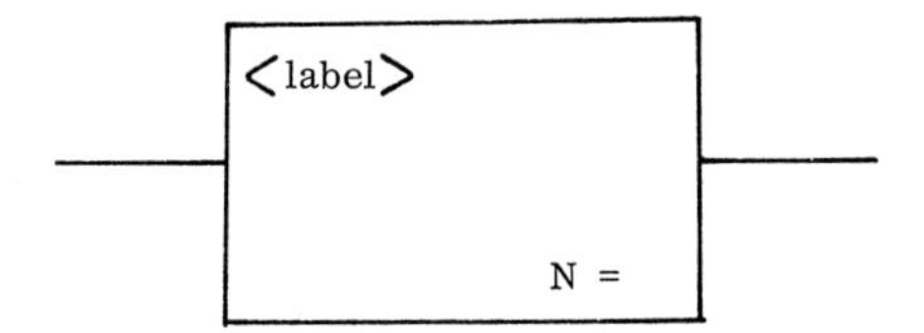

2. <label> DO FOR <set>

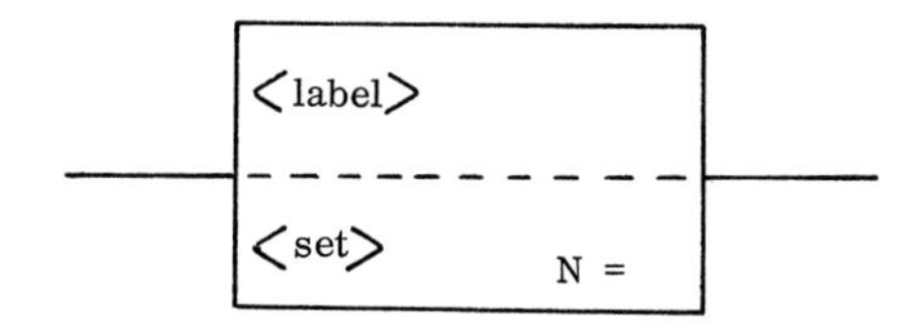

3. <label> IF <exp> THEN

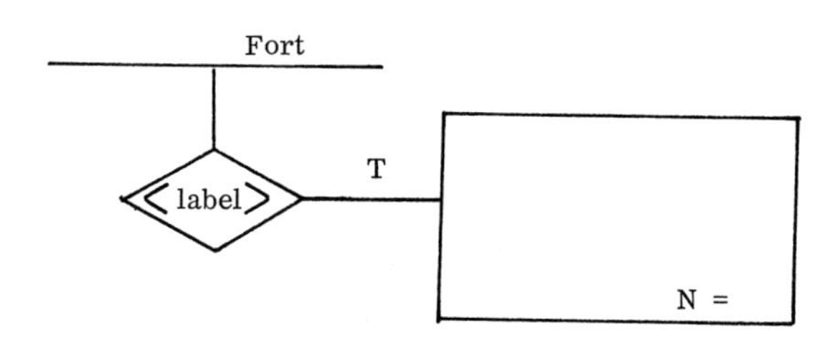

4. <label> ELSE IF <exp> THEN

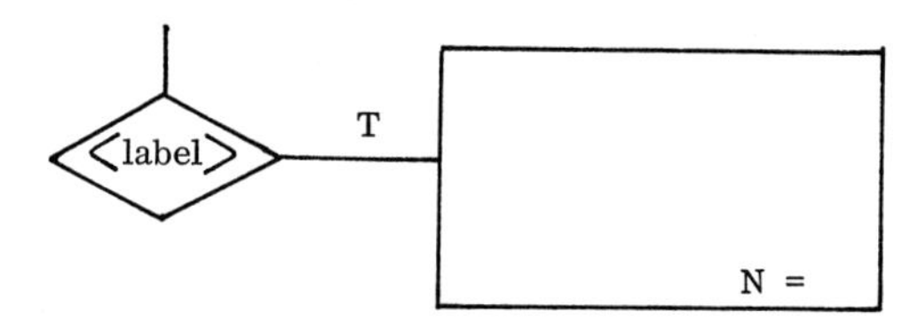

Fig. 7 Flow Diagrams

is specified in terms of its level, relative to the operator being defined and the number of DO levels within the DO. A third level DO with two levels lower could show as N = 3, 5 .

The convention on the IF statement is somewhat non-conventional. The actual branch is not shown on a flow line but is shown below the flow line. The branch for 'true' always goes to a DO block sequence and the last DO block in the sequence has no exit. It is assumed that the last DO block always goes back to the flow line at the point where the conditional test was attached to the flow line. The absence of GO TO statements and the use of the DO nesting convention makes the generation of flow diagrams very simple. Only one level of nesting needs to be shown at a time, and that only for the DO sequence the user is interested in. The ELSE IF diagram is simply connected on to the button of the IF diagram.

9. CONCLUSIONS

Fundamental to the system is one concept - that by using an interpretive processor to control all input and output from rather conventional subroutines, the concepts of conventional programming can be merged with graphic techniques to get the best of two methods - the ease of using graphics, with the flexibility of programming. The methods that have been developed for handling interactive compilers should be directly applicable to a graphics system of this type. In the paper, little effort has been given to describing specific operators, because the objective has been to illustrate the major concepts of concern:

(a) the continuous use of syntax and semantic information to control the input-output data flow from operators.

(b) the recognition that in graphics, variables must be defined as expressions, with a token displaying a variable being only one of many forms possible for the variable.

and (c) the merging of graphic operators with simple programming statements to achieve iterative control over sequences of graphic operators.

10. BIBLIOGRAPHY

DODD, G.G. (1966), "APL - A Language for Associative Data Handling in PL/1"; Proc. AFIPS Fall Joint Comput. Conf., Vol. 29, pp 677-684.

KNUTH, D. E. (1968), "The Art of Computer Programming"; Vol. 1, Addison-Wesley Publishing Company.

11. DISCUSSION

Bobrow: Does the system remember the entire sequence of operations that you have done in a console sitting?

Jacks: Yes, that is inherent in the system. As I mentioned, we use a mode switch, so that we can do it. You can turn on the sequence or turn it off; that is, you remember or you don't remember internally.

Narasimhan: In relation to making an association to, say, 'point' when the designer chooses this, when does the system start checking for types?

Jacks: It does that on every association.

Narasimhan: Then it extracts information on 'point'?

Jacks: There is a table set up for this variable, and in that it says, that is a 'point-type' verb. I mean by that, its an expression that must mean 'point'.

Narasimhan: But I am pointing, with the light pen I am pointing to a token. So it'll have to check that the token is of the proper type?

Jacks: Yes.

Narisimhan: That it does because the token is appropriately labelled, or what?

Jacks: The programs that control the variable display set up a completely independent set of sets, one of those sets contains all points, another set contains all lines, and when he points at the token, it means, 'your turn'. The set that it is in, is checked. Now for display purposes, we do not use the ring structure. In fact, we use essentially an incident array, which is just a simple matrix, which one way is set-type - it may be 'point'; and the other way across is variable. The variable at type 'point', will have a bit saying it is type 'point', and under the set for a line, will be a zero. So for every token on the screen, we have a bit pattern which identifies the sets that it belongs to for screen purposes. Besides the variable types, there are other things you want to display, for instance for control of the image, if you want to move it, rotate it, shift it; if you want to do that independently of such things as the list of names you have on the screen for your own purposes, you may have a list of set names, and you will find that those variables which are in the actual image that you want to control by rotation are in one of these sets, and the other variables are in different sets. This is done with the incident array. The vector for each variable is kept with the variable in the display buffer, so we can actually do the check in the display buffer.

One of the things I mention relative to that, the current system accepts, "knows", that 'point' is the next variable, and it actually disables all tokens coming out of the box. This turns out to be the wrong way to do it. We think it is going to be better to simply check them, and if you are getting the wrong one, if the man is pointing at something and he's coming back with the wrong variable at that point, put up a little message saying the wrong variable has been selected and then disable everything. The reason he is going to get the wrong variable is that you have a denser area in the screen with the points, lines and surfaces all packed together, and he can't select the right one. So then we will disable and make sure he can get only the right type of variable.

Rutter: Having done 'point', how do you get to doing 'line'?

Jacks: What happens is that we follow the sequence; as soon as he's done the point, we'll bring up the number '2' . But we don't say that this is dead, so if, you know you are not doing what you want, and if he points at this item, 'line', we change the background.

Macleod: This is where your ring structures come in handy, is it - to be able to get back to where you were, readily?

Jacks: Yes, the system is built for ring structure approach.

Macleod: If you just went straight through, there wouldn't appear to be any advantage in a ring structure - if you just went straight through the code, that is, followed one path.

Jacks: Except that a separate ring structure is keeping track of all the input to this thing too, and if you just do a normal programming, you don't keep track of data flow. This method shows two things at the same time, data flow and sequence flow, and the ring structure helps doing that because you have got an entirely separate ring set up for that.

Macleod: You've got a ring for your data and a ring for your control?

Jacks: Yes. With the data ring being tied to whatever subroutine calls are going on, and the control ring being this sequence.

Narasimhan: You talked about various forms of operations. There seem to be so many details possible to variance of the operations on surface normal for example.

Jacks: Yes, in our current system we have 106 geometric operations in the written descriptive geometry language for operations at the surface normal level. Most of them have been brought up to the screen, in some way or other; programs have been written which will interface most of these 106 operations. You get chaos if you do this, because each operation gets treated a little differently; there is no consistent approach. To give you a feeling for how complicated are these 106 operations, one of them is: 'define a circle or a segment of a circle', and within that one there are 13 different ways to define a circle or a segment of a circle. So each of the 106 operations has in addition, many different forms. What you want is to get these as consistent as possible so that the designer can work on them easily. We hope eventually that we can do better than the 106 - by better selection of primitives we hope we can reduce the number, but I do not think we will come out with less than 50, say.

Clowes: How do you define and actually construct all the 106 schema? Can you use this same apparatus again, can you treat that as a variable?

Jacks: You can treat it as a variable, but we are no longer in this same problem I have been talking about. We have two approaches - we are setting up a written language version that will map to this thing; the other approach is we intend to bring it completely to the screen as a variable.

Narisimhan: During your talk, you have said that in your process you don't use variable names at all. Actually you do use variable names, but use them implicitly; one, two, three, four are in fact names.

Jacks: The man at the console, the user, at this level need not use names.

Narasimhan: In going from the pictures to a sequential operation, gathering together the input and the output parameters and ignoring the internal parameters - you called this pruning in your talk - how do you make sure that when you do the pruning you do not have a name flag? Will not local names clash?

Jacks: There won't be a clash. The point is though there is really closure on the name also, because this is simply kept as a ring, and if you start it over, you see one, two and three, associations, there wouldn't be a clash.

Narasimhan: No, but when you put them both together?

Jacks: But this is what stops it, the fact you have an executable point. This is strictly a mechanism which is used within one operator, and the scope of these things is just within one operator. We don't even build up names - we don't even generate the 'two' internally, because what we are doing is hanging the data blocks on the set. (There is a scope problem, however).

Stanton: Do you in fact use sequences to disambiguate this situation? Say you point at 'point'. Say 'point' brightens up, and that's the next thing to be in. Now you say, 'well the rest of it is not dead, he can change his mind and perhaps get 'line''.

Jacks: If he does this, he automatically is making something else active and the former is no longer active. Only one variable is going to be active at a time.

Stanton: Say he does the following operation - points to 'point', he points to 'line' and then it swaps over. Then he says "now I do that" and he points back to 'point' again. If there are two points in the ring, as input to the routine, which one is he now referring to?

Jacks: If there are two points there, there is probably different meaning to them for the construction job. If there are not, in the situation where he wants a sequence of points going in, then that is different from the one point going in as far as these operators are concerned.

Stanton: You would see a separate name on there for every point?

Jacks: No, at one time (I haven't looked lately at our current system), he might see the word 'point set', and in that mode of operation, this would stay active until he selected 'surface': but then the class of variables which you are getting is a little different too, and it has to be kept track of. There is a collection of them being gathered together, and that is not treated as two separate points. The key point here is that by careful structuring, you get control over programming, and then the programming is the easy part.

Stanton: Can we go back to the sets? Are these sets predefined by the system or can the user define his own sets?

Jacks: Both.

Stanton: Then can we have a variable which is a member of more than one set at a time?

Jacks: Yes. In one case the user is doing it because he wants sets. We have for variables, predefined sets that we use to keep track of these things that they are members of. So there are two different levels here, internal level and external level.

Stanton: Before you carry out any remember procedures, operations, what do you start with in the machine, what is involved?

Jacks: In the system we are designing the first thing when you walk up to a console is an operator fixture here, everything else is blank. There is a standard sign-on procedure, and you have to define it within the operator's structure: for sign-on, you start out with name, password, and then you get the system - that is what it amounts to. You want to design and do any other operations that you may have, then you branch off into the particular systems that you want and you start attaching records.

Stanton: What are the primitives which are built into the system at the beginning? Are they points? How do you get off the ground, what set of routines do you have initially before the user can build up these other routines?

Jacks: Within the system, there will be a complete descriptive geometry design package. Now, for the people that are writing that descriptive geometry package, defining the primitives, there is going to be a version of PL/1 - there is in the interim system.

Stanton: You said you can build up these routines from the console.

Jacks: I was addressing as a designer you see, as a draftsman you can build these up.

Stanton: But you didn't sit down and build up a set for him, you actually wrote it in code, you didn't build them up from the consoles.

Jacks: The current set is being built up with the written language, which has the same properties, but we are trying to implement the graphics so that it is consistent; this is one of the things we are shooting for, a consistent interface from each operator through to a sequence of operators. And at this level, because we think it is fairly simple to understand, and we are looking at the draftsman on the board, he will understand it. Some of those fellows, once they get through this, they will want to go on to programming - that would be fine.

Macleod: Would you say a few words about the interface with the keyboard and how it can be used. I noticed on the film (in your talk "Design Principles for Graphics Software Systems"), that there didn't appear to be any need for the keyboard; that you were pointing at numerals on the right hand side. Is this used in preference to a keyboard?

Jacks: If there is a different way of doing it, someone will want to do it so - you can always go to the keyboard if you want to.

Macleod: I was wondering about constructing a diagram, would you be keying in the letters with the keyboard?

Jacks: No, we put these in the variable space up above, so when you are doing an association where you need it, you pick anything you want in it, and you can take and use Holerith characters, because they are legitimate variables of the right mode.

Macleod: If you wanted to put in a fairly long word, say for the operator, it would be a matter of constructing it a character at a time?

Jacks: The keyboard will be used on that. That is the basic idea, you always make sure you have the alternatives there by recognising different expressions which have the same variable type when you have done it. So far we have not used a 360 keyboard on a screen for any of this work, but it has been used on some other work in the 360.

DATA STRUCTURE AND QUESTION ANSWERING

C.J. BARTER

Department of Electronic Computation
School of Electrical Engineering
University of New South Wales

1. INTRODUCTION

A brief discussion of the "question-answering" problem will be given. Certain aspects of the problem of system design will be discussed; in particular, we will examine what is meant by "the structure of a data base". A survey of existing systems will not be attempted, but a few well known systems will be used as examples to illustrate various points. (See Simmons 1965, Woods 1967, Schank 1968, Bobrow 1969).

The essential tasks of representation of the information content of questions and data in the data base both require a theory of the structure of information. An adequate theory of information structure should provide a descriptive mechanism (a language) in which to represent information. All question-answering systems assume such a theory, although the descriptive mechanisms may be hidden in ad hoc programming code. Some existing systems will be compared in this respect, and contrasted with the thesis developed in this paper.

A central theme of the paper is concerned with the representation of relationships between entities in the data base. The point has been made (Fraser, 1967), that some questions are difficult to answer, not because the information is not in the data base, but because the data base is not organized for such questions; questions are most easily answered when the relationships referred to in the question are explicitly represented in the data base. This leads to an examination of the concepts of "explicit representation" and "restructurable" data bases.

It has also been suggested (Bobrow, 1964), that a question-answering system will require some representation of the problem area which is the subject of the system. This component of a system has been called "the model of the world", and can be identified at least implicitly, in all systems.

The system's description of the problem area will be, in general, statements about entities and their interrelationships in the problem area; they will be considered as statements in a problem language. Accordingly, some descriptive mechanism is needed in order that statements in the problem language may be overtly characterized, (Clowes, 1968). All of this will be subsumed by an adequate theory of information structure.

The latter part of the paper makes some suggestions towards the design of a picture language machine with a data base containing map information.

2. QUESTION-ANSWERING (Q.A.) SYSTEMS

The question-answering problem has been an attractive applications problem, because it has held out the hope that semantic analysis of natural language expressions might be at least operationally successful. To the information retrieval worker, short questions allow full linguistic analysis, whereas the economics of document retrieval systems force the use of statistical techniques. To the linguist, it is the restriction of the subject area of interest which is attractive, and makes the problem more tractible, than, say machine translation.

The usual model of a Q.A. System has three components, a syntax analyzer, a semantic interpreter, and a data base.

The syntax analyzer embodies some natural language grammar, and is used to recover syntactic relationships between the symbols of the string representing the query.

The semantic interpreter has the task of translating the structural relationships uncovered into a formal statement of the content of the question. This formal statement of content only represents the "meaning" of the query in an operational sense, in that it is a formal procedure for processing the data base to produce an answer to the question.

The data base is sometimes organized to anticipate the kind of query most expected in the problem domain, and it is often claimed that appropriate data structuring simplifies the task of semantic analysis.

There is, currently, no adequate grammar for natural language; indeed, it may well be the case that no problem-independent grammar will ever be written.

Consequently, the systems which have been reported either abandon the attempt at natural language analysis, and interrogate the data base through a formal query language, or use a grammar which describes a subset of natural language.

The BASEBALL system (Green, et al., 1963) uses a very simple grammar to uncover a few grammatical relationships (such as subject and object of main verbs), in queries expressed in a highly restricted subset of English. The bulk of the analysis consists of key word or phrase identification by lexical match, grammatical relations being used in the main to disambiguate active/passive voice constructions, and subject/object relationships predicated on non-symmetric verbs (viz. "beat").

A system for the interrogation of airline schedules (Woods, 1967) assumes the existence of a transformational grammar for English, which will make a deep structural analysis available as a starting point for semantic interpretation (Chomsky, 1965). The system uses a pattern matching technique to find a small number of syntactic relationships in the deep structure of the query, which are identified with particular grammatical relations. (E.g. S dominates NP and VP dominates (V), identifies with "subject of verb".)

The grammatical relationships between terminal items are equated with expressions in a formal language of a predicate calculus nature, by means of a translation rule of the kind:

"syntactic pattern (satisfy lexical restrictions)"
$\Longrightarrow$ formal procedural expression.

Although Woods stresses the importance of the structure of a data base, he does not mention the representation of data. The structure of the data is contained in the rules for semantic interpretation, and in point of fact, his system would decide on the semantic well formedness of a query without a data base.

In contrast, the DEACON system (Thompson 1966, Craig et al. 1966) relies heavily on the representation of data for semantic analysis.

In the BASEBALL program, the problem knowledge needed for semantic interpretation is represented in the structure of both the data base and the analysis component.

3. LIMITATIONS OF THE MODEL

The conventional model of a Q.A. system has led to some confusion in thinking and terminology. The "structure of the data" not only means the structural representation of factual items in the data base, but also the representation of problem knowledge, which need not reside in the data base. Alternatively, the component of semantic interpretation rules which describes problem knowledge might be thought of as data; and we now could recognize two kinds of data - specific factual data, and general problem descriptive data. This is recognized in the CONVERSE system (Kellog, 1967), where specific data is organized in a "representative" data base, and general problem data, together with syntactic and semantic rules, in an "interpretive" data base.

To discuss these issues at any deeper level we should recognize that the three components of a system are not clearly separable, and more to the point, do not seem to constitute an adequate paradigm for Q.A. systems, in that it does not suggest many insights valuable to system research. It seems to have the status of a loose model which is useful for a surface discussion of the problem.

4. PARADIGM FOR Q.A. SYSTEMS

A more substantial way of looking at the Q.A. problem, is to consider the process as one of language translation. For this comment to be meaningful, we must identify what is meant by "language" in a Q.A. context, and what is involved in the process of "translation".

Consider the following problem (Clowes, 1968). The question "What is the voltage across R2?", is addressed to a simple circuit diagram, with a table of component values.

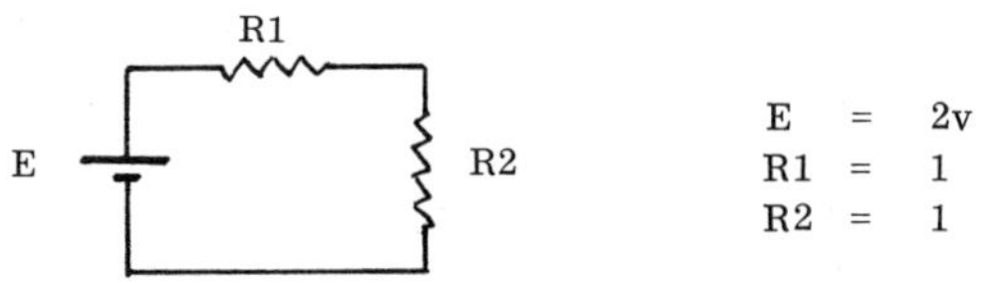

Three groups of information are represented, the query, the diagram and the table. Each of these are expressions in a language. The query language is English. The diagram is an expression in a picture language, with a syntax which describes relationships (such as near, join, parallel) which exist between a finite set of symbols (line, character, -WW- etc.).

Excluding the pictorial layout (which is a second picture language) the table is an expression in a tabular language, whose syntax describes the notion "association" between symbols which stand for "names of objects" and "numeric values". However, competence in these languages is not sufficient to answer the question. Even to interpret the question requires that the three expressions be related.

Interpretation of the question might be thought of in terms of the set of real world objects (or abstractions of) being referred to by the three kinds of expressions. If these abstract referents underly the representations in symbol form in the three languages identified, then they also provide a base through which expressions in each language may be related.

Further, the problem knowledge (such as Ohm's law) which is required to answer the question is concerned with the relationships which exist between such abstract referents as current, voltage, resistance, frequency, etc. A statement of some problem relationships might be thought of as an expression in a problem language, and a description of the relationship, the syntactic structure of the expression. Problem language statements will usually be expressed in some other language. For example, if the electrical relationship "in parallel" holds between two resistors, this may be expressed in a picture language,

or in English "the two resistors are connected in parallel", or in tabular form, as in an interconnection table. These alternative characterisations of the relationship are essentially paraphrases of the base problem language expression. But if we are to use problem language statements to uncover paraphrases, then we must be able to characterise them; and this requires that some other language be used as a medium.

For this purpose, it is proposed that a descriptive mechanism be devised, which may be used to characterise problem relationships in a way which approximates as closely as possible the structure of abstract problem languages. The design of a problem descriptive mechanism would thus seem to be a difficult task, since its structure is to be modelled on that of an abstract language, with no overt examples of sentences to examine. However, there are many examples of problem expressions in other languages; by an examination of the properties of ambiguity, anomaly and paraphrase in these languages, we may be able to develop an adequate descriptive mechanism for problem knowledge that may be useful as a base for recovering the semantic content of expressions in other languages. This concept mirrors the transformational grammar treatment of linguistic paraphrase.

It is clear that the syntax of the descriptive mechanism will be concerned with such abstract entities as relationships, functions, expressions, which are usually regarded as meta-syntactic entities. Statements in the descriptive system will however be concerned with direct problem entities.

5. DESCRIPTIVE SYSTEMS

It is interesting to examine existing Q.A. systems in terms of the paradigm outlined above. In doing this we will identify expression in a problem language with the structure of data (general and specific data); and the recovery of a base problem expression with semantic interpretation. This will be done by a close examination of the descriptive mechanism used to represent information in the system.

There is a class of Q.A. systems which assumes that an appropriate descriptive mechanism is to be found in the description of natural language grammars. One grammar based system (Rosenbaum, 1968), includes the data base as a set of restrictions in the categorial component of a transformational grammar. The answer to a question is either yes or no, depending on whether it can be parsed.

This approach assumes that the descriptive mechanism used by the transformation grammar is adequate for the representation of problem knowledge. However, the inclusion of data restrictions in the grammar in no way improves its adequacy in accounting for those examples of paraphrase which are termed "nonlinguistic". The descriptive mechanism of the grammar used has no way of representing the problem knowledge needed to identify this important class of paraphrase situation.

Another proposal has been made where data is represented in the form of English kernel sentences, and question-answering is an attempt to identify a paraphrase relationship between the query expression and data expressions (Simmons, 1966). It is interesting to note that an attempt to extend the descriptive mechanism was made, based on an identification of syntactic relationships in a kernel expression with a set of abstract logical relationships (viz. set membership).

A more recent paper by the same author (Simmons, 1968), proposes an improved natural language grammar based system. The grammar for English used is more concerned with the representation of grammatical function (e.g. agentives, instrumentals, locatives, datives) rather than syntactic relation (Fillmore, 1966, 1967). The motivation behind this grammar would seem to require a descriptive mechanism close to that which we have specified. For this reason, Simmons is able to relate the structures recovered by this grammar, with the notion of a "conceptual base", which is a direct parallel to "base expressions in a problem language".

Woods' airline system is based on a descriptive system which characterizes relationships which hold between objects or sets of objects, functions predicated on objects which address other objects, and the logical combination and embedding of predicate expressions. This is a very powerful system for representing problem knowledge. But if, as is the thesis of this paper, the descriptive mechanism should account for the structure of both the data base and the queries, it is interesting to examine what Woods has done. He has not discussed the representation of data, but instead related his descriptive system to the structure of English. But to show how such a system underlies a grammar for English is an immense problem, so Woods has taken the entire syntactic apparatus of a generative grammar and related it to his system. But the whole concept of the generative base component in the grammar for English used is deeply opposed to the "performance" issue of basing an English sentence on a problem language expression. This is made evident firstly by the observation that few syntactic relations in the base phrase marker are useful to Woods' system; and secondly many relationships expressible in Woods' system cannot be represented in the base phrase marker. The second point accounts for why such a system would seem to require so many rules for semantic interpretation.

6. DESCRIPTIVE SYSTEMS AND NATURAL LANGUAGES

The descriptive system developed by Woods was based on an informal theory of information structure described by Bobrow (1964). It was informal in the sense that the structure of the descriptive system was given in English. Bobrow's theory of coherent discourse is a theory of information structure; the need for problem language statements is identified (speaker's model of the world), and some syntactic categories of a problem descriptive system are given. The concept of a base representation of problem relationships underlying other overt expressions is presented as a discussion of "generative semantics", supported by a set of examples of English phrases.

But as is the case with Woods' system, Bobrow's algebra question answering scheme does not fully exploit the base properties of his descriptive system. The algebra system has no specific factual data base (the question contains the data), and very little general data (e.g. there are twelve inches in a foot), so there is little opportunity to see how the descriptive system might underlie the structure of data. Because of the choice of subject area, only a few functions and relations are used (arithmetic operations). This is by no means adequate to account for English, but is sufficient for statements about algebra, and consequently, a large number of linguistic forms in the English query are simply ignored by the (heuristic) parsing routine.

We might agree that a descriptive system should provide a base for natural language analysis; and conversely, examples of natural language forms (grammatical relations) provide a rich source of experience for the design of a theory of information structure. However, to demonstrate how natural language analysis could be accomplished in this way is tantamount to the task of specifying a complete grammar for English with explanatory adequacy, and this has proved to be extremely difficult.

7. DESCRIPTIVE SYSTEMS AND DATA STRUCTURE

A class of question-answering systems has been identified previously, where the choice of a descriptive system was heavily influenced, and sometimes dominated by, natural language considerations. There is another class where the motivation seems in part to arise out of consideration of machine representation of data. These systems use descriptive mechanisms which might be called associative programming languages and in general propose various ways to represent the associative relation of attribution. A meta-statement of this relation is "ATTRIBUTE of OBJECT is VALUE", and an example is

"FATHER of BILL is JOHN"

and has been expressed in some convenient notation, for example:

FATHER (BILL) = JOHN

or FATHER. BILL = JOHN

JOHN. FATHER (BILL)

This relation may be used functionally in a machine system where specifying the attribute of an object returns the associated (or computed) value. Alternatively it may be used as an associative triplet, where any element of the triplet is addressable through the other two. Recursion and embedded relationships are expressed by treating "value" as an "object".

The BASEBALL program is based in part on this kind of language. The problem area is concerned with the results of all games played in a year. The only object (excluding values of attributes) known to the system, is "GAME". The attributes of GAME are participants, venue, time and outcome of the game. Because only one object is known to the system, the associative triplet may be represented as an attribute-value pair, the object being known to the system implicitly. The attribute-value pairs which represent a game are the set:-

MONTH	=	month name
DAY	=	day number
PLACE	=	place name
TEAM	=	score
TEAM	=	score

Thus the descriptive system used describes a set of attribute value pairs, and one object class, "GAME". (GAME is also the name of an attribute, with a value which is the game's serial number, being the name of the game; this confuses the issue and is left out of the current discussion.) The descriptive system has direct influence on natural language analysis. Because only one object class is known, and represents a particular event, only one linguistic relationship is recognized. This is the relation "play". (The relation "beat" is recognized as a variant of "play" where teams are tagged as "winning" or "losing" by an ad hoc routine.)

The remainder of linguistic analysis is directed towards identifying the set of attribute value pairs in the question. The paradigm example statement of a game event allowed by the descriptive mechanism might be represented in English as:

"team, scoring runs, plays team, scoring runs,
at a place, on a day, in a month".

The linguistic routines attempt to analyze a query, and represent its content in a formal descriptive language. This representation in the BASEBALL system is called a "specification list".

The descriptive system also directly underlies the structure of the data; game information is organized into sets of attribute-value pairs. We have remarked before that the structure of data is not exactly the same thing as the structure of the data base in a machine. The actual representation of data in a machine also needs to be described. In BASEBALL, the descriptive system for this purpose is a list processing language, IPL-V. Thus the structure of the data (sets of attribute-value pairs) is represented in a machine data base by a list structure, where "set" is represented by a "path" through the list structure, and attribute value pairs as "atoms" at list nodes.

Questions are answered by exploiting the attribute-value associations within a set. A set (or sets) is identified (and by implication, a game event) by the known attribute value pairs in the specification list. The remaining set members are then available to allow named attributes to be given values, which are answers to questions. More complex questions are discussed later.

Several observations are worth making. Firstly, it is clear that the descriptive system used underlies the major components of the program, and that this is the main reason for its simplicity and success.

The descriptive system limits the kind of question which may be asked. For example, questions involving embedding or logical conjunction cannot be handled because the descriptive system cannot represent relationships between sets of attribute-value pairs.

Secondly, very little general problem data is represented. What there is is either implicit in the descriptive system (e.g. teams have scores) or represented in ad hoc code (e.g. a place name is ambiguous and may represent a venue name or a team name).

This has been possible because of the highly restricted nature of the problem. The fact that only one object and one relationship are involved allows them to be implicitly represented without ambiguity. In more complex situations, the descriptive mechanism would have to be expanded to allow general problem knowledge to be stated explicitly.

Thirdly, we can now give concrete support to earlier remarks about the relationships between the base descriptive system and surface languages. The structure of the BASEBALL data base in machine memory is in part specified by the syntax of

IPL-V. Further structure is imposed by the rules (outlined previously) which map between problem expressions in the base descriptive system and IPL-V expressions. Both sets of rules of syntax, in an appropriate compiler would constitute a procedure for translation between base problem expressions and machine memory contents. The status of IPL-V is that of a convenient intermediate language in this translation procedure. It is convenient to isolate (and use), as its syntax is mostly concerned with the medium of representation (machine memory), and of course had already been written.

Including IPL-V, we see an hierarchy of three languages for the representation of data expressions. At the highest level, problem expressions have the most overt structure and the information content of an expression is characterized by the expression together with the structure of the problem descriptive system. At the lowest level, expressions in machine memory have most of this structure removed and placed in the syntax of the mapping rules. Ultimately, these mapping rules are defined on the structure of the machine. IPL-V expressions mediate between these expressions.

It is in this sense that some high level languages (e.g. the associative programming language LEAP, described later) may claim to be suitable languages for writing question-answering systems. Their adequacy is dependent on the problem descriptive power of their base descriptive mechanism.

Precisely similar comments apply to other surface languages. This has been demonstrated for a class of picture language (Stanton, 1969) where the surface languages involved are represented in machine memory, and also in two dimensional line drawings. Mapping between the two surface expressions is achieved by basing the structure of each language on a description of the base problem (topology of region maps). As before, the procedures which map between problem base expressions and surface expressions contain the structure implied by the surface expression and made overt in the base expression. The surface expression is mapped through an hierarchy of languages in the graphics language for convenience (two dimensional line drawing to point coordinates to vectors).

That the same comments apply to English expressions can only be conjecture. But if they do, then it is clear that the medium of expression (viz. the restriction of word tokens in sequence), is the problem which has been the concern of a large part of conventional linguistics. In the present paradigm, the question whether certain linguistic structures and transformations are "meaning bearing" is meaningless. All that can be implied is an observation on how much structure has been overtly characterized, and how much is assumed by the structure of the descriptive mechanism of the grammar which uncovered the structure. An active/passive transformation may require little overt structure, since it is largely definable within those aspects of English concerned with medium of expression. Other transformations, say paraphrase, will involve to a varying degree, more overt structure. The distinction between linguistic and non-linguistic paraphrase would then seem to be a comment about the adequacy of a descriptive mechanism which attempts to recover an overt structure which reveals paraphrase.

Two other systems are interesting in the way they approach query-data translation. The DEACON system provides a structural description of query and data surface expressions, and attempts to recover a base expression from each. However, the base expression is not made explicit. Instead, structure is recovered from the query and the data in parallel, by the application of a rule pair (one from each language). When the query parsing is complete (i.e. no more rules apply), the structure of the data base reveals the answer to the query. Otherwise, the system fits within our framework: the descriptive system implicit is associative, and underlies the surface languages; data structure is represented by a ring structure language, and English by a phrase structure model which exploits a few association-like kernels.

The S.D.C. time shared data management system (T.D.M.S.) which uses a formal query language, is particularly interesting (Bleier, 1967). The descriptive system uses only two relationships; "dominance/subordinance" and "dominated by the same node", between nodes in a list structure. (A ring structure representation would do precisely this.) This is also the structure of the data base. The interesting aspect of this system is, that problem relationships are overtly stated in the system in the form of a list structure as well, called "the data base description". The data base description being a list, directly represents the relation "dominate", and represents a grouping on the same level by a device called "a repeating group", a recursive notation for varying the number of branches from a node. The description list allows the naming of data. Normal nodes represent direct data name, repeating group nodes represent generic class names. The description list names and defines the structure of the data base, and is used, once specified by a user, to actually structure named string data. It is then used in the inverse way to retrieve data by name from the data base. Of particular interest is the way the data description list directs the formulation of questions. Basically, only two kinds of questions may be asked, involving the relations "dominate", or "common parent node". However, these relationships are made transparent to the user by framing questions in terms of the description list. This in turn assures that questions are framed with a structure which corresponds to organization of the data base.

8. ASSOCIATIVE PROGRAMMING LANGUAGES

The associative language LEAP (Feldman, 1968), and an extended version of it, TRAMP, (Ash and Sibley, 1967) provide a descriptive mechanism for the "attribute of object is value" relationship. The relation is extended to allow embedding by defining "value" to be an object, and a "set of objects" to be an object. Hash coding techniques and pointer systems account for the machine representation of data with this structure, and the descriptive system is included with the syntax of ALGOL to form a powerful associative programming language.

It has been proposed that such a language is useful for writing question-answering programs. It is no doubt useful for some problem areas (e.g. family relationship data) where the attribution relation is sufficient. The attribution relation seems necessary to describe the attributes characteristics and associates that an object _has_; it seems less clear that it is adequate to describe what an object _is_ in terms of its parts and the way in which they are related.

9. RESTRUCTURABLE DATA BASES

A data set, based on an attribution model, which addresses a value object in terms of an attribute-object pair, is well organized for questions which ask for the value object, but not for those which ask for either the attribute or the main object. This is overcome in associative languages by storing all three organizations of the data.

Fraser (1967) has pointed out that "some questions .... cannot be easily answered, not because the information is not available in the data base, but because the information is not organized for such questions. And the variations on the data base organization are practically indefinite, depending on where the emphasis is being placed". He suggests that the criterion used to structure data (e.g. airline schedules) in a fixed organization, is based on that of the most common or useful kind of question. What should we do when the same data is to be accessed from different points of view; especially when all the different data organizations are too many to store?

"An interesting but yet unexplored question is how to dynamically reorganize the data base on the basis of reorientation of user questions. But actually we cannot expect this problem to be solved until some theory of information structure is developed. Such a theory must provide us with a metalanguage for talking about information, its structure, for stating the types of relationships possible within the structure, and must provide a way of automatically and systematically interpreting the data structure with respect to its content and the type of relationships which are explicitly and those which are implicitly stated."

Descriptive systems, for example those identified previously, are metalanguages for talking about information, and have influenced data base organization.

We may now identify what is meant by "dynamic restructuring" of a data base. Previously, the comments made about data organization were implied to concern a fixed data base. An analysis of a question produced a specification of the relations implied by the question, stated in a problem language. This formal expression, together with the structure of its descriptive mechanism directly influenced the explicit representation of data. All questions, if they are not anaphoric, contain some description of what is being queried; that is, they specify some problem situation. In BASEBALL, this situation is restricted to segments of the "team play team at place, etc." situation, and is represented by the specification list, which together with the structure of the attribute-set descriptive mechanism, defines the data organization needed to answer the question. For simple questions about games, a single data organization is adequate.

We will call this statement of relationships implied by the question a situation specification; it is a major part of that which is sometimes called the formal representation of the content of a question. The situation specification represents the "orientation" of user questions mentioned by Fraser. So far we have considered the case only where one basic situation could be specified by a question, and the organization of the data base appropriate to it. If this could be done systematically, then different situation specifications could also specify different, but appropriate data organizations

This in fact, would achieve dynamic reorganization of the data base in the way Fraser outlines, according to the point of view of the question.

10. QUESTION DIRECTED RESTRUCTURING

The formal representation of a question will have two parts: the main part is the situation specification which contains a description of the objects being referenced and the way they are related. The second part is the actual question part, which determines the kind of question and thus the answering procedure involved. There are three basic kinds of question:

1. Those which ask whether a situation specification can be verified in the data base. This question component underlies such linguistic forms as "Is it true that ...... ?", or "Does ...... ?" or "Is ...... ?". Such questions involve a data check, and a yes/no answer.

2. Those which ask for the name of an object (or perhaps a relationship) identified indirectly through the data base, by the situation specification. These are "who, when, where, which, what" questions, where interrogative pronouns act as adjectival modifiers on objects (which are usually generic, as in "which University ...... ?", and sometimes understood as in "which person ...... ?" being implied by "who...... ?") known to the situation specification. Such questions may be regarded as "which" questions, and require data retrieval.

3. Those which require numeric computation. Excluding those questions where the numeric value is already computed and stored (and are thus "which" questions), these questions in general imply calls to arithmetic processing routines. The most common question of this kind is the "how many ...... ?" question, and involves a count of the members in a set identified by the situation specification. Other general arithmetic operations such as comparison of set counts, have corresponding linguistic forms, for example "is ...... more than ...... ?".

Questions may be complex through embedding. The question component of a query is actually a procedure for defining an object, and these indirectly defined objects may be embedded in further situation specifications to form complex questions.

Such questions are really multiple questions, and need to be answered in stages. For example, the question -

"Which company owns the flight which leaves Canberra for Sydney at 7.00 a.m.?"

has an embedded situation specification:

"flights leave Canberra for Sydney at 7.00 a.m."

and a "which" question:

"which flights?"

The answer to this question is a flight, F (which is an object or set of objects). The entire question is now:

"A company owns a flight F."

and a "which" question:

"which company?"

This breakdown of the question has a natural parallel in the anaphoric form of the question:

"A flight leaves Canberra for Sydney at 7.00 a.m.: Who owns <u>the</u> flight?"

Anaphora is a linguistic device which is an alternative to question embedding. It uses an inter-sentence marker (usually "<u>the</u>") in place of embedding markers (pronouns) to construct complex situation specifications. It is normally used to simplify the linguistic expression of a complex situation. Otherwise the formal question situation specification is the same for both.

The BASEBALL program provides an example of the way in which a situation specification provides structural information for the reorganization of a data base. Although only one basic situation exists, complex situation specifications arise out of the indirect specification of objects by numeric computation.

Two numeric operations are used; they are "count set member" and "check set member count against a value". An example of this kind of query is -

"On how many days in July did eight teams play" which breaks down into a sequence of situation specifications:

1. "teams play on days in July" - "Which teams?"
   which organizes the data by month, then days then teams. The answer to "which teams?" is a set of team names for each day in July. A set count routine, counts the number of elements in the set (teams) for each day. A count check compares this with "eight", leaving a structure organized by days, then by team sets with a count of "eight".

2. "There are days in July (those on which 8 teams played) - "How many days?"
   A set count routine counts the members of this "day" set.

More complex questions with many stages of searching and processing can be handled in this way. Unfortunately the BASEBALL program does not make this process of restructuring explicit, but the intention that question-answering be a process of data restructuring is clear from the examples given of the way in which the "specification list" and "found list" interact for complex questions.

The BASEBALL system does not handle complex questions which involve embedding due to indirect object definition by means of "which" questions. An example of this type of question would be:

> "Which team played in the park, (which park was) used by the Yankees on July 7, on July 14th?"

Kuno (1968) has suggested how the BASEBALL program might be extended to handle this type of embedding, and essentially his comments involve extensions to the basic descriptive mechanism of the system. This was discussed previously.

We might now extend Fraser's remarks about data structure. The situation specification should direct the restructuring of the data base in order that the relationships implied by the question are explicitly represented in the data base. This makes questions easy to answer (i.e. the question component of the query can be a general retrieval or processing routine). Further, we note that restructuring operations on the data base are essential to the process of answering complex (multi stage) questions. It is attractive to view question-answering as a process of data restructuring under the control of a situation specification. When we consider that some questions require not simply an object (or set) as an answer, but a structured set of objects (or sets), then the answer itself could be considered as the data base in a suitably restructured form. This defines restructuring to include reclassification and subsetting by either direct match (retrieval by name) or by more complex processing (arithmetic routines).

11. DATA REPRESENTATION

Given that the structure of the data is determined by a situation specification, what is now meant by "data structure" and how is data to be represented? In the earlier discussions, "problem expressions" specified only one class of situation and a fixed data organization was possible, and moreover because it was fixed, a large amount of the structure of the data could be invested in the structure of the representation of the data base.

What does it mean to have different situation specifications defined on the same data base? In general it will mean that different objects will be talked about, but they will be (sometimes) high level objects defined on primitive objects common to the data base. (If this is not the case, then the data base would be separable, and two systems exist.) In BASEBALL, the high level object known to a situation specification was "GAME", defined on primitives such as team, day, place, etc. A weak example of another object perhaps useful in BASEBALL might be "PERFORMANCE", which is an opponent independent view of a game from a particular team's point of view. "PERFORMANCE" would be defined on the same data primitives as "GAME". A better example would be the two views of an airline schedule, one being the view of movement of individual flights, useful to booking agents, another the view of an individual airport concerned with traffic through the port. Each view would inspire different questions, but use the same data organized differently.

Similar to this is the different situation specifications which arise out of taking different aspects of the same relation. If a binary relation between two objects is stored consistently, and the retrieval of objects (or sets) is easier for one object in the relation than the other, the two kinds of object accessing questions might well require different situation specifications. This might be the case in an account system recording transactions between departments of a company (say a retailer) and other companies (say manufacturers).

Two points of view are required. One from the point of view of departments (what they each bought from other companies, and is a classification of purchases first by departments then by manufacturers), the other the point of view of other manufacturers (what manufacturers sold to departments, and is the inverse classification of the first view).

These of course may be used as arguments for separate well structured data bases, admitting the cost of multiple storage of data items. As has been remarked, this was the choice taken by the associative languages mentioned before.

The alternative is to store a data set with a minimum amount of structure (viz. that structure which is useful - and in common to all structures erected by the various situation specifications). There is a strong argument for doing this: if the structures that are to be imposed, (or already exist in multiple fixed storage) are complex then they will consume a large amount of storage space. If there is any chance of partitioning the data set (e.g. separating baseball games into years), and obtaining a useful subset, then not all of the data base needs to be structured to answer a question. This is of particular importance in a storage system with a cost/speed hierarchical structure. The decision here will be made by considering the usual time/space trade off. The region map program (Stanton, 1969) which treats picture point data from two points of view decides heavily in favour of erecting structure dynamically rather than storing separate complex structures.

12. SUMMARY OF THE Q.A. SYSTEM PARADIGM

Treating the Q.A. process as one of translation between a query language and a data base language through a base problem language, has led to a set of observations on the nature of the languages involved.

The base problem language, being abstract, is approximated by a descriptive system, which characterises general problem knowledge, the semantic content of which is contained in the implicit structure of the descriptive system. In simple systems, restrictions on the problem area are implicit in the scope of the descriptive system; in more complex systems which cater for "reorientation of user points of view", the descriptive system needs to be extended in scope, but is restricted by a "situation specification" at question-answering time. The descriptive system directs the question-answering process in two ways. Firstly by directing and specifying the data structure required to answer questions (either determining the fixed organization of data by its own implicit structure, or by doing this dynamically from the structure of situation specifications). Secondly, by basing retrieval and processing operations on the structure of the descriptive system; such processing and retrieval will be simple operations based on complex structures erected by the situation specification.

The surface languages (e.g. query data) which map between surface and base expressions may be specified in stages. It will probably be convenient to make use of already existing languages (list and ring structure, and associative languages) as intermediate languages.

Finally, considering the structure of the data to be the explicit structure of the data base together with the structures of the data base language and the base descriptive system, constructing an answer to a question can be seen as a process of transforming the structure of the data base through the structure of the data, directed by the structure of the question.

13. PICTURE LANGUAGE MACHINES

In some problem areas, the most convenient way to express some problem relations is in pictorial form. A Q.A. system based on such a problem area would require the specification of a picture language, for surface expressions, and the representation of pictorial information in a graphics data base. The nature of the base descriptive system would be the same as outlined before, because the abstract objects and relations characterized by it will not be treated differently just because they are to be represented pictorially. This is the concern of the picture language which maps abstract problem relations into pictorial relations.

Systems which relate pictorial expressions to other forms of surface expression have been called Picture Language Machines. One of the earliest of these systems (Kirsch, 1964) was a Q.A. system which addressed natural language queries to a graphic data base. The system used a predicate calculus descriptive system as a base for uncovering paraphrase situations between pictorial expressions, and natural language descriptions of pictures.

Because of the limitations on the number of objects and relations represented, it was possible to design a picture language on a simple pattern recognition device, and recover sufficient structural information from the picture to construct adequate problem expressions. The natural language expressions were simple enough to allow rules of syntax to be associated with the structure of the descriptive system used.

The structural analysis of general pictorial expressions has proven to be a difficult task, and as with other linguistic expressions, seems to require knowledge of the problem area being described for the process of semantic interpretation (Clowes, 1968). However, success has been had with restricted classes of pictorial expressions, in specifying languages which uncover some useful pictorial relationships. One such language (Stanton, 1969) describes plane region (map) line graphics, and uncovers a set of useful topological relations. An examination of this system will reveal that the analysis scheme reflects the concepts for general linguistics analysis outlined earlier in this paper; and in this particular system, the base problem expression describes the problem area of two dimensional topological regions.

A problem area for a Q.A. system will now be introduced, where the data base includes both graphical and non-graphical data. The data base will contain information about the surface of the earth, which is normally represented on topographical maps. The natural restrictions on the nature of this information make it particularly suited to the descriptive systems available for both pictorial and non-pictorial languages.

14. GEO-INFORMATION SYSTEMS

Two applications can be considered which make use of geo-information (information about and measurement of the surface of the earth). One is the actual production of maps (cartography), and the other, the interrogation of such a data base in a question-answering fashion. Both aspects have been implemented in a computer based land inventory system in Canada (Tomlinson, 1967). Considered as a picture language machine, it is not particularly interesting as it is based on a keyword addressing scheme, and the small amount of picture data processing involved is done by ad hoc routines. It does however, point the way to a very interesting and useful application area, and offers evidence that the two applications of automated cartography and question-answering are quite similar in their data structural requirements.

A Q.A. system may require an answer expressed in pictorial form (i.e.

a map). If the answer is complex, the picture language may need to be as complex as that needed for normal map production. Conversely, various specialized maps are now produced to meet the information needs of individual organizations (e.g. land use maps, weather maps, road maps, etc.). Automated cartography offers the flexibility to allow even more specialized map production. In this sense the system would be used to represent information in highly specialized maps, which would be the answers to highly particular questions; the questions themselves would be in terms of the objects to be represented on the map. The difference between the two applications is only a difference in the generality of questions. Maps are usually general answers to general (multiple) questions, and they are useful as an intermediate (subset) data base to which more specific questions may be put. These specific questions are the kind directly handled by a Q.A. system approach, where the detailed recovery of relationship between specific map objects will be required. Map production on the other hand, will require just the display of general classes of map objects. For this purpose a specialized command language will probably be most suitable, and the specification of multiple sets of map objects will be reflected in the display language by an overlay mechanism. But as has been suggested, both applications overlap, and some questions may require both facilities.

Before we can discuss the data structural requirements of a geo-information system, we will supplement our model of Q.A. systems with an examination of the map making process.

15. DATA ORGANIZATION FOR MAP PRODUCTION

The organization of data files has to take into account (and currently does in manual systems), three stages of map production:-

1. The gathering and compilation of primary data files.
2. The organization of primary data into useful map "objects".
3. The representation of map objects on a map.

These stages correspond with the functions of:-

1. The conventions and techniques of surveying, and data logging.
2. The <u>use</u> of the map.

3. The conventions and techniques of map making, i.e. printing, layout, aesthetics.

And in a Q.A. system, would correspond to:-

1. The input and representation of raw data.

2. The statement of problem descriptions in a descriptive system, and the structuring of data.

3. The specification of a surface pictorial language.

Each of these stages may vary greatly in their demands on file organization, and conflict with each other. Examples of difficulties in each phase are:-

1. The amount of structural information available with input data may vary. For example, land height data may be available in point height/coordinate form. Alternatively it may be available in contour form (as for example as the output of a stereo plotter, or copied from another map). Rainfall data may be available as long term average measurements from named weather stations, or as short term sets of data from rainfall grids at coordinate points. It is important to accept and store primary data in the form in which they become available. This will, however, complicate the problem of structuring data.

2. The objects which are to be created from primary data are specified from considerations of map use (which is the question-answering process). For example, from a data set containing point/rainfall measurements a useful map object to construct might be isohyets (lines of equal rainfall). Or similarly, the construction of contours from point/height data - or even if contours are already available, but not all needed.

3. Alternative representations of objects (image and name) are a characteristic of picture languages for maps. A region may be represented by a coloured area, a boundary or both, with a "name" inside, or even simply a name and no distinct boundary. Some clear rules exist for the pictorial representation of map objects. Other rules are difficult to articulate, and it is not clear that this is because the situations are complex, or pure aesthetics are involved. However, one would expect that it would be possible to characterize the aesthetic component of picture languages if the aesthetic component is

based on some concept such as "the ease of recovery of pictorial relations".

In summary, each phrase corresponds to the process of:-

1. Data gathering and gazette compilation, which includes little in the way of picture data, except for that data taken from air photos or other maps.

2. Re-organization (structuring) of primary data and the creation and naming of new objects which will be useful to the map user. It is at this point that primary data with coordinate references are used to construct pictorial objects.

3. The representation of map objects in an easily interpretable way using the art/techniques of conventional mapping. In an automated system, pictorial expressions for display may be restricted by the difficulty of characterizing some display language techniques, and the limitations of some display media (e.g. colour on oscilloscope displays). In general, the display language component which overlays different map objects on a coordinate framework (the abstract grid system) is straightforward. Two exceptions will prove difficult to characterize. The first is the necessary distortion of some picture objects (e.g. the thinning of rivers and roads, the artificial separation of superposed or adjacent but near objects such as road alongside a railway line), to achieve clarity at the expense of accuracy. The other is the introduction of non-graphic file data (e.g. names) into a picture as picture objects.

This last point invokes a comment about the way in which general alphanumeric file data is related to picture data. It is fortunate in the case of the map problem, that the coordinate grid system is already available and useful as a mechanism for relating gazette and image data. Thus each item of general file data is defined on a named point, line or area in the image file, and these objects are ultimately defined on the point coordinate system. It is clear that the base descriptive system will have to describe point coordinates as a basic problem "object".

16. PICTURE LANGUAGE AND DATA REPRESENTATION

As part of the task of specifying a picture language for maps, we will assume that we have available a picture language such as that described by Stanton (1969) for the description of topological relations between pictorial objects in a plane region map. This language describes the objects point, line, and region and will be useful as an intermediate language for both data representation and display languages. It will prove to be a useful intermediate language because it can directly characterize some of the important pictorial objects and relations required by the base descriptive system. For example:

point inside region.
region adjacent to region.
line joins line, etc.

And also allows arithmetically computed attributes:

length of line.
area of region.

And may be extended to create objects such as:

"region whose boundary is some distance from a point".

Another particularly useful extension would be the ability to overlay two region maps and define intersection regions by some logical operations. We would, for example, require conjoint and disjoint pictorial operations.

A set of mapping rules will be required to map between pictorial expressions in the base system and expressions in the topological language.

The plane regions language already includes a display component. This will have to be supplemented by other display rules (including specification of perhaps shading, line thickness, and colour) determined by the nature of objects represented in the base. We have noted that this will be difficult for full map production.

17. THE BASE DESCRIPTIVE SYSTEM FOR MAP DATA

In a problem area with mixed graphic/alphanumeric data, the object NAME needs careful definition. Real world objects (e.g. a city called Sydney) are represented by a collection of data about each object. This data will be both alphanumeric (e.g. population of) and graphic (e.g. pictorial boundary data of the region). The name of the object (e.g. SYDNEY) stands for the object in that it will be the external representation of the object in some surface expressions (e.g. English, tables) and also address the object data when used in problem expressions. It also addresses object picture data when problem expressions imply pictorial relations (e.g. inside (the region) Sydney).

Finally, it is also the name of a pictorial representation of an object, (e.g. a display of a region may be named SYDNEY). Providing the metalanguage of the descriptive system distinguishes between pictorial and non-pictorial relations, it will always be clear how a name is to be used.

Because of the natural restriction that map data always refers to objects <u>at</u> a specific location, we may adopt a simple descriptive system. The data associated with a map object (addressed by name) is amenable to attribute-object-value representation (and although picture information is an exception, this too could be described by a "location of" attribute). This simplification is possible because all of the objects implied by such data, are not map objects. For example, we may have population data on cities, but "people" are not map objects, and are only indirectly implied by the attribute "population of".

The descriptive system will need to handle <u>sets</u> of objects. One way of defining sets is through the relation "has the same attribute-value". (E.g. the set of all points with "height above sea level" of 1000 feet). Logical (and, or, not) and arithmetic (more than, equal to, less than) operations between attribute-value pairs of an object will be useful (e.g. the set of all points with height 1000 feet <u>and</u> more than 10 inches annual rainfall). These operations will of course include pictorial information in the sets created, so that picture sets will exist as well.

One special kind of set is the generic set (e.g. the set of all cities), where the set has a name (generic name). It will be convenient if these sets already exist as objects in the system, and not have to be overtly specified by questions. Whether explicit problem expressions will be used to dynamically create these sets, or whether they will be known implicitly to the metalanguage of the descriptive system based on some classification structure, is an economic consideration. For simplicity, we assume the latter representation.

We have indicated that the relation "has the same attribute-value" is useful between objects, and that primitive logical relations are useful between attributes of single objects. This is essentially a way of defining subsets, by restriction relations on larger sets (including the whole data set).

A second way of defining sets is by defining relationships (based on attributes again) <u>between</u> sets (e.g. the region which has more than 10 inch rainfall <u>and</u> above 1000 feet). This second kind of operation will involve a merge of sets into a set upon which a logical restriction is placed. For non-pictorial data, this operation is best represented by the "has the same attribute-value" procedure of the first kind mentioned above. For picture data, the merge will correspond to a pictorial overlay, and logical operations will be of the graphical kind, creating new graphical objects (or sets of).

Sets will generally contain at least one attribute in common, and perhaps the same value of that attribute, which is simply a reflection of the way sets are formed. An extreme example of this is seen in sets which represent only one attribute, for objects which have no particular name. For example a useful file to store as a set would be a set of rainfall readings at points on a rainfall grid (coordinates). These points are not worth storing as individual objects, and storing them as a set reflects the way in which primary data becomes available.

18. PICTURE LANGUAGE MACHINE EXAMPLES

Many questions will not use graphics data, in particular those which are represented by the attribution relation on non-pictorial attributes. For example:-

"What is the population of Sydney?"

"Which cities have a population over 10, 000?"

But many questions involve graphics processing:

"What is the area of Sydney?"

"What is the length of the N.S.W. coastline?"

involving arithmetic computation on region and line data.

Some involve interaction between attribute data and graphic data:

"What cities in N.S.W. have population over 10, 000?"

This involves an evaluation of the graphical relation region (city) <u>inside</u> region (state), to create the set of cities inside N.S.W., followed by an attribute evaluation to create the subset restricted by population.

Suppose we wished to answer a very complex question, such as, "Where are the potential rice growing areas in N.S.W.?". The requirements for a rice growing area need to be articulated, and probably would not be specified in single question form. Either a command language or anaphoric questioning would be more useful.

Nevertheless we will examine the process involved in answering the question. We might specify:

"A region within 50 miles of a large river."

We need to create the set of "large rivers" either by naming them individually, or finding them through some attribute such as flow: from computations (within 50 miles) on the pictorial representation of this set of rivers (lines) we generate the appropriate regions.

"within 10 miles of transportation"

We define transportation on road or rail systems, and generate another region set as before.

"of particular soil type"

Regions accessed from soil map data.

"flat, cleared, open farm land"

Regions accessed from land use map data.

These regions would be overlaid and a set of regions created from their logical conjunction, as being potential rice growing areas. We might further restrict these areas now by specifying that each set member exceed some acreage, which is again a picture data operation. These regions could then be displayed on a map, perhaps with other objects (such as towns <u>in</u> or <u>near</u> these regions).

Apart from being useful for display, picture data is largely used for the representation of spatial information not represented in the alphanumeric attribute file. While the attribute information will be explicitly stored (because it can be done simply), most spatial information (such as length, area, points inside regions) is only implicit, and needs to be recovered by restructuring operations. The structures imposed on the picture data by base problem expressions, and question situation specifications, can be seen as imposing structure on the whole data base, because of the way picture data is linked to alphanumeric file data. For example, the relationship that the state of N.S.W. has a city called Sydney is not explicitly represented in the data base; but it is recoverable through graphical operations on the picture data linked to these object names.

19. CONCLUSION

This paper has attempted to provide a paradigm for question-answering systems which we believe will be useful in the evaluation of problems revealed by past work, and point out the direction in which further research should proceed. We have tried to show how it provides a basis for the integration of various concepts of data structure, which we have found to be, at times, conflicting.

Early systems effectively demonstrated the advantages of appropriate data organization, to both semantic interpretation and retrieval. Their success was limited to problem areas which were highly restricted, and were not easily extended. Woods pointed out that this was precisely because they depended too heavily on data base structure for semantic interpretation, and that the structural devices used were inadequate for the representation of any but the most simple semantic models. It was not possible to use such systems for wider (or even different sometimes) problems, because no mechanism was available for the representation of new (or different) kinds of problem relationship.

This led Woods and others to direct their attention away from data base representation, and concentrate on the representation of problem relationships, reflecting at the same time Bobrow's concern with "knowledge of the world". But the point made by early attempts, that data base organization is important, still remains valid.

Fraser indicated the way in which the relationships inherent in a problem area should influence data organization, but showed that any rigid organization might not be appropriate for all questions. His remarks about "dynamic reorganization" suggested that data base structure should be based on an analysis of questions, and to do this we require a "theory of information structure".

The paradigm presented assumes part of such a theory. A theory of information structure will be the metalanguage of a descriptive system used to characterise abstract problem relations and situations, which characterisations will underlie surface language expressions in the problem area. The descriptive system, together with a description of surface languages, if it is to be directly useful, will manifest itself as a language in which we may write question-answering programs.

The specification of surface languages will prove to be difficult. It does seem clear that two aspects of surface language exist: the representation of relationships which exist in the base; and those aspects concerned with the conventions and restrictions of the medium of representation.

Most of our remarks have been made about surface languages for machine representation of data. For obvious reasons of difficulty, little has been ventured with respect to natural language. Picture language specification seems tractable for restricted classes of expressions (e.g. plane region line drawings). But even for picture expressions which seem to involve artistic components, such as aesthetics, have yielded to quantification. Some of these considerations which apply to the production of quality maps have been formalized (Williams, 1968). Although the study of complex pictorial expressions perhaps belongs to the theory of art, we may here even find encouragement.

"Styles, like languages, differ in the sequence of articulation and in the number of questions they may allow the artist to ask ...... But what matters to us is that the correct portrait, like the useful map, is an end product of a long road through scheme and correction. It is not a faithful record of a visual experience, but the faithful construction of a relational model." (Gombrich 1960).

ACKNOWLEDGEMENTS

I would like to acknowledge the invaluable help and time given to me by my colleagues M.B. Clowes, R.B. Stanton and D.J. Langridge, and the assistance of Miss N. Walsh, in the preparation of this paper.

20. REFERENCES

ASH, W. and SIBLEY, E. (1968), "TRAMP: A relational memory with an associative base"; CONCOMP Project Technical Report No. 5, University of Michigan

BLEIER, R.E. (1967), "Treating hierarchical data structures in the SDC time-shared data management system (T.D.M.S.)"; Proc. ACM Nat. Meeting, pp. 41+

BOBROW, D.G. (1964), "A question-answering system for high school algebra word problems"; Proc. FJCC, pp. 591-614.

BOBROW, D.G. (1969), "Natural language interaction systems"; This Conference.

CHOMSKY, N. (1965), "Aspects of the theory of syntax"; M.I.T. Press, Cambridge, Mass.

CLOWES, M.B. (1968), "Transformational grammars and the organization of pictures"; Seminar Paper No. 11, C.S.I.R.O. Div. of Computing Research, Canberra. Presented to the Nato Summer School on Automatic Interpretation and Classification of Images, Pisa, Italy.

CLOWES, M.B. (1968), "Pictorial relationships - a syntactic approach"; Presented to the Fourth Machine Intelligence Workshop, Edinburgh, Scotland. To be published.

CRAIG, J.A. et al. (1966), "DEACON: Direct English Access and Control"; Proc. FJCC, pp.365-380.

FELDMAN, J.A. and ROVNER, P.D. (1968), "An ALGOL-Based Associative Language"; Stanford Artificial Intelligence Project Memo AI-66.

FILLMORE, C.J. (1966), "A proposal concerning English propositions"; Monograph Series on Languages and Linguistics, No. 19, Georgetown University Institute of Languages and Linguistics.

FILLMORE, C.J. (1967), "The grammar of hitting and breaking"; In 'Studies in English Transformational Grammar', R. Jacobs and P Rosenbaum (Eds.), Ginn-Blaisdell.

FRASER, J.B. (1967), "On communicating with machines in natural language"; In "Computer and Information Science II", Julius Tou (Ed.), pp. 315-335.

GOMBRICH, E.H. (1960), "Art and illusion"; Phaidon Press, London, p. 90.

GREEN, B.F. et al. (1964), "BASEBALL: An automatic question answerer"; In "Computers and Thought", Feigenbaum and Feldman (Eds.), McGraw-Hill, N.Y., pp.207-216.

KELLOG, C.H. (1967), "On-line translation of natural language questions into artificial language queries"; SDC Document SP-2827/000/000.

KELLOG, C.H. (1967), "CONVERSE: A system for the on-line description and retrieval of structured data using natural language"; SDC Document SP-2635/000/000.

KIRSCH, R.A. (1964), "Computer interpretation of English text and picture patterns"; Trans. IEEE, Vol. EC13, (August).

KUNO, S. (1967), "Computer analysis of natural language"; In "Mathematical Aspects of Computer Science", Proc. Symposia in Applied Mathematics, Vol. XIX, pp. 52-109.

ROSENBAUM, P.S. (1967), "A grammar base question-answering procedure"; Comm. ACM 10, 10, pp.630-635.

SCHANK, R.C. (1968), "The use of conceptual relations in content analysis and data base storage"; Report No. 68-347-U to TRACOR Inc. Austin, Texas.

SIMMONS, R.F. (1965), "Answering English questions by computer: A survey"; Comm. ACM 8, 1, pp.53-70.

SIMMONS, R.F., BURGER, J.F. and LONG, R.E. (1966), "An approach toward answering English questions from text"; Proc. FJCC, pp.357-373.

SIMMONS, R.F. (1968), "Linguistic analysis of constructed student responses in CAI"; Report No. TNN-86 of the Computation Center, University of Texas at Austin.

STANTON, R.B. (1969), "Plane regions: A study in graphical communication"; This Conference.

THOMPSON, F.B. (1966), "English for the computer"; Proc. FJCC, pp.349-356.

TOMLINSON, R.F. (1967), "An introduction to the geo-information system of the Canada land inventory"; Pre-publication report.

WILLIAMS, N.L.G. (1968), "Partial anatomy of a map"; Presented to the Automated Cartography Study Group, Canberra, Australia, (December).

WOODS, W.A. (1967), "Semantics for a question-answering system"; Harvard University Ph.D. dissertation and report No. NSF-19 to the National Science Foundation.

21. DISCUSSION

Narasimhan: In discussing the electric circuit problem in your paper, you introduced the notion of four languages. Supposing I say that there aren't in fact four languages. There is only one problem language, and these are just expressions in the same problem language. Would you deny that?

Barter: No.

Narasimhan: Is that what you are saying?

Barter: There are two ways I can say it. Obviously I am influenced by this Chomsky based surface expression concept - I find that a convenient fiction. But whether or not you have sets of surface expressions, they are all just related to (and I would like to put some kind of model up here), these are expressions all in the same problem language, and you say that's one language. Alternatively, you say you are just going to relate them all as para-phrases of the same thing, and I just find it more convenient to think about it that way - particularly when finally I am up for writing down a problem language. Since I am going to write down the problem language by looking at all of the ways one can say things, I like to think of it as one level up.

Clowes: Could I make a comment? In expressing the situation there are four ways of doing this, there are four languages and the relation between them. I can't think of what else you can say but that three of the languages represent something about the fourth, specifically that an expression, or expressions in the fourth language (which Barter calls a problem language), can be related to expressions in any one of the other three. Now the question that you are asking here is, should we regard the relation of representation as being an expression in a language? If we do, then clearly all four are in that language.

Narasimhan: Yes, the point is these three expressions talk about three different aspects of a particular situation. That situation is embedded in the problem world, and so it is not correct to say that these three expressions are three different ways of saying the same thing. They say three different things. All three put together defines a particular situation.

Clowes: We could have three things which all talked about the same aspect of the problem situation.

Narasimhan: They don't do so in this picture (of the circuit).

Clowes: No, but they could. The point is, it is no longer relevant to talk about different aspects now, its a question of talking of how we talk, that is different. We talk in English, we talk in pictures, we talk in tabular form, algebraically: we talk differently, and there are three different ways of talking, not three different aspects of talking. You have to characterize the 'how' not the 'what'. How is it possible to talk differently in three different ways?

Narasimhan: I don't think we are arguing about the same problem. What I say is this problem is just an elliptic way of expressing the following statement - (He ought to actually write it out, in order to be completely unambiguous in what to say) - 'What is the voltage across R_2 in the picture where, so and so?' You just drop in and where, so it appears like three different sentences.

Clowes: Following the 'where' we will have an English expression, what we have expressed pictorially. What else would we have?

Narasimhan: I can embed all kinds of languages in English, but I don't want to do that.

Clowes: I understood your intention was, 'what is the voltage across R_2 in a circuit in which R_2 appears in series with R_1?'

Narasimhan: I don't have to paraphrase the picture to do that. What I am saying is I embed a picture language in English.

Bobrow: It is not the usual thought that people have when thinking about what English is, to think about pictures in it. If you want to characterize these things, you may say, well that's one part of one super language, but if you get it in the squiggly lines and things, they will have one syntax, and maybe we ought to recognise it's there. The mapping is very very different if it is two dimensional than it was if you had this one dimensional string, which doesn't make any difference to 'what is the voltage across R_2' if there are two lines or one line, but it does make a difference when R_1 is near that first squiggly line and near some place else, and the tabular form has a completely different kind of syntax, and you would like to be able to handle that differently. Now you can say, 'I'll subsume these and recognise that there were three parts, and I want to call them all the same language, because they are dealing with the same problem, but that's a different orientation - the semantics is what determines the language, not the syntax.

Narasimhan: This happens all the time, for example in ALGOL we have Arithmetic expressions, Boolean expressions and so on; each expression is defined by its own syntax and semantics. You put them all together and you call it a block, and that block is an ALGOL, an evolved sub-language is an ALGOL.

Bobrow: That's a sub language of English too, that's a slightly different way of representing the same problem. I think it more useful to talk about separate languages if you are going to worry about separating out the definitions.

THE MODIFICATION OF THE ASPECTS-BASE AND INTERPRETIVE SEMANTICS

R.J. ZATORSKI

Languages Section
Faculty of Science
University of Melbourne
Victoria

1. INTRODUCTION

It is the intention of the author to argue on purely linguistic grounds that the representation of the referential domain is essential for adequate description, that such representation involves the specification of the situation, or event underlying an utterance, in terms of its primitive objects, parts, attributes and relations, and that mapping procedures must be evolved to translate such specifications into linguistic structures over which transformational rules operate to derive certain kinds of paraphrases.

The paper first discusses the devices recently introduced into the Aspects-base, then proposes an alternative approach to linguistic description in line with the statement in the opening paragraph.

2. THE MODIFICATION OF THE BASE

Since the publication of Aspects (Chomsky, 1965), a massive effort has been devoted to the validation or otherwise of the Aspects-system and to a sharper definition of the status of Katz' semantic theory within it. The results to date are interesting and instructive. On one view (Postal, 1967:2), the Aspects-system represents a wholly premature attempt, undertaken "in the absence of explicit awareness of the (full) range of facts (which a theory must handle)" - a strange judgement considering that without the guidance of a theory one would have no idea what the "facts" of language might be and how to go about discovering them. Other views hold the Aspects-system to be correct in outline but requiring a degree of elaboration. In particular, two rival conceptions have come to light within the latter school, defined respectively as transformationalist (Lakoff, Chapin) and lexicalist (Chomsky) which take contrasting attitudes on the issue: "whether certain descriptive problems can be handled by enriching the lexicon and simplifying the categorial component of the base, or conversely; or by simplifying the base at the cost of greater complexity of transformations, or conversely" (Chomsky, 1967: 3). In general, whereas the investigations of the "transformationalists" revealed serious shortcomings of the Aspects-base, the solutions proposed appear to be counter-intuitive as long as the construction of a syntactic base capable of handling paraphrase, cooccurrence, conjunction, coordination, entailment, etc., remains a possibility. Accordingly much of the research of the "lexicalist" group has concentrated on broadening the scope of the existing base, Chomsky himself contributing a number of radical proposals.

Additional stimulus to this activity has come from the investigations of problems lying on the so-called boundary between syntax and semantics, as that boundary is defined in Aspects. In particular, there has been some discussion of the role of the grammatical relations in the assembly of semantic readings. It is my intention in this paper to consider briefly some of the devices recently introduced in the base and their effect on the functioning of the semantic component in the sense of Katz (1967).

2.1. Rule Schemata

The admissibility of rule schemata in the syntactic base was discussed in Aspects in connection with conjunction and coordination (Chomsky, 1965: 98-9, 155-6, 224 n. 7), and no objections in principle were raised to their incorporation. Concrete proposals for their inclusion were subsequently made by a number of workers, including Postal, Kayser and Dik, the latter, in the context of a "functional" grammar which attempts to assemble surface structures without employing transformations. The least developed among these proposals seem to be those of Kayser and Postal. Kayser (1968) notes that adverbs have considerable positional freedom in sentences and introduces the so-called "transportability convention" which (without the formal apparatus having been specified anywhere) labels all nodes suspended from a single dominating node as "sisters". The selection of a category lexically specified as [+ transportable] in any "sister" position results in the derivation by the transformational component of a set of structures, one of which is selected for further transformational "processing" and all others suppressed.

Thus given a deep structure:

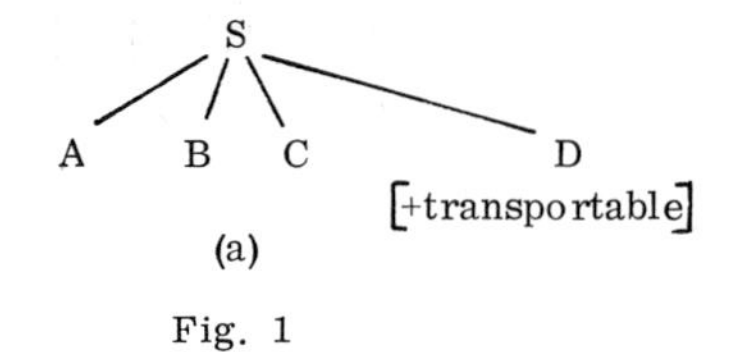

Fig. 1

the transformational component derives, in addition, the set:

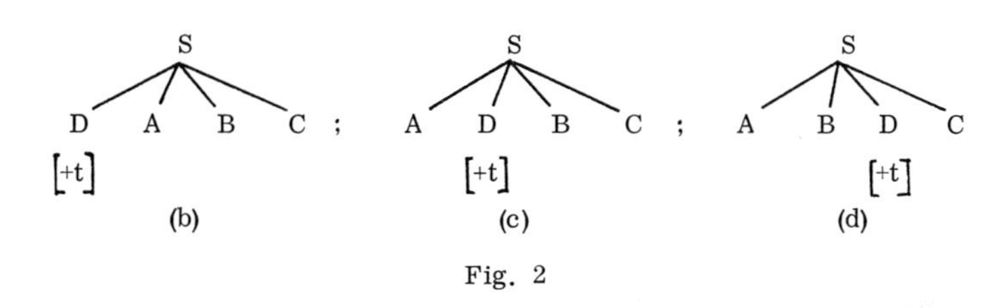

Fig. 2

allowing (a), (b), (c) or (d) to be selected to account for such sentences as:

(1) i Immediately, John mailed the letter home;

ii John immediately mailed the letter home; or

iii John mailed the letter home immediately.

Postal (1967) raises the question of rule schemata in his discussion of McCawley (1967) who rejects the derivation of certain plurals by the process of conjunction. As counter-argument, Postal asserts that the sentence:

(2) Approximately 300 men volunteered.

has the structure:

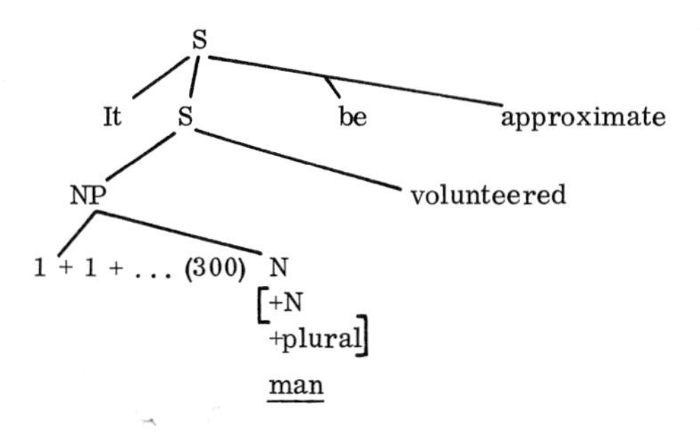

Fig. 3

For sentences with terms such as infinite, boundless, etc., Postal suggests deep structure schemas generating phrase markers of the above kind in which appropriate notation is used to represent infinite sets.

Dik (1968: 92-115; 185-6; 204-208) accounts for the coordination of multiple subjects and the conjunction of multiple elements of subjects, illustrated respectively by

(3) i John, Bill and Harry came, and
ii John, Bill and Harry met,

by the use of a rule schema

(4) $c \rightarrow F_n (n > 1)$ where c = category
F = function (e.g. subject, predicator, modifier)

which generates an infinite set:

(5) $c \rightarrow F$
$c \rightarrow F^\frown F$
$c \rightarrow F^\frown F^\frown F$
.
.
.

in each application, without any restriction on the number of F's. Assuming that such a restriction can be specified, the inclusion of schema (4) in the base would allow it to assign to (3)i and (3)ii respectively, the phrase markers (a) and (b) in Fig. 4 below.

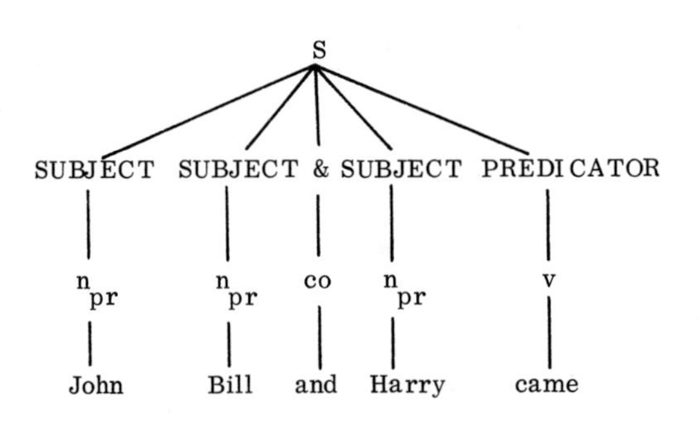

(a)

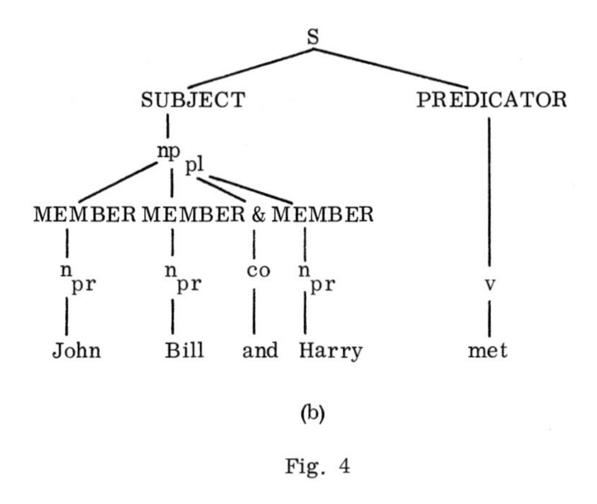

(b)

Fig. 4

Unlike Kayser's schema, which forms part of the transformational component and generates a finite number of phrase markers (this number being a function on the number of "sisters" at the given level) the rule schemata proposed by Dik and Postal are designed to ensure unrestricted generation of subordinate nodes in the base phrase marker and we are given no clues as to the manner of selecting a rule with precisely the number of elements on the right of the arrow required for lexical items such as several or many, e.g. in sentences

(6) i several of my 12 students suffer from asthma
ii several among the crowd of students were injured
iii several of my many students were absent.

A related question posed by the inclusion of such schemata is whether grammars so elaborated remain finite systems and whereas Postal's proposals uphold the general condition of finiteness, Dik extends the Chomsky-Schützenberger (1963: 133) position on rule schemata in context-free grammars, viz.: "the grammar, though still specified by rule schemata, consists of an infinite number of rules", to a context-sensitive "functional" grammar which "must be said to be a finite set of rules and rule-schemata [my emphasis - R.J.Z.] determining an infinite set of rules, determining an infinite set of linguistic expressions together with their structural descriptions" (Dik, 1968: 185).

2.2. Transformations in Phrase Structure

To account for the assembly of derived nominals such as (7), several of John's proofs of the theorem, in the base, Chomsky (1967), in accordance with the lexicalist hypothesis, dispenses altogether with the Aspects-base and introduces one from which the notion of "category" is eliminated. In place of categorial nodes, Chomsky proposes complex symbols specifying sets of features which allow lexical entries to be substituted. These are introduced by rule schemata (8) ii and iii in the grammar base below, reconstructed from his 1967 paper.

Given X = N, A, V

(8) i $S \rightarrow \bar{\bar{N}} \frown \bar{\bar{V}}$

ii $\bar{\bar{X}} \rightarrow [\text{Spec}, \bar{X}] \frown \bar{X}$

iii $\bar{X} \rightarrow X \frown (\bar{\bar{X}})$

iv $[\text{Spec}, \bar{N}] \Rightarrow$ Predeterminer$\frown$of, $[\pm\text{Def}, (\bar{\bar{N}})\ (S)]$

v $[\text{Spec}, \bar{A}] \Rightarrow [(\text{Comparative}), (\text{very}), (\bar{\bar{A}})]$

vi $[\text{Spec}, \bar{V}] \Rightarrow [(\text{Adv}_t), \text{Aux}, (\bar{\bar{V}})]$

Accordingly, the sentence

(9) several of John's proofs of the theorem appeared in 1968

is assigned a phrase marker:

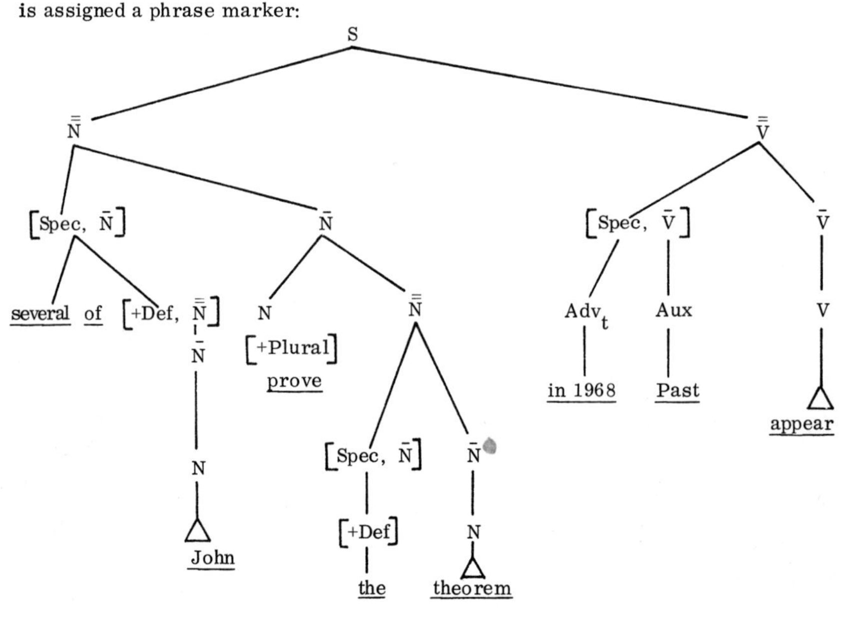

Fig. 5

In the grammar (8), N, A and V are lexical categories in the sense of Aspects (1965: 166); double bars over symbols indicate "phrases"; $\bar{\bar{N}}$ and S are recursive elements; rules (8) iv, v and vi are transformational in the "local" sense (Chomsky, 1965: 215).

A different scheme for the introduction of transformations into the base has been suggested by Fillmore who maintains that since subject and object are not "among the syntactic functions to which semantic rules must be sensitive" (1966: 21), we must "abandon the notions of deep structure subject and .... object, and must .... accept a model of grammar in which the subjects and objects that we see in surface structure are introduced by rules" (1967: 15). The relations which Fillmore accepts as relevant to semantic interpretation are ergative, dative, locative, instrumental, agentive, etc.; they assume a categorial status in the fragment of his grammar (9) shown below:

(10) i S → Modality⁀Aux⁀Proposition

ii Modality → { Adverbial sentence / Adverbial time / Q / Neg }

iii Aux → { Pres / Past }

iv Proposition → V(Erg) (Dat) (Loc) (Inst) (Ag) ....

v { Erg / Dat / Loc / Inst / Ag } → NP

vi NP → P(Det) (S) N

vii P → [+P, +Act]

viii Det → [+Det, +Act]

ix N → [+N, +Act]

(vii–ix) where Act = actant, i.e. Erg, Dat, Loc .... dominating the NP along the same branch.

x [+N, +Ag] → [+Animate]

xi [+P, +Erg] → for/Mod⁀Aux⁀<u>blame</u> ___

xii [+P, +Dat] → on/Mod⁀Aux⁀<u>blame</u> ___

Mod⁀Aux⁀<u>depend</u> ___

xiii [+P, +Dat] → to/Mod⁀Aux⁀<u>object</u> ___

.

.

.

.

A substantial number of conventions not specified in the form of rules, especially with respect to prepositions other than those in rules (9)xi-xiii above, further constrain sentence generation. Rules (9)v-x specify "local" transformational rearrangement of the base phrase marker on the pattern:

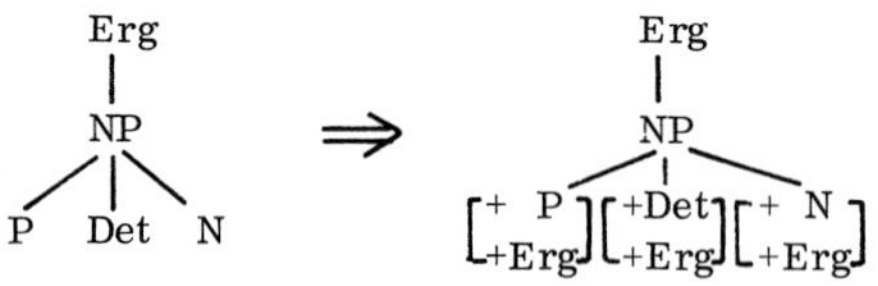

Erg | NP — P Det N ⇒ Erg | NP — [+P, +Erg] [+Det, +Erg] [+N, +Erg]

Fig. 6

Thus, while Chomsky is essentially concerned with increasing the scope of the noun phrase branch of the generalized phrase marker in such a way as to account for mutual derivation of certain categories of words and is careful not to upset the relata of Katz's semantic component where projection rules combine the readings of modifiers with those of heads, those of verbs with those of objects, those of subjects with those of predicates, etc., Fillmore substitutes for the notions subject, object, modifier, head, etc., a set of functional notions: ergative, agentive, locative, etc., based on case systems, which enter directly, as labelled nodes, into the base phrase marker. In this way (and also through the use of conventions and complex symbols) Fillmore (like Dik) does away with the need for interpretive derivation of the elements: subject, object, main verb, etc. He does not however introduce any changes into the operations of projection rules. Specifically, Fillmore does not equate projection with selection (see Langendoen, 1967: 100, also Clowes et al., 1969: 23, for an opposing view), and steers clear of the notion of generative semantics. In this way he continues to accept the fiction of a "boundary" dividing generative syntax from interpretive semantics and fails to perceive the mutually exclusive character of systems one of which assembles only "meaningful" strings and the other accepts for interpretation any random PS-derived string with its structural description and endows it with a semantic classification.

In the latest work elaborating the Aspects-base, Jacobs & Rosenbaum (1968) use transformations in the base, in the manner of Fillmore, to accomplish a great variety of tasks, for example the derivation of the determiner from a feature on N, as follows:

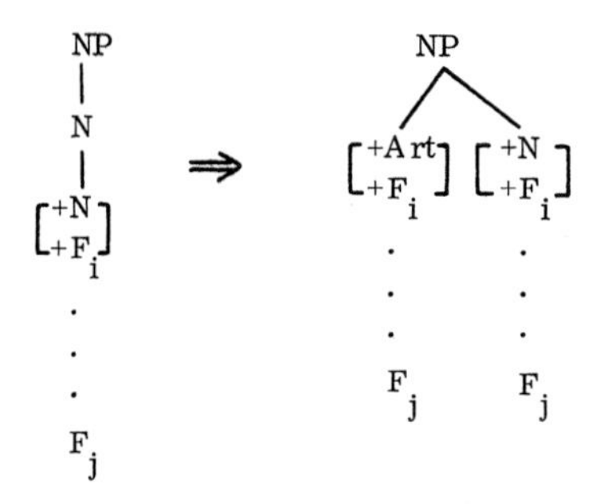

where F_i, ... F_j include such features as ±Def, etc.

Fig. 7

3. THE INADEQUACY OF INTERPRETIVE SEMANTICS

Allowing that the modifications discussed above increase to a considerable degree the weak generative power of the Aspects-base, they make no contribution whatever to the task of specifying the structural relationships among the sentence pairs, triples, etc., involved in any of the more complex types of paraphrase, inference, entailment, etc., illustrated below:

(11) i John helped Mary <=> John assisted Mary*;

ii John's crying caused Mary to be upset <=> Mary was upset because John cried;

iii John might do it <=> It is possible that John will do it;

iv If John blushes then he is embarrassed <=> John's blushing indicates he is embarrassed;

* The sign <=> designates an unspecified relation.

v John's diligence led to his promotion<=>John's promotion stemmed from his diligence;

vi John committed a crime and was duly arrested<=>John's arrest followed his committing a crime;

vii John owns a Holden, a Ford and a Peugeot<=>John owns motorcars;

viii John believes that Mary is a good cook. All good cooks enjoy entertaining<=> John believes that Mary enjoys entertaining.

If we interpret the Katz and Postal hypothesis on the non-introduction in the transformational component of meaning-bearing elements as implying that all the paraphrastic variance in sentence sets, such as those listed in (11)i-viii, must share a common deep structure, the "new" words or constructions introduced into whichever we assume to be the derived variant must be treated either as non-meaning-bearing *in the context of the given deep structure* - a notion defying useful formalization, or the Katz and Postal hypothesis must be abandoned. This last alternative is clearly counterintuitive in view of the well-known empirical utility of the transformational component operating under the Katz and Postal restriction. We are thus forced to conclude that in terms of the present *Aspects*-base, the sentence sets in (11) above are not in fact paraphrastic variants; having said this, we are left with no means whatever of describing the obvious relationships observed among such variants.

The deep reason for the failure of the *Aspects*-base in such cases lies in the conception of deep structure as a *verbal* construct, required to obey simultaneously the rules of word adjunction mediated by the single relation "followed-by", and the specification of the underlying referential domain, currently packaged in a set of subcategorial and selectional rules which do not even begin to come to grips with the range of relations obtaining among objects, events, entities, their parts, attributes and values, in that domain. Of the two arguments offered by Chomsky for the present base (1965: 123), the first which asserts the need for the specification in deep structure of the subject, objects, etc., functions, has been shown to have no deep relevance (Fillmore, 1967: 15; Zatorski, 1968: 17); the second, concerned with the need for the ordering of elements, is satisfied by any principled string-generating device (Zatorski, 1968: 17, 18). We are therefore obliged to reconsider our conception of generative syntax and explore systems in which:

(a) the deep structure of a sentence will comprise a symbolic specification of the structure of "fact" describing the objects, events, etc. underlying the verbal manifestation of the sentence, and the relations governing their mutual interaction;

(b) mapping rules will "translate" the referential specification in (a) above into a deep structure of a sentence, e.g. in the form of a bracketed verbal string;

(c) the transformational component will operate on such verbal deep structures in a normal way to produce surface structures.

4. REFERENTIAL SPECIFICATION

Current work in picture syntax (Clowes, 1969; Stanton, 1969) proposes a conceptual and notational framework for the description of geometric figures and there are hopes that it may prove possible to broaden that framework to enable it to handle certain fundamental problems associated with the formalization of referential descriptions essential to a system outlined in the previous section. As an illustration, consider the simple sentence:

(12) the boy hit the ball.

Leaving aside the question of <u>the</u>-specification, we must begin by defining the different referential "configurations" of <u>boy</u>, <u>hit</u> and <u>ball</u> within their particular domains; having derived such <u>referential deep structures</u> we must find a way of mapping from their formulae to verbal strings. The different senses of <u>boy</u> may be determined from sentences:

(13) i boys have proportionately shorter limbs but larger heads than adult men;

ii the boy spoke to his teacher;

iii even at eighteen, he was still a boy;

iv after the victory at Simla, the English colonel ordered his boy to fetch some champagne.

In (13)i, we are concerned with <u>boy</u> in the anatomical sense, i.e. as a member of the subdomain {anatomical objects} whose referential specification must account for the primitive parts on which the object is defined and the relation(s) which govern the assembly of such parts. We can provide such an account in the formula:

(14) BOY$_{\text{anatobject}}$ →> HUMAN.MALE<HEAD, TRUNK, LIMBS>[join<2, 3>, join<3, 4>],
(numbering: HUMAN.MALE = 1, HEAD = 2, TRUNK = 3, LIMBS = 4)

where: anatobject designates the referential subdomain;

HUMAN.MALE specifies inherent, "inalienable" or list properties;

→> "is defined as";

< > enclose sets of parts, attributes or values;

[ ] enclose relations governing the assembly of parts or attributes.

In (13)ii and iii, however, the differentiation of the two senses of boy rests on the insight that "boyhood" in the subdomain of human entities is a function of both the physical and mental age, characterized respectively by the referential formulae:

(15) i BOY$_{\text{entity}}$ →> HUMAN.MALE<BODY, MIND>[age<2>, age<3>, low<4>, low<5>, equal<4, 5>]
(numbering: HUMAN.MALE = 1, BODY = 2, MIND = 3, age<2> = 4, age<3> = 5)

ii BOY$_{\text{entity}}$ →> HUMAN.MALE<BODY, MIND>[age<2>, age<3>, ~low<4>, low<5>, greater than<4, 5>]
(numbering: HUMAN.MALE = 1, BODY = 2, MIND = 3, age<2> = 4, age<3> = 5)

The terms "~low<4>", "low<5>" constitute abbreviated forms of fully specified relations with numerical values defined on human-specific criteria.

The situation is quite different in the case of boy in (13)iv which is merely a synonym for servant, predicated on a service relation between humans; assuming that this relation can be exhibited as

(16) serve<MASTER, SERVANT>[ ], without, for the time being,
(numbering: MASTER = 1, SERVANT = 2)

specifying the conditions in square brackets, we must associate (16) with the word boy via the definition of serve and via a mapping such as

(17) serve<MASTER, SERVANT>[ ] map<1, BOY>
(numbering: MASTER = 1, SERVANT = 2)

where we ascribe the appearance of boy in (13)iv to a substitution of a verbal token for an entity servant with an appropriately defined referential structure. The exact form of (16) or (17) is highly tentative, nevertheless the need for both the definition of service as a relation and for treating boy in (13)iv as a mapping of SERVANT into BOY, seem to me beyond question.

The description of hit in (12) must be based on the discussion in Fillmore (1967: 14) which suggests the following two referential definitions:

(18) hit <OBJECT, {HUMAN / OBJECT}> [travel <1> [t=-1], direction
                 1   2

<towards <2>[ ]>, touch <2, 3> [t=o]]

where 1 designates an instrument, in Fillmore's sense,

(19) hit <HUMAN, {HUMAN / OBJECT}, OBJECT> [propel <1, 3>[ ],
                 1   2   3

travel <3> [t=-1], direction <towards <2>[ ]>, touch <2, 3> [t=o]]

where 1 represents Fillmore's agent.

Thus, the formula (18) specifies hit for sentences such as

(20) one train hit another;

while the formula (19) specifies it for such sentences as

(21) John hit the snake with a stick.

In specifying ball, we may formalize the three readings investigated by Katz (cf. extensive discussion and references in Clowes, et al., 1969) respectively as

(22) i BALL$_{obj}$ → SOLID <MATERIAL, GEOMETRY> spherical <3>[ ]]
                 1   2   3

map <METAL, 3> [rigid <2>[ ]]
                 2

ii map <RUBBER, 3> [elastic <2>[ ]]
                 2

iii BALL$_{event}$ → SOCIAL <EVENT> [dance <PARTNER, PARTNER>[ ],
                 1   2   3

together <2, 3>[ ]]

To simplify further discussion, we can abbreviate the referential formulae above as follows:

TABLE 1

		Abbreviation
Formula	15i	(child) BOY
"	15ii	(man) BOY
Mapping	17	(servant) BOY
Formula	18	HIT 1
"	19	HIT 2
"	22i	(cannon) BALL
"	ii	(cricket)BALL
"	iii	(dance) BALL

It emerges with some clarity from the discussion on the preceding pages that the assembly of deep structures of complex entities or events (such as that underlying (13)) consists in the substitution of the specifications of objects, events, etc. with given inalienable properties, such as HUMAN. MALE in (15)i, ii or SOLID in 22i, ii, for the equivalent arguments in the description of parts such as 1, 2 in (16) and (18), 1, 2, 3 in (19), etc., in relations. Thus, on HIT 1 we can define a set of referential structures:

(23)	(cannon) BALL	HIT 1	(cannon) BALL
	"	"	(cricket) BALL
	(cricket) BALL	"	(cannon) BALL
	"	"	(cricket) BALL

A predictably larger set is defined on HIT 2:

(24)	i	(child)BOY	HIT 2	(cannon)BALL	with	(cannon) BALL
		"	"	"	"	(cricket) BALL
		"	"	(cricket)BALL	"	(cannon) BALL
		"	"	"	"	(cricket) BALL
	ii	(man)BOY	"	(cannon)BALL	"	(cannon) BALL
		.	.	.		.
		.	.	.	as in (24) i	.
		.	.	.		.
	iii	(servant)BOY	"	(cannon)BALL	"	(cannon) BALL
		.	.	.		.

(iii ctd. ...)

. . . as in (24) i .

. . . .

Mapping from the referential domain into language involves as the first step the assembly of such complex deep structures from primitive specifications. The resulting formula constitutes the symbolic description of a complex event, entity, etc. which can now be mapped in part or in full into sets of paraphrastic variants ranging from simple to most complex in cases where every single element and relation inherent in the referential formula underwent mapping into words. We can illustrate by substituting into (19) the formula (15)i and (22)ii, leaving one argument place unfilled and introducing an additional convention to the effect that unsubstituted argument places in relations are either specified in language as unknowns or are deleted. Thus,

(25) map ref. domain hit<BOY.HUMAN.MALE<BODY, MIND>,
1 2 3

5. age<2>, 5. age<3>, BALL.SOLID<RUBBER,
5

GEOMETRY> spherical, elastic, OBJECT< >>
6

[propel, travel [t=-1], direction, towards, touch [t=o]]

⇒ lang

boy Past hit ball

boy Past hit rubber ball with unknown OBJECT

.
.
.
.

boy human male age(d) 5 (in) body (and) mind Past hit solid, rubber, spherical, elastic ball with unknown object (which (same) boy) propelled (so that (same) unknown object) travelled .... (until it) touch (ed) the (same) ... ball.

5. CONCLUSIONS

Difficulties inherent in the notion of "verbal" deep structure and the descriptive inadequacy of the *Aspects*-base compel the exploration of systems which map referential specifications into sets of verbal strings; one such system has been considered in the previous section. The principal difficulty involved is the systematic construction of referential formulae for individual objects, events, relations, attributes, one obviously contingent on an exhaustive investigation of the language-independent structure of such objects or events; in the technical sense, the problem is rather analogous to that of assembling the lexical entries in the present Aspects-system and no generative framework need be postulated for primitive formulae. Sentence-generation in such systems is viewed as mapping from referential specifications to sentential deep structures which in turn form input strings to the transformational component;

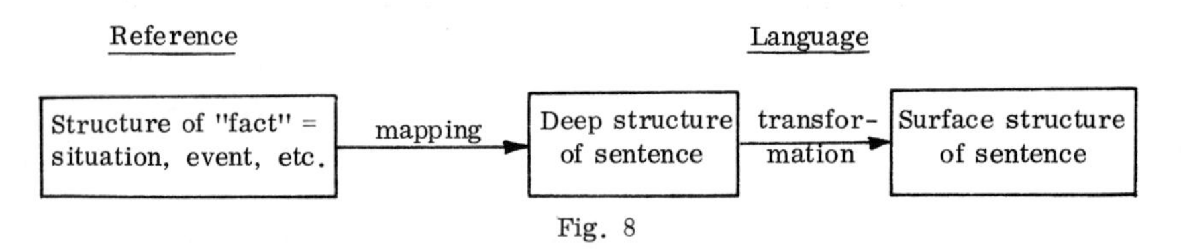

Fig. 8

The semantic component in the sense of Chomsky and Katz, is transformed into a reference-specifying, i.e. generative system, explicitly characterizing the structure of objects, events, etc., which underlie verbal communication.

6. REFERENCES

CHAPIN, F. (1967), "On the syntax of word derivation in English"; Ph.D. Dissertation, M.I.T., Cambridge, Mass.

CHOMSKY, N. and SCHÜTZENBERGER, M.P. (1963), "The algebraic theory of context-free languages"; In Braffort and Hirschberg (Eds.), "Computer programming and formal systems". Studies in logic series, Amsterdam.

CHOMSKY, N. (1965), "Aspects of the theory of syntax"; M.I.T., Cambridge, Mass.

CHOMSKY, N. (1967), "Remarks on nominalization"; Department of Modern Languages and Linguistics, M.I.T., Cambridge, Mass.

CLOWES, M.B., LANGRIDGE, D.J. and ZATORSKI, R.J. (1969), "Linguistic descriptions"; Seminar Paper No. 14; Division of Computing Research, C.S.I.R.O., Canberra, A.C.T.

CLOWES, M.B. (1969), "Picture syntax"; Seminar Paper No. 15; Division of Computing Research, C.S.I.R.O., Canberra, A.C.T.

DIK, S.C. (1968), "Coordination"; North-Holland Publishing Co., Amsterdam.

FILLMORE, C.J. (1966), "A proposal concerning English prepositions"; In Dinneen (Ed.), "Problems in Semantics, History of Linguistics and English". Monograph series on Language and Linguistics, No. 19, Georgetown University.

FILLMORE, C.J. (1967), "The grammar of hitting and breaking"; In "Working Papers in Linguistics". The Ohio State University Research Foundation, Columbus, Ohio.

KATZ, J.J. (1966), "The philosophy of language"; London and New York, Harper & Row.

KAYSER, S.J. (1968), Review of JACOBSON, S., "Adverbial Positions in English"; 1964, Stockholm, AB Studentbok. Language 44: 357-374.

LAKOFF, G. (1965), "On the nature of syntactic irregularity"; Ph.D. Dissertation, Harvard University, Cambridge, Mass.

LANGENDOEN, D.T. (1967), "On selection, projection, meaning and semantic content"; In "Working Papers in Linguistics", The Ohio State University Research Foundation, Columbus, Ohio.

McCAWLEY, J. (1967), "How to find semantic universals in the event there are any"; April 1967 Texas Conference Papers.

POSTAL, P. (1967), "Linguistic anarchy notes"; I.B.M. Yorktown Heights, N. York.

STANTON, R.B. (1969), "Plane regions: a study in graphical communication"; Seminar Paper No. 16, Division of Computing Research, C.S.I.R.O., Canberra, A.C.T.

ZATORSKI, R.J. (1968), "Grammatical relations and the structure of the categorial Aspects-type base"; To appear in Linguistics, International Journal, The Hague, Mouton & Co.

7. DISCUSSION

Clowes: In Figure 8, the right hand box I think I understand, but I am not quite sure whether when you use the phrase 'deep structure' it is referred to by that on the left - is that Chomsky?

Zatorski: No it is not Chomsky at all any more. I can call it a referential deep structure.

Clowes: But surely there is some sense in which anything in which T rules turn into surface structure ought to be called deep structure, if we are to retain some semblance of the Chomskian notation.

Zatorski: Well that's a notational point. I can accept that. Just to cover that point I can talk about referential specification or something of that kind.

Narasimhan: In note 13, the different senses of 'boy' - this specification that you give, does that go into the last box or into the middle box of Figure 8 ?

Zatorski: Into the first box.

Narasimhan: So you visualize that as being linguistic?

Zatorski: Yes, why not? I probably would not argue that this is semantic, any more, but in a sense semantics now mean something different to me from what they meant six months ago.

Narasimhan: I'm just trying to find out what is the meta symbolism there?

Zatorski: I'm trying to use in some way the same symbols that Clowes et al. have used, but I am finding great difficulty.

Bobrow: You really feel after working on this for a while, that what you really want to do is to separate out these various separate meanings, because it seems somehow there are a whole lot of properties that you are talking about - about boys - and sometimes you are talking about one, and sometimes another. It seems that probably you can construct sentences which focus attention on any number of these properties - the first three are all the same kind of boy, somebody's young, somebody's small physically, and somebody's not very smart yet, and maybe not very sure, and a few other such things for his age range. Somehow it seems that by just separating them out in this way, you are getting yourself into terrible combinatorial problems.

Zatorski: I agree. I am at this stage quite unsure; this is exactly the problem that Fillmore has when he asks himself, when he talks about break one, break two, break three, break four, or hit one, hit two, hit three, and he says that 'it can't be like this in language': we only have one word in language, so it can't be like this. I recognise the problem you are posing, but at this stage I cannot say whether it is a better procedure to get hold of relations first, and discuss this all on the basis of relations with argument places and then fill these argument places in, or whether I have to characterize each object and include every relation possible assignable over that object included in this description - I just don't know. I am exploring this.

Narasimhan: Can I make the suggestion - it may not be acceptable to you - but it seems to me that one can get around this problem, if one is willing to accept that the structuring of the referential deep structure is in fact done in terms of expressions rather than in terms of units like this. It seems to me to capture the whole lot of all these three cases.

Zatorski: No, I am after a neutral universal description.

Clowes: I am rather confused, you agree with Narasimhan when he says that the left hand box was linguistic.

Zatorski: No, not at all.

Clowes: I thought that was what you agreed.

Anonymous: It is referential.

Zatorski: As I said at the beginning, all I want to say is there's an insight here - the insight is that there are these varieties of boys and they have to be accounted for in different ways. There is a different insight when the similar boy is mentioned, that is, the similar boy is really a synonym on servant and servant has to be built upon a relation, servant or service, and the relation 'servant-service' presumably is defined on master and servant, and there are some conditions which specify this relation which I can't write down. Therefore I postulate here that one possible way of handling synonyms might be that if I had a relation like this, I can map it in such a way that I will substitute 'boy' in this situation for 'servant' in the specification.

Narasimhan: In this referential structuring that you have given an example, you use labels as specifiers?

Zatorski: The list property referring to human mind or something?

Narasimhan: Yes things like that. These are object labels, not meta-labels. Would you just use labels like object, attribute?

Zatorski: I would like to do that yes. I wouldn't want any list properties of that kind to be there - I used it solely as a hand hold to experience.

Langridge: Surely you can introduce attributes as values?

Narasimhan: So then they will be, as values that will belong to the specific object worlds and the question is, whether one can at all think of having a universal set of attribute values, that will cover all kinds of object worlds. Would you not think that it is necessary in this particular kind of language system - unless you allow yourself to let the language expressions themselves function as values? Then it is open-ended and you can do all kinds of things, when you have constructed some language, some portion of the language, you can use the expressions as attribute values of other portions of language, you can keep on doing it.

Bobrow: You don't need just the expressions, if you have any mapping for expressions into these things, you can use the compound objects also. So it's not a matter of staying on the level of expression. But you do want to let these compound things exist as values as well as simple ones.

Narasimhan: What of other compounds?

Bobrow: Suppose you have an expression which can in fact be analysed, like 'crushed Florida oranges'. I know its attributes, and we can now have a thing, which maybe its only English expression is 'crushed Florida oranges', so we can have inside a structure, which represents those attributes of 'crushed Florida oranges' that we want to replace.

Narasimhan: What I am talking about as expressions are themselves expressions represented as networks, so they are structured entities, not strings. I am quoting Lamont, that is all.

Bobrow: That wasn't clear to me before. I missed where you differentiated between the generation process and the analysis process. They seem to me to be so very separate kinds of things, and if you are worrying about the generation processes having something in the structure of fact reference, you don't worry about all these different kinds of 'boys' and 'hits' and 'balls' - you know what you are trying to say, so you can say it in any number of different ways, but you don't have to worry about all these multiple ambiguity problems. If you are going to try to analyse the sentence 'the boy hit the ball' it seems to me that in general you're not going to get that alone or in isolation, you can get it in a context in which this particular boy, the boy, (and that means the one you can be referring to when you know what kind of boy he is), you know the context of hitting the ball. So you probably won't get too much of that other kind of ambiguity. I understood you as saying you were in the middle and you were going both ways.

Zatorski: You want to look at this from the point of view of the hearer as against the speaker, this analysis and synthesis.

Bobrow: Isn't that what your arrows intend to imply? Chomsky says, 'start with the syntax and go down to the semantics and go out to the surface structure'.

Zatorski: I mean it in exactly the same sense in which Chomsky does, that is, I see generation as a sort of mirror process on analysis and vice versa. I would say I am neutral within the speaker and the hearer, in the same sense that Chomsky says he's neutral within the speaker and the hearer.

Bobrow: Well, then, do you feel that this is not a statement of a process which in fact people might carry out, rather than some kind of formal syntagmatic statement about this process? How do I know if I wish to describe some of the information there if it doesn't tell you how to carry the process through? Narasimhan has made a distinction between paradigmatic, which is the processes that we go through in some sense, and a formalization (syntagmatic) of those things which may state something about some of the information that's used in these processes. Now are you working on the latter formalization of the information?

Clowes: I think that in this business of neutrality which you have just expressed; clearly the idea of a neutral description as essential is that we are trying to characterize something about what you need to know in order to go either way. Now, among the kinds of things you'll need to know, are what sort of objects exist in what sort of domains, what sort of relations exist in these domains? Specifically you will need to know things like 'near', how is 'near' computed? It is inconceivable to me that irrespective of which way you are going, how in this synthesis you wouldn't need to know these things, and know something about their correspondences. Now, I'll grant you that in fact to go procedurally in either direction, you will almost certainly need to add to this base line of knowledge. The base line of knowledge which is what I think we have been trying to find, perhaps illadviseably, this base line is what I would want to think of as neutral description. I don't know whether Stanton would agree with it; I would guess that is what Zatorski is talking about when he is trying to take this neutral position which he characterizes as being that of Chomsky. Although I agree that Chomsky hasn't (taken a neutral position), you know, Chomsky's activities belie many of his words.

Bobrow: Then you are saying that this is not enough information yet, and to say we have problems about this ambiguity, somehow says that you're going beyond the stage of saying we have this information. You have to talk about techniques for adding information and worry more about the ambiguity problem. If you are going to stay neutral, then you want to talk about it neutrally and you don't want to worry about these processes.

Clowes: I would reiterate that there is a sense in which you can talk about neutral descriptions - it's exactly the one I have tried to clarify. It is inconceivable to me that this current knowledge would not appear in both the characterizations of both directions.

Narasimhan: It seems to me that one can in fact dispute whether one can meaningfully talk about what is to be known, except in the context of how it is actually made use of.

Clowes: As it is perfectly clear, Zatorski's gear for example is going one way, generating, it has been pointed out to me that my analysis of LETER is going the other way, is analysing - my point still stands. What has been teased out in terms of these two procedures is a common element, which will figure in a characterization of the reverse direction. That common element I have again underlined, is a neutral component.

Narasimhan: Yes, but the very fact that you have been trying to arrive at these common elements later in terms of performing analysis or in terms of performing a generation, already you see contradicts your stand. You are not arriving at anything in any noncontextual framework.

Clowes: We don't have a neutral discovery procedure, that I grant you, but part of what we discover we would claim to be neutral, in so far as it must figure in both directions.

Bobrow: I think that you can only claim that it really is neutral provided that you show that it is used in both directions and in the same form and that that is a reasonable form for both directions.

Clowes: I think that we are rubbished on the grounds that this neutral thing is meaningless - I would simply think of it as an open question.

Narasimhan: If you don't know how to assign meaning to it, it seems to me that it's meaningless.

Clowes: Well actually no, I think that's an open question. You don't have to assign meaning - if you can clearly show that no meaning can be assigned, then it is meaningless.

ON THE DESCRIPTION OF BOARD GAMES

M. B. CLOWES

Division of Computing Research
C.S.I.R.O., Canberra

1. INTRODUCTION

Game playing programs have occupied a central role in attempts to develop Artificial Intelligence. Among the considerations which may have prompted researchers to devote so much effort to this area we might list the following.

Situations in games like Chess and Go, can be, and usually are, very complex, calling for the capacity to interrelate many factors in deciding upon a play; dealing with complex situations effectively (i.e. promptly and flexibly) is generally thought to require intelligence. Bound up with our intuitions about intelligence is the concept of learning: different games provide graded environments in which learning strategies may be studied. Finally (in this list), there is a certain dramatization inherent in communicating and competing with a machine: the language mediating this interaction - discrete states of the board - is trivially automated by comparison with situations in which communication is expressed in a natural language like English, or some pictorial language say flowcharts or topographic maps.

Our interest here lies in the possibility of illuminating games by formulating them in syntactic terms. The idea that this might be possible and even rewarding arises from the observation that the basis of a game, specifically (but not exclusively) a board game, is relational. Thus in Chess and Checkers, opposing pieces may be juxtaposed so as to attack (i.e. threaten to capture) one another. Similarly, like pieces may defend one another. Defence or attack are not necessarily symmetrical between the two pieces involved. These relations belong to the game domain, and are essentially primitive relationships. They are expressed pictorially in terms of more or less complex geometrical relations and objects which may differ according to the piece (e.g. Pawn vs Knight, Man vs King) which is attacking or defending. In evaluating a board state it seems clear that relationships are grasped between groups of pieces as well as single pieces, indeed in Games such as Go Moku (in which 5 collinear contiguous pieces of the same colour defines a win, relations involving pairs and triples of pieces are crucial to any evaluation. A syntactic account of a game would therefore be in terms of relations between these more or less complex objects. We would expect that the recovery of these syntactic structures would be crucial in guiding play, in much the same way as the recovery of sentential structure is crucial to communicating in a natural language.

This 'linguistic' viewpoint has not been an overt feature of game playing programs, indeed there are some grounds for believing that the descriptive techniques employed in many programs are close to those employed in that approach to pattern recognition which treats it as a partitioning of a signal space. Recently however, Elcock and Murray (1967, 1968) and Koffman (1968) have developed an approach to a restricted class of games (including Go Moku and Noughts-and-Crosses) which appears to deploy an underlying syntactic model.

In the paper we will characterize the principal elements of the 'traditional' approach to game playing programs with a view to substantiating the above. The more recent developments will be outlined and couched in the syntactic notation introduced earlier (Clowes, 1969). The areas of deficiency and the possible extension to Draughts will complete the paper.

2. THE 'TRADITIONAL' APPROACH TO GAME-PLAYING PROGRAMS

In a board game like Draughts the complete history of a game may be characterized as a sequence of board states related by legal moves made alternately by each player. Since any board state will in general present many possible legal moves, this sequence is essentially a path through a tree whose nodes are board states and whose branches (or links between nodes) are legal moves. This *moves tree* is the central representational device employed in traditional game-playing programs and also figures in a number of cognate Artificial Intelligence topics, e.g. theorem proving (Newell and Ernst, 1965). For Draughts a terminal board state is one in which no legal move is possible for the player whose turn it is to move (neglecting drawn games). That player *loses*: the situation may arise because the player in question has no man, or there are no squares to which he can move them. Given a representation of the board and some means of recognising legal moves in terms of that representation, the problem for a game playing program may be expressed as that of finding some path through this tree which terminates on a winning board state. This exploration is termed look ahead. It has been estimated (Newell & Simon, 1964) that in an average Draughts game there are 10 legal moves available for each board state in such a tree and about 70 moves in a game. Thus the tree is very large (70^{10} paths) and an exhaustive examination of every path (complete look-ahead) is not feasible. Procedures are required which may be applied by the program so as to limit the amount of the moves tree which has to be explored. To the extent that these procedures restrict the search to parts of the moves tree which are advantageous to the program's play they are useful. It is a characteristic of heuristic procedures that they do not in general guarantee this - they are at best rules of thumb. Restricting the look-ahead distance is the tree-pruning heuristic employed by Samuel (1959) in his Draughts-playing program. Briefly the program (when it is the program's turn to move) develops all legal moves to a depth of three, i.e. Program-Opponent-Program, and makes no further development from any board state unless one of the following conditions occur: (1) The next move is a jump; (2) the last move was a jump; or (3) an exchange is possible. Where any one of these conditions is met, development continues subject to increasing strictures and an eventual limit for any one development of 20 moves. In a subsequent paper (Samuel, 1967) additional and more sophisticated tree-pruning heuristics have been added, In general look-ahead always terminates well short of a board state which is terminal for the game. The program must then make some comparison of the board states it has looked at in order to choose between the alternatives open to it. At the same time it must take into account the opponent's likely response(s) to any move or series of moves, since we must assume that the opponent will in general oppose any line of play advantageous to the program.

Comparison of non-terminal board states is effected through the application of an evaluation function. In Samuel's (1959) case this is a polynomial whose terms (or parameters) reflect what are considered to be significant features of the board. Thus

"CENT (Centre Control I)

The parameter is credited with 1 for each of the .... (centre) squares which is occupied by a passive man". The passive side is the one which moved to reach the board state which is being evaluated. Terms have coefficients which may be set by the programmer or manipulated by a 'learning' procedure so as to give more or less weight to different aspects of the board state.

Thus the program is able to assign values to those board states at which look-ahead terminated. In order to select a move it is necessary to assign values to the board states which are just one move ahead of the state from which look-ahead was initiated. That is if S_{oo} (Fig. 1) is the present state, from which look-ahead is initiated, states

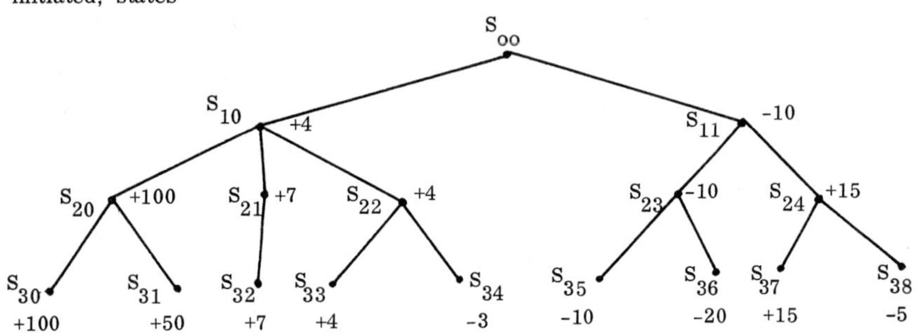

Fig. 1

S_{30} - S_{38} will have values assigned to them. However, move selection will have to be based upon a comparison of state S_{10} with state S_{11} : between them and S_{30} - S_{38} the opponent has a move. If we assume that he deploys the same form of the evaluation function then we should adopt a pessimistic view of his reply namely that he will select a board state to which we assigned a minimal value. Where we make the selection we opt for our most highly valued state. This assumption about our opponent provides us with a minimax procedure by which to 'back up' the scores assigned to the terminal states according to

(a) if $(\text{board state})_{p+1}$ is attained by the opponent's move then $(\text{board state})_p :=$ $\min\,(\text{board states})_{p+1}$; or

(b) if $(\text{board state})_{p+1}$ is attained by own move then $(\text{board state})_p := \max$ $(\text{board states})_{p+1}$.

Thus in the illustrative fragment of the look-ahead tree depicted in (Fig. 1) we see that S_{10} is preferable to S_{11} since their minimaxed values are +4 and -10 respectively.

The introduction of the evaluation function is in a sense a reflection of an underlying heuristic namely that where some goal, i.e. complete look-ahead, cannot be achieved, the problem is restated in terms of a different, but hopefully equivalent, goal. The rules of the game provide only evaluation criteria for terminal board states; where we cannot reach these, we must substitute alternative criteria (an evaluation function). In so doing we are in effect changing the game. The problem of course is to devise alternative criteria which accurately reflect the true goals as defined by the rules of the game.

3. A PATTERN-ORIENTED APPROACH

Go Moku is played on a 19 x 19 square mesh. Player b (w) has a supply of indistinguishable black (white) pieces. The players take it in turns to play a piece on a lattice point. The winner is the first to complete a (horizontal, vertical or diagonal) line of five and only five adjacent pieces of his colour. (From Elcock & Murray, 1967). Elcock and Murray illustrate the essential features of the game with an example which we reproduce in Fig. 2. White wins with move 21. Black's last four moves 14, 16, 18, 20 are all forced upon him as he tries to avoid defeat. They respond to different patterns of white pieces: the diagonal of three created by 13, the horizontal of four created by 15, the vertical of three created by 17, the (winning) diagonal of four created by 19. Elcock and Murray (1968) illustrate forcing patterns of considerable complexity. In general, the significant forcing patterns are those formed by the overlap of two potential 5-patterns. These configurations present the possibility of creating two forcing patterns with a single move because the piece played in that move is an element of both patterns. 17 is just such a move in Fig. 2; typically Black is now unable to block both the diagonal through 17 and the vertical. The lattice point at which 17 is played is a common node of the two potential 5-patterns. Just as a winning pattern may be created by playing at a common (own) node, it may also be destroyed by an opponent's play at that point. Thus both defensive and offensive play may be formulated in terms of recognition of common nodes as well as the simpler types of forcing pattern (e.g. the horizontal pattern created by 15). Elcock and Murray (1968) have developed a program which analyzes the board to recognise a wide variety of forcing patterns and selects moves in terms of the patterns it finds. The program is claimed to play at the level of 'expert'.

Thus the program would make the sequence of forcing moves 15, 17, 19 by recognising these as positions in forcing patterns; equivalently it would be com-

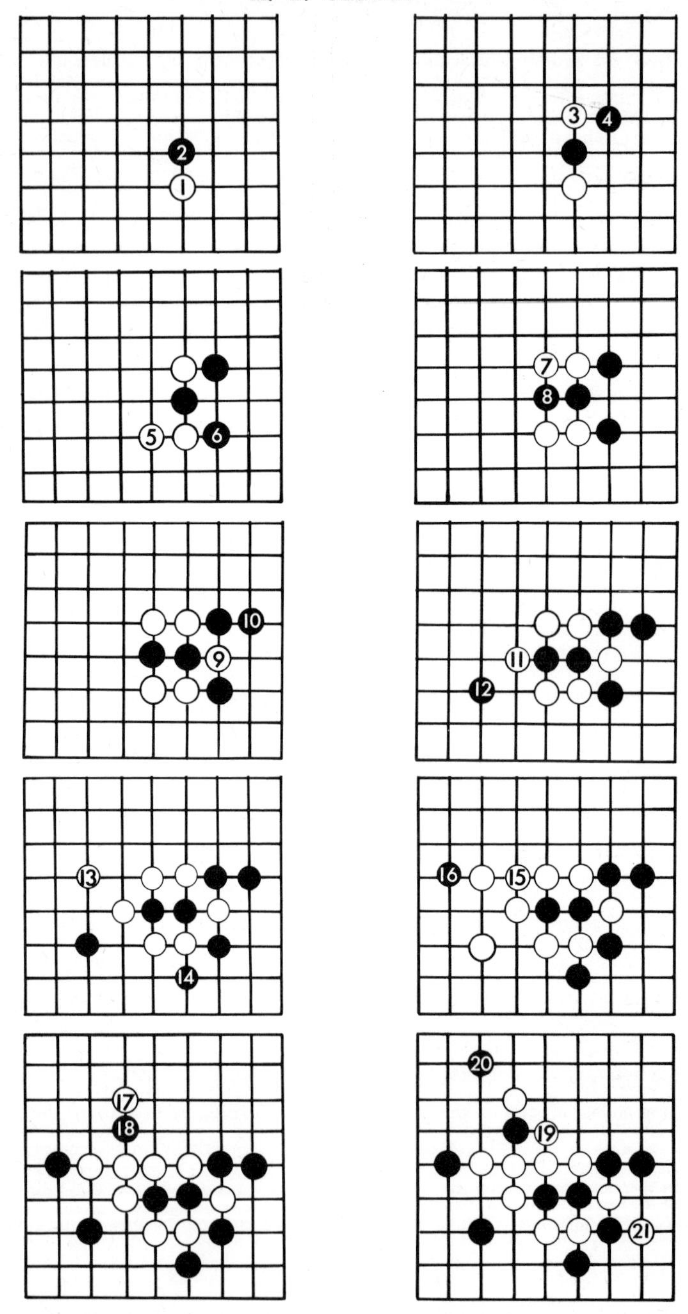

Figure 2. Example from GO - MOKU.

petent to defend at 14, 16, 18, 20.

This is however a purely analytic view of the play. The crucial aspect of play is surely the synthesis or creation of forcing patterns. The development by white in the example above does not bring this out because the moves which lead to the creation of the winning force are themselves components of other forcing patterns. The account given by Elcock and Murray (1967) of move selection where no forcing patterns (self or opponent) have been recognised is: "A default procedure is called to select a move. The default procedure may use quite different criteria or no criterion at all (e.g. selection using a random number generator)". Thus they fail to characterize the creative phase of the game.

4. THE SYNTHESIS OF FORCING PATTERNS

One rather obvious 'default procedure' would utilise the knowledge of the geometry of forcing patterns to select moves which would facilitate their synthesis. In the absence of any such constructive approach, forcing patterns can arise only by accident although when they do we would expect the program to be inhumanly efficient at detecting them. Such a program might be likened to Dickens' famous character Mr. Micawber - 'always waiting for something to turn up'. We can examine the difficulties involved in designing a creative 'default procedure' by studying play in a restricted version of Go Moku namely Noughts and Crosses (Tic-Tac-Toe). The sequence of moves leading to the board state illustrated in Fig. 4 was X: 3; 0:6; X: 1; 0: 2; X:7 (using the numbering convention of Fig. 3). 0 has lost because he cannot block both X's forcing patterns on 7, 5, 3 and 7, 4, 1. Thus the move X: 7; is, like move 17 in Fig. 2, play at a common node. The forcing pattern of which 7 is a common node is comprised of a pair of triples (3, 5, 7; 1, 4, 9). There are two minor variants of this 'fork' pattern namely those in which the common node is 5, or one of the set 2, 4, 6, 8. That is the fork patterns are pairs of triples overlapping (i.e. having a common node) at a corner, (corner fork) in the centre (centre fork) or on the side (T fork). Noughts and Crosses is 'simple' by comparison with Go Moku not only because the game is

Fig. 3

1	2	3
4	5	6
7	8	9

Fig. 4

X	0	X
		0
X		

limited to a maximum of 9 moves but also because the variety and complexity of forcing patterns is so limited.

0's loss of this illustrative game seems due to the weakness of his first response 0: 2. Not only does it leave most of the board 'open' for X to develop his winning force, it also provides 0 with little potential to do any counter-forcing. Thus the only possible development of his move is as the line 4, 5, 6. By contrast 0: 5; would have given 0 three possible lines: 4, 5, 6; 1, 5, 9 and 2, 5, 8 as well as 'cramping' X a good deal.

While X is clearly in a good position at this stage X: 3; 0: 6; in that with his next (second) move he can create a common node of a corner fork pattern - indeed he has four corner fork patterns to choose from - judgement is required. Specifically the ordering of his second and third moves reflects the fact that in general 0's possible defenses include both play at X's common nodes and/or preventing X from playing at them. The latter prospect arises whenever 0 can mount a simple force such as he would create, X permitting, were his next (second) move to be 0: 5; or 0: 4. Had X's second move been X: 7; 0 would have been forced to reply 0: 5; in order to block X's line 7, 5, 3. In so doing he would force X to make his next move at a position other than his own common node. The move X: 1 which creates a common node at 7 and a simple force on 0 at 2 avoids this dilemma. Thus this simple game illustrates that moves may have more than one role simultaneously e.g. 0:5; would have been both 'cramping' to X and a useful development for 0, while the move X: 7; if made too early would have made 0's response (0:5;) both an adequate defence of the simple force and an adequate defence of the force represented by X's corner fork.

This partial analysis of the game brings out some of the distinct types of move, e.g. developmental, defensive, forcing, as well as the fact that move and response may be interrelated very strongly. In general the significance of a move is to be seen only in terms of the patterns that it develops and/or disrupts. Development is another way of talking about creative play. Thus those moves which are developmental belong to the creative phase of play, i.e. to the 'default procedure'. If a program were able to grasp the play in this game in a way which exemplified the features of our partial analysis, and specifically those features having to do with 'development', then it might (given other facilities too) be able to play 'creatively'. To provide the program with such a grasp it is necessary to show that our partial analysis can be expressed in a formal language. Such a language can then become the vehicle through which the program grasps the play.

In the next section we attempt such a representation of board patterns and their development. The language used is, as in characterizing picture organization (Clowes, 1969) one which distinguishes between objects, relations and attributes. The objects, relations and attributes introduced are restricted to those necessary to characterize the particular choice open to X for his second move, which was discussed above.

5. TOWARDS A FORMAL NOTATION FOR DESCRIBING PLAY

In general terms we see the game as a moves tree in which the program will carry out look-ahead. The nodes of this tree are not however simply board states, but analyses of board states in terms of the program's knowledge of forcing patterns. Similarly the branches of the tree are not merely legal moves but moves viewed as relating these analyses of the board states.

In analyzing a board state we will distinguish between the geometrical or spatial relationships between positions on the board, and the location of pieces at those positions. Thus denoting a position as a primitive object P we will designate the occupancy of such a position by, say, X in the form of a relation:

(1) $\underline{\text{at}}\langle P, X\rangle$

There appear to be two types of spatial relationships necessary to the specification of useful geometrical objects. These are the nearest neighbour (Nn) relation, and the four directional relations: 'horizontally to the right of' r, 'vertically below' b, 'diagonally below and right' dr, 'diagonally below and left' dl - all regarded as primitive relations. We will treat the latter as values of the relation direction.

Thus positions 1, 2 in Fig. 3 enjoy the following relations:

(2) $\underline{\text{direction}}\langle P, P\rangle[\underline{\text{r}}]$ and

(3) $\underline{\text{Nn}}\langle P, P\rangle$

We may define the relation collinear in terms of direction:

(4) $\underline{\text{collinear}}\langle \underset{1}{P}, \underset{2}{P}, \underset{3}{P}\rangle \subset \underline{\text{Eq}}\langle \underline{\text{direction}}\langle 1, 2\rangle, \underline{\text{direction}}\langle 2, 3\rangle\rangle$

With these relations we may now specify the varieties of pictorial object necessary to the characterization of board organization. Let us call the pictorial object which underlies a win, a TRIPLE:

$$(5)\quad \text{TRIPLE}\left\langle \underset{1}{\text{P}}, \underset{2}{\text{P}}, \underset{3}{\text{P}}\right\rangle \left[\underline{\text{collinear}}\langle 1, 2, 3\rangle, \underline{\text{Nn}}\langle 1, 2\rangle, \underline{\text{Nn}}\langle 2, 3\rangle\right]$$

Notice that we may define an attribute of a TRIPLE such as end:

$$(6)\quad \text{end}\left\langle \underset{1}{\text{TRIPLE}}\right\rangle \left[\underset{2}{\text{P}}\right] \subset 1\langle \text{P}, \text{P}, 2\rangle, \text{ or } 1\langle 2, \text{P}, \text{P}\rangle$$

The positions occupied by a corner fork form an ANGLE.

$$(7)\quad \text{ANGLE}\left\langle \underset{1}{\text{TRIPLE}}, \underset{2}{\text{TRIPLE}}\right\rangle \left[\underline{\text{Coinc}}\langle \text{end}\langle 1\rangle, \text{end}\langle 2\rangle\rangle\right]$$

Its apex may be defined:

$$(8)\quad \text{apex}\left\langle \underset{1}{\text{ANGLE}}\right\rangle \left[\underset{2}{\text{P}}\right] \subset 1\left\langle \underset{3}{\text{TRIPLE}}, \underset{4}{\text{TRIPLE}}\right\rangle, \underline{\text{Eq}}\langle 2, \text{end}\langle 3\rangle\rangle, \underline{\text{Eq}}\langle 2, \text{end}\langle 4\rangle\rangle$$

TRIPLE and ANGLE are patterns of positions. Patterns of pieces, e.g. cornerforks, arise when pieces are placed <u>at</u> specific positions <u>on</u> certain patterns. The definition of position patterns like TRIPLE and ANGLE requires us to specify in detail their component positions; all such definitions have the general form:

$$(9)\quad \text{OBJECT}\langle \{\text{P}\}\rangle \left[\text{relationships}\right]$$

i.e. as a set $\{P\}$ of positions, between which specified spatial relationships hold.

We may utilise this generalization to define <u>on</u>:

$$(10)\quad \underline{\text{on}}\left\langle \underset{1}{\text{OBJECT}}, \underset{2}{\text{X}}\right\rangle \subset 1\left\langle \underset{3}{\{\text{P}\}}\right\rangle, \underline{\text{at}}\left\langle \underset{4}{\text{P}}, 2\right\rangle, \underline{\in}\langle 4, 3\rangle$$

where $\underline{\in}$ is the relation of set inclusion.

Thus the piece X is <u>on</u> the object 1 if the position (subscripted 4) <u>at</u> which X is located is a member of the set of positions (3) belonging to the object.

The pattern immediately preceding a win which we will call a FORCE is a TRIPLE with two pieces played upon it:

$$(11)\quad \text{FORCE}\left\langle \underset{1}{\text{TRIPLE}}, \underset{2}{\text{X}}, \underset{3}{\text{X}}\right\rangle \left[\underline{\text{on}}\langle 1, 2\rangle, \underline{\text{on}}\langle 1, 3\rangle\right]$$

The empty position of a FORCE is an attribute of FORCE:

$$(12)\quad \text{empty}\left\langle \underset{1}{\text{FORCE}}\right\rangle \left[\underset{2}{\text{P}}\right] \subset 1\left\langle \underset{3}{\{\text{P}\}}, \text{X}, \text{X}\right\rangle, \underline{\in}\langle 2, 3\rangle, \underline{\text{at}}\langle 2, \text{NIL}\rangle$$

The cornerfork pattern takes the form:

(13) $\mathrm{CNRFK}\langle \underset{1}{\mathrm{ANGLE}}, \underset{2}{X}, \underset{3}{X}\rangle\left[\underline{\mathrm{at}}\langle \mathrm{apex}\langle 1\rangle, \mathrm{NIL}\rangle, \underline{\mathrm{on}}\langle \mathrm{triple}\langle 1\rangle, 2\rangle, \underline{\mathrm{on}}\langle \mathrm{triple}\langle 1\rangle, 3\rangle\right]$

where triple is a structural attribute of CNRFK identical with a part (a TRIPLE) of the ANGLE on which the CNRFK is based.

Piece patterns (and specifically forcing patterns) thus take the general form:

(14) $\mathrm{OBJECT}\langle \underset{1}{\{P\}}, \underset{2}{\{X\}}\rangle\left[\text{relationships between elements of 1 and 2}\right]$

i.e. objects comprising a set of positions $\{P\}$ and a set of pieces $\{X\}$.

We have suggested that creative play is concerned with the development of forcing patterns such as 11 and 13. The pattern from which a forcing pattern (or any other piece pattern) is developed will be called its predecessor. We can give a very general characterization of predecessor using the generalized definition of piece pattern stated in (14). Let us treat the pattern and its predecessor as the arguments of a relation dev (meaning development) whose value is taken to be the played piece which creates the pattern from its predecessor*:

(15) $\underline{\mathrm{dev}}\langle \underset{1}{\mathrm{PREOBJ}}, \underset{2}{\mathrm{OBJ}}\rangle\left[\underset{3}{\underline{\mathrm{at}}}\langle P, \underset{4}{X}\rangle\right] \subset 1\langle \underset{5}{\{P\}}, \underset{6}{\{X\}}\rangle\left[\underset{7}{\{\text{rels on 5 and 6}\}}\right],$

$2\langle 5, \underset{8}{\{X\}}\rangle\left[\underset{9}{\{\text{rels on 5 and 8}\}}\right],$

$\underline{\mathrm{Eq}}\langle 4, \mathrm{difference}\langle 8, 6\rangle\rangle,$

$\underline{\mathrm{Eq}}\langle 3, \mathrm{difference}\langle 9, 7\rangle\rangle,$

Thus $\underline{\mathrm{at}}\langle P, X\rangle$ in 15 is the play needed to develop the OBJ labelled 2 from the PREOBJ labelled 1. Given any definition of an OBJ, e.g. CNRFK, 15 characterizes the immediate predecessor of that OBJ.

Specifically, by substituting the definition of the OBJ into the r.h.s. of 15 (as the string beginning $2\langle 5 \ldots\ldots)$, the admissible values of 1 and 3 may be obtained. The board can then be analyzed to find instances of 1, thus identifying possible developmental plays (the related values of 3). Such a board analysis is then merely an extension of the analysis procedure outlined in Section 3, in which

* Alternatively, regard dev as a predicate and $\underline{\mathrm{at}}\langle P, X\rangle$ as an attribute of one or other piece pattern. The formulation given here is just a convenient abbreviation.

only forcing patterns and not their predecessors are searched for.

Selecting the 'best move' from the set of possible developmental plays thus involves a knowledge of

(a) the forcing patterns of the game;

(b) the relation dev; and

(c) the two forms of defensive move:

(i) playing at an opponent's forcing positions;

(ii) forcing him to play at other than his own forcing positions.

As characterized earlier, play at forcing positions develops (or blocks) two forces simultaneously. That is it is a special case of dev where the PREOBJ is a forcing pattern and for at least one value of $\underline{\text{at}}\langle \text{P}, \text{X}\rangle$, OBJ has two values which are both forcing patterns or where OBJ has the single value: a winning pattern.

Thus we might define forcing position for the first case as an attribute:

$$(16)\quad \text{force}\underset{1}{\langle \text{OBJ}\rangle}\underset{2}{[\text{P}]} \subset \underline{\text{dev}}\langle 1, \text{OBJ}\rangle[\underline{\text{at}}\langle 2, \text{X}\rangle]$$

which may be read: 'a force is a position in an object play at which develops that object'.

An adequate defence of type 1 is a move which blocks such a development and we can treat it as a relation - def 1:

$$(17)\quad \underline{\text{def1}}\langle \underset{1}{\underline{\text{at}}}\langle \text{P}, \text{X}\rangle, \underset{2}{\underline{\text{at}}}\langle \text{P}, 0\rangle\rangle \subset \underline{\text{dev}}\langle \text{PREOBJ}, \underset{3}{\text{OBJ}}\rangle[1], \text{force}\langle 3\rangle[2]$$

The second type of defence (c(ii)) can also be characterized as a relation between two moves:

$$(18)\quad \underline{\text{def2}}\langle \underset{1}{\underline{\text{at}}}\langle \text{P}, \text{X}\rangle, \underset{2}{\underline{\text{at}}}\underset{3}{\langle \text{P}}, 0\rangle\rangle \subset \underline{\text{dev}}\langle \text{PREOBJ}, \underset{4}{\text{OBJ}}\rangle[1], \underline{\text{dev}}\langle \text{PREOBJ}, \text{OBJ}\rangle[2], \underline{\text{Eq}}\langle 3, \text{force}\langle 4\rangle\rangle$$

We may now use this apparatus to give a formal characterization of the ordering of X's second and third moves, given in Section 4. Fig. 5 sets out the relations which are evaluated by X and exhibits the positions identified by each evaluation. Thus (a) and (b) in Fig. 5 illustrate positions satisfying two types of dev, and (c) those positions common to both. Considering only such positions, restricts 0's defences to be of type (ii) only, and hence we need examine only possible developments of FORCE's by 0. This is exhibited in (d). Only one of X's moves identified in (c) can be defended by 0 with the dev of (d). This is given in (e). This leaves X with plays at X: 1 (as discussed in Section 4) and also at X: 5 illustrated in (f) and (g) respectively. For each of these 0 has no play for which def2 is true.

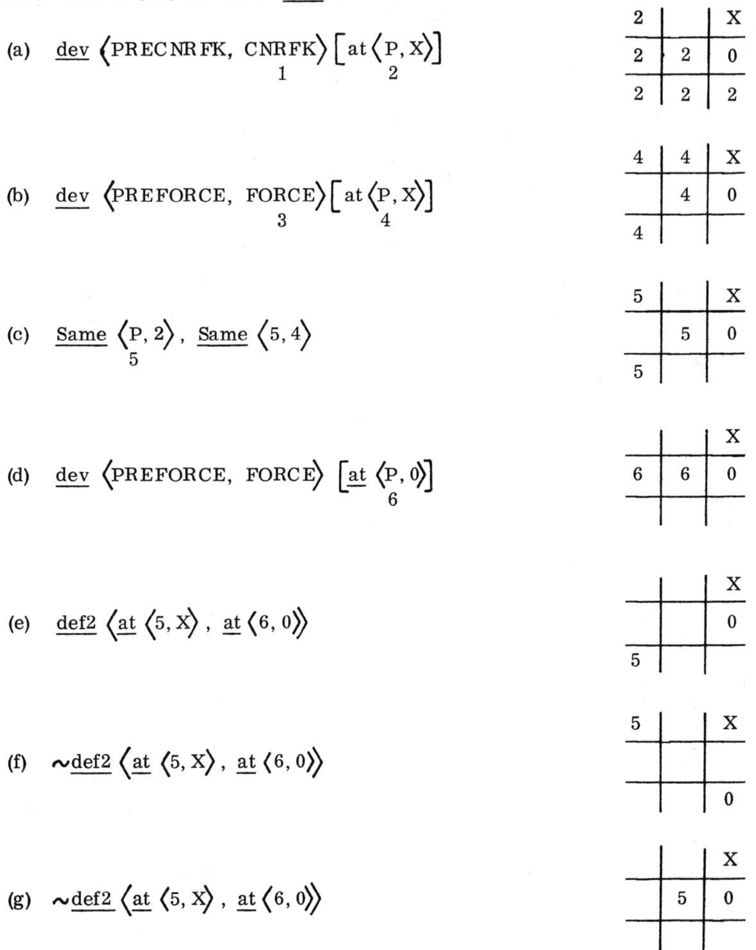

Fig. 5 - Formal characterisation of X's second move.

This is of course only a partial formalization since it omits the problem of selecting the relevant relations to be evaluated and we would require this characterization to be embedded in some procedural description. Among the additional knowledge which would be embodied in such a procedure we might list:

(1) A characterization of the value of moves achieving more than one development (X:1, X: 5, X: 7).

(2) The relevance of depth of development, i.e. ANGLE is PREPRECNRFK.

(3) Related to 2 the restriction of evaluations of the opponent's def2 responses to those which achieve development in fewer moves than the own development under consideration.

What we have done here is to identify the way in which the program should 'look' at the board to achieve 'creative' play. (Refer also Appendix, Section 9.)

6. OTHER GAMES AND OTHER APPROACHES

It is hard to draw conclusions of a general kind from this fragmentary characterization of a game usually regarded as trivial. Our objective is not however to design a universal game playing program, but to use the characterization developed in Section 4 and Section 5 to illuminate other games and other approaches. Noughts and Crosses is a special case of the class of board games (see Koffman, 1968) of which Go Moku is a member. The recovery of forcing patterns is, as noted previously, the basis of an existing 'expert' program: our account above differs only in that it described simpler situations and uses a notation which explicitly identifies objects, relations and attributes. The notation used in the 'expert' program by Elcock and Murray (1968) provides a highly edited description of objects and no characterization of relations (either pictorial or game relations). Use of attributes is restricted to (1) number of pieces played in each line of the pattern; (2) number of plays required to create a win with the pattern. One weakness of their notation lies in its inability (because most of the object is omitted) to provide for the expression of interrelations between objects, e.g. an OBJ and its predecessor. Insofar as their approach to play does not call for the identification of such relationships, this weakness is irrelevant. To the extent that such relationships are crucial to effective play, the weakness may be crippling. We conclude therefore that there is no reason to believe that the account of Noughts and Crosses given above would not 'transfer' to Go Moku, where the deployment of a weaker version of it has already yielded an 'expert' program.

The account (in Section 2) of 'traditional' approaches to game playing centred upon Draughts. What evidence is there for thinking a syntactic approach may illuminate that game? Hopper (1956) gives many illustrations of forcing patterns in

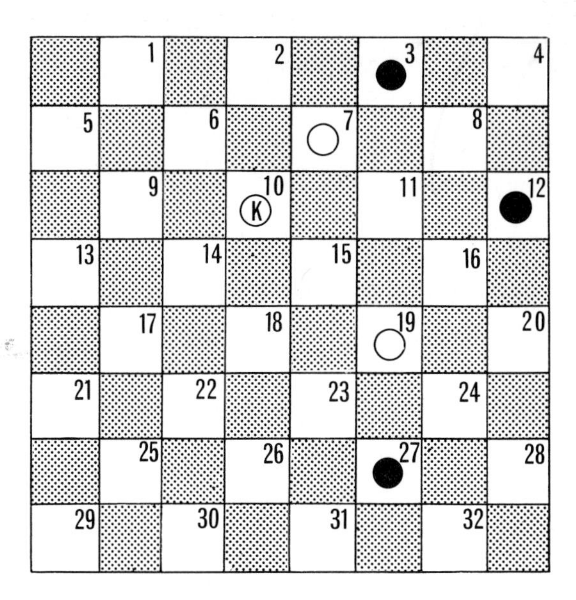

Figure 6. White in a winning position.

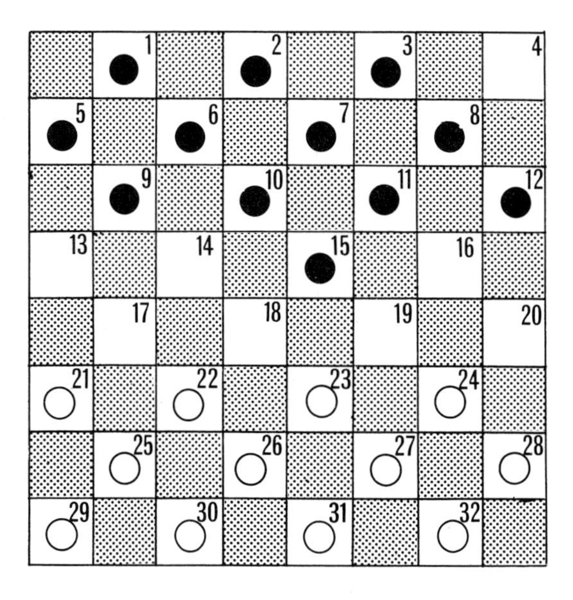

Figure 7. Formation of a centre wedge.

Draughts. Thus in Fig. 6 (Hopper 1956, Diag 36) White playing up the board is in a winning position. He makes the two forcing moves (no 'huffing' allowed) 19-16 and 10-6 to 'clean up' his opponent. Developing these profitable exchanges is a common feature of play.

Elsewhere Hopper, in discussing the opening game, outlines patterns of a strategic nature, e.g. formation of a centre wedge, (Fig. 7), which provide a base from which to develop forcing patterns. The principal distinction between Draughts and Go Moku may be said to lie in the fact that achievement of a forcing pattern in Go Moku guarantees a win, in Draughts it may give no more than a temporary advantage.

If we make these tentative identifications it is possible to compare the 'traditional' with the 'syntactic' approach.

6.1 Searching the Moves Tree

In Section 5 we characterized moves as relating - developing or defending - structured objects. Given that we see the current board state as PRECNRFK, only certain moves are relevant to its development. For the purposes of development all others may be ignored. Thus, if we recognise the spatial relation in Fig. 6 of the King on 10 to the man on 27, we need only consider those (forcing) moves by white necessary to create the intervening two jumps between the King and the man (i.e. 6-15, 15-24) implicit in this spatial relationship. White has four other legal moves at this point most of which go unnoticed.

The idea of restricting the moves to be evaluated according to some plan was described by Newell, Shaw & Simon (1958). The analyses of the board situations carried out to initiate the proposal of moves-to-be considered was however restricted to broad numerical and positional evaluations, into which all the material on the board would contribute, e.g. 'Material Balance'. Some indication that searches of the moves tree should be constrained by structural analyses of the board is one interpretation of the introduction of "plausibility analyses" as a heuristic to prune the growth of the moves tree in the evaluation phase (Samuel 1967, Greenblatt et al. 1967). The absence of relevant constraints on the development of the moves tree shows up in an aimless pattern of play. Samuel (1967) acknowledges this deficiency (op. cit p. 616) and accepts that the program needs 'deep objectives', a phrase readily interpreted from the standpoint developed in Section 5.

6.2. Evaluation Functions

The device which replaces board structure in 'traditional' approaches is the evaluation function. Even here we find an implicit acknowledgement of the importance of relationships in that approximately half of Samuel's parameters have to do with the identification of relationships between pieces, e.g.

"FORK

The parameter is credited with 1 for each situation in which passive pieces occupy two adjacent squares in one row and in which there are three empty squares so disposed that the active side could, by occupying one of them, threaten a sure capture of one or the other of the two pieces".

In general it seems fair to conclude that the evaluation function is deficient in much the same way as 'signal space' is defective as a basis for recognising patterns. Namely, that it fails to provide that articulation of the object (or pattern) necessary for its manipulation.

6.3. Minimaxing

This expresses the program's assumptions about its opponent. However, what we know about an opponent is often quite specific - that he appears to be developing a certain kind of attack. It is hard to see how the 'traditional' apparatus could cope with this. The analysis given in Section 5 however is wholely concerned with its characterization.

7. CONCLUSIONS

These comparisons are of course speculative in the sense that their validity could be tested only by developing programs along the lines indicated in Section 5. It does seem however that a certain aspect of the problem of complexity - specifically of board games - has been neglected and that this omission can be seen to be related to the central ideas (Moves tree, evaluation function, minimaxing) around which these programs revolve. Crucial to any reformulation will be the definition of a powerful descriptive language having much in common with the languages used to describe syntactic structure in other domains, e.g. English (Clowes et al., 1969) and Pictures (Clowes, 1969). The overlap may arise through the necessary use in a board games language of relationships, e.g. Nn, developed as part of a more general picture language. More generally however we would assume that the overlap would occur in the type of notation used. Thus all languages descriptive of syntactic structure must provide for the characterization of objects, relations and attributes.

Formulating games in terms of syntactic models provides a new standpoint from which to discuss complexity, learning and man/machine dialogue. These three topics were singled out earlier (Section 1) as among the primary motivations for work on game playing.

Newell, Shaw and Simon (1958) relate human mastery of complexity to linguistic abilities " ... given a more powerful language, we can specify greater complexity ... ". Elsewhere Newell & Simon (1964) imply that complexity is linked with having to choose between vast numbers of alternative actions or sequences of actions. This, formulated as a tree-search problem, is discussed primarily from the standpoint of ways of reducing the tree. In terms of board games this moves tree is a characterization of the very large number of board states addressed by legal continuations. Heuristics tend to say what we can ignore. A creative approach on the other hand visualises only those board states which exhibit desired patterns and generates only that fragment of the moves tree containing them. We have shown how the description of the board state in terms of desired patterns permits an interpretation of a move as having more than one role. It appears that this requires a considerable degree of articulation in the description of board patterns, which rules out the use of evaluation functions as pattern descriptions.

The stock of board patterns a player has is used both creatively and analytically. In this sense it is a representation of the player's 'competence' analogously to Chomsky's (1957) characterization of linguistic ability. Thus the idea of dialogue implicit in game playing achieves a new significance. The communication is in terms of patterns realised as board states: a board state may, and in general will, contain many patterns not all of which are intentional or active. We may liken 'deep objectives' to patterns which are effectively disguised by the player: to this extent the dialogue has diametrically opposed purposes to that normally assumed in verbal behaviour - it is intended to obscure meaning rather than to illuminate it.

Elcock and Murray (1968) describe a variety of learning in which the program acquires a 'stock' of patterns by analyzing the games it loses. Crucial to this process is the existence of an appropriate language representing 'nodes', 'line patterns', etc., in which to formulate descriptions of the board. Narasimhan (1966) has suggested that 'real intelligence' consists in devising this descriptive language. This evidently has two components: specific and general. The latter has to do with the form of the metalanguage. Thus we have utilized certain kinds of expression, e.g. name⟨ ⟩[ ], and more complex ones too, within which we formulate descriptions. The specific issues on the other hand have to do with what particulars appear in the metalanguage, i.e. the particular objects, relations and attributes that apply to a particular game.

Chomsky (1968) believes that the form of a syntactic metalanguage is innate: acquisition of language has to do with the particular objects (syntactic categories, e.g. S, NP) and attributes (features, e.g. ± Animate) which characterise a particular language. There seems some ground for thinking that a metalanguage which provides for the characterization of objects, relations and attributes perhaps along the lines developed here and elsewhere (Clowes, 1969) may apply over several domains of interest. Such a metalanguage is an essential component in Chomsky's (1968) idea of universal grammar.

The framework in which this analysis of a board game has been presented has been that of devising a computer program to play that game. There is however a different standpoint from which to view this attempt at formulation. Namely that as competent players of any game, we know a great deal about the meaning of various moves, the significance of various configurations on the board and so on. Such knowledge appears on the face of it to be of the same general character as that knowledge of a natural language for which Chomsky has attempted to give a formal characterization. Such a characterization of our intuitions about a game in terms which are neutral as to whether it is a machine or a man being characterized must yield some insights about the topics outlined in the Introduction. Whether these insights lead directly to the development of programs either as implementations of the characterization or as rejections of it would be an issue for subsequent decision. If experience with the characterization of natural language is anything to go by, we may expect a formal description of our grasp of games to reveal a subtle and very powerful system. The relative failure of existing computer programs to play games is compelling if indirect testimony to that prophecy.

ACKNOWLEDGEMENTS

This paper has profited greatly from lengthy discussions with my colleagues R.B. Stanton and C.J. Barter. I gratefully acknowledge their contributions while absolving them from any responsibility to defend the result.

8. REFERENCES

CHOMSKY, N. (1957), "Syntactic structures"; The Hague : Mouton.

CHOMSKY, N. (1968), "Language and mind"; Harcourt, Brace and World.

CLOWES, M.B., LANGRIDGE, D.J. and ZATORSKI, R.J. (1969), "Linguistic descriptions"; Seminar Paper 14, Division of Computing Research, C.S.I.R.O., Canberra.

CLOWES, M.B. (1969), "Picture syntax"; This conference.

ELCOCK, E.W. and MURRAY, A.M. (1967), "Experiments with a learning component in a Go-Moku playing program"; Machine Intelligence 1, pp. 87-103, Collins & Michie (Eds.), Edinburgh : Oliver & Boyd.

ELCOCK, E.W. and MURRAY, A.M. (1968), "Automatic description and recognition of board patterns in Go-Moku"; Machine Intelligence 2, pp. 75-88, Edinburgh : Oliver & Boyd.

GREENBLATT, R.D., EASTLAKE D.E. III, CROCKER, S.D. (1967), "The Greenblatt chess program"; AFIPS Conference Proceedings 31, pp. 801-810.

HOPPER, M. (1956), "Win at checkers"; New York : Dover.

KOFFMAN, E.B. (1968), "Learning games through pattern recognition"; IEEE Trans. System Sciences & Cybernetics, SSC-4, 1, pp. 12-16.

NARASIMHAN, R. (1966), "Intelligence and artificial intelligence"; Tech. Report No. 16, Computer Group, Tata Institute, Bombay.

NEWELL, A., SHAW, J.C. and SIMON, H. (1958), "Chess playing programs and the problem of complexity"; IBM J. Res. and Dev. 2, pp. 320-335. Reprinted in 'Computers and Thought'. Feigenbaum and Feldman (Eds.), McGraw-Hill

NEWELL, A. and SIMON, H. (1964), "Problem-solving machines"; International Science and Technology, 36, pp. 48-62.

NEWELL, A. and ERNST, G. (1965), "The search for generality"; Proc. IFIP Congress 65, pp. 17-24. Washington : Spartan Books.

SAMUEL, A.L. (1956), "Some studies in machine learning using the game of checkers"; IBM Journal of Res. and Dev. 3, pp.211-229. Reprinted in 'Computers and Thought'. Feigenbaum and Feldman (Eds.), McGraw-Hill, 1963.

SAMUEL, A.L. (1967), "Some studies in machine learning using the game of checkers. II - Recent progress"; IBM J. Res. and Dev. 11, 6, pp.601-617.

9. APPENDIX

Objects, relations and attributes defined to characterise the board situation of Fig. 4.

(1) $\underline{\mathrm{at}}\langle P, X\rangle$

(2) $\underline{\mathrm{direction}}\langle P, P\rangle\,[\underline{r}]$

(3) $\underline{\mathrm{Nn}}\langle P, P\rangle$

(4) $\underline{\mathrm{Collinear}}\langle \underset{1}{P}, \underset{2}{P}, \underset{3}{P}\rangle \subset \underline{\mathrm{Eq}}\langle \underline{\mathrm{direction}}\langle 1, 2\rangle, \underline{\mathrm{direction}}\langle 2, 3\rangle\rangle$

(5) $\mathrm{TRIPLE}\langle \underset{1}{P}, \underset{2}{P}, \underset{3}{P}\rangle\,[\underline{\mathrm{Collinear}}\langle 1, 2, 3\rangle, \underline{\mathrm{Nn}}\langle 1, 2\rangle, \underline{\mathrm{Nn}}\langle 2, 3\rangle]$

(6) $\mathrm{end}\langle \underset{1}{\mathrm{TRIPLE}}\rangle\,[\underset{2}{P}] \subset 1\langle P, P, 2\rangle$

$\qquad\qquad\text{or } 1\langle 2, P, P\rangle$

(7) $\mathrm{ANGLE}\langle \underset{1}{\mathrm{TRIPLE}}, \underset{2}{\mathrm{TRIPLE}}\rangle\,[\underline{\mathrm{Coinc}}\langle \mathrm{end}\langle 1\rangle, \mathrm{end}\langle 2\rangle\rangle]$

(8) $\mathrm{apex}\langle \underset{1}{\mathrm{ANGLE}}\rangle\,[\underset{2}{P}] \subset 1\langle \underset{3}{\mathrm{TRIPLE}}, \underset{4}{\mathrm{TRIPLE}}\rangle, \underline{\mathrm{Eq}}\langle 2, \mathrm{end}\langle 3\rangle\rangle, \underline{\mathrm{Eq}}\langle 2, \mathrm{end}\langle 4\rangle\rangle$

(9) $\mathrm{OBJECT}\langle \{P\}\rangle\,[\mathrm{relationships}]$

(10) $\underline{\mathrm{on}}\langle \underset{1}{\mathrm{OBJECT}}, \underset{2}{X}\rangle \subset 1\langle \underset{3}{\{P\}}\rangle, \underline{\mathrm{at}}\langle \underset{4}{P}, 2\rangle, \underline{\in}\langle 4, 3\rangle$

(11) $\mathrm{FORCE}\langle \underset{1}{\mathrm{TRIPLE}}, \underset{2}{X}, \underset{3}{X}\rangle\,[\underline{\mathrm{on}}\langle 1, 2\rangle, \underline{\mathrm{on}}\langle 1, 3\rangle]$

(12) $\mathrm{empty}\langle \underset{1}{\mathrm{FORCE}}\rangle\,[\underset{2}{P}] \subset 1\langle \underset{3}{\{P\}}, X, X\rangle, \underline{\in}\langle 2, 3\rangle, \underline{\mathrm{at}}\langle 2, \mathrm{NIL}\rangle$

(13) $\mathrm{CNRFK}\langle \underset{1}{\mathrm{ANGLE}}, \underset{2}{X}, \underset{3}{X}\rangle[\underline{\mathrm{at}}\langle \mathrm{apex}\langle 1\rangle, \mathrm{NIL}\rangle, \underline{\mathrm{on}}\langle \mathrm{triple}\langle 1\rangle, 2\rangle,$

$\qquad\qquad \underline{\mathrm{on}}\langle \mathrm{triple}\langle 1\rangle, 3\rangle]$

(14) $\mathrm{OBJECT}\langle \underset{1}{\{P\}}, \underset{2}{\{X\}}\rangle\,[\text{relationships between elements of 1 and 2}]$

$$(15)\quad \underline{\text{dev}}\langle \underset{1}{\text{PREOBJ}}, \underset{2}{\text{OBJ}}\rangle [\underline{\text{at}}\langle \text{P}, \text{X}\rangle] \subset 1\langle \underset{5}{\{\text{P}\}}, \{6\}\rangle \left[\underset{7}{\{\text{rels on 5 and 6}\}}\right]$$

$$2\langle 5, \underset{8}{\{\text{X}\}}\rangle \left[\underset{9}{\{\text{rels on 5 and 8}\}}\right]$$

$$\underline{\text{Eq}}\langle 4, \text{difference}\langle 8, 6\rangle\rangle$$

$$\underline{\text{Eq}}\langle 3, \text{difference}\langle 9, 7\rangle\rangle$$

$$(16)\quad \text{force}\langle \underset{1}{\text{OBJ}}\rangle \underset{2}{[\text{P}]} \subset \underline{\text{dev}}\langle 1, \text{OBJ}\rangle [\text{at}\langle 2, \text{X}\rangle]$$

$$(17)\quad \underline{\text{def 1}}\langle \underset{1}{\underline{\text{at}}}\langle \text{P}, \text{X}\rangle, \underset{2}{\underline{\text{at}}}\langle \text{P}, 0\rangle\rangle \subset \underline{\text{dev}}\langle \text{PREOBJ}, \text{OBJ}\rangle [1], \text{force}\langle 3\rangle [2]$$

$$(18)\quad \underline{\text{def 2}}\langle \underset{1}{\underline{\text{at}}}\langle \text{P}, \text{X}\rangle, \underset{2}{\underline{\text{at}}}\langle \text{P}, \underset{3}{0}\rangle\rangle \subset \underline{\text{dev}}\langle \text{PREOBJ}, \underset{4}{\text{OBJ}}\rangle [1], \underline{\text{dev}}\langle \text{PREOBJ}, \text{OBJ}\rangle [2],$$

$$\underline{\text{Eq}}\langle 3, \text{force}\langle 4\rangle\rangle$$

10. DISCUSSION

Macleod: In attempting to devise a development analysis for developing forcing patterns, does this preclude the possibility that the opponent can be forcing you at the same time as you are forcing him, except that he's got a pattern that forces you in one move less than yours?

Clowes: Essentially I think you must do these two things simultaneously, and what you are up for is saying, as you say, if the opponent has a forcing pattern which he can win from you in 5 moves, and you have a forcing pattern in which you can force a win in 6 moves, then clearly you are all for blocking in, not playing at yours. This idea of number of moves, is an essential component - it has to be, and I think that is of course where in a sense we are doing some kind of look ahead, because we are really predicting something about what the pattern game is going to be, assuming, of course, that both players know about forcing patterns, and both players know how to do that. So in a sense what we are doing can be characterized as a restriction of the moves tree. We do not have to bother about the moves tree because we know that given that this pattern exists, we will inexorably follow the particular path, and of course the particular quality of Go - Moku and other similar games apparently is this inexorability. Games like draughts and chess don't have this particular inevitability about them, but nevertheless I think that there are components, which we can tease out in common between these, and this is what I have tried to do.

Overheu: In your example of the Go - Moku game, I would have said that game was lost at step 10.

Clowes: Yes that is true. The difficulty with that particular example is that black is fighting off white much earlier, and that's really why he can't block that force.

Overheu: It seemed to me that black could have made a better move at that step - he doesn't make a move which is consistent with fighting off white and getting himself a development.

Clowes: Yes, if you look at the actual short commentary on that play in the Machine Intelligence I paper, this supports what you have said.

Bennett: Would it be reasonable to write a program that could learn some patterns? You have discussed writing them into the program.

Clowes: I think that such a thing already exists for Go - Moku in the program that Elcock and Murray have written, because this had a back track procedure which in fact recovered these things. Its even more interesting as we come to speculate as to whether, if you know the rules of the game, there isn't some way in which you can deduce these patterns. Isn't that what a lot of programs need to play these games, given the rules?

Rutter: You could set two programs to play each other in a game.

Bobrow: That kind of thing is being worked on, but I don't recall where it has been published.

Overheu: In support of the theme of this paper, I recall a near end game in chess that I read about. The objective of the game had been made deliberately the checking of the king, irrespective of anything else. The final goal was to check the king and this particular game playing program did very much better - it does depend on this kind of pattern development.

Clowes: This is the sign of great playing.

Macleod: The strategy of seeing where you are and what you can develop seems to be just a different way of looking at the principle of searching ahead and seeing what is going to come out, and taking notice.

Clowes: But it is a very different way of looking at it, that's exactly my point.

Macleod: You are looking at the same processor.

Clowes: Well, ultimately you'll play a game, if that's the processor you are talking about, yes.

Macleod: I am thinking here of the evaluation of the next move.

Clowes: Essentially what it means is that if you do the look ahead without any constructive idea about what you are going to make, you cannot say anything about the particular moves of relevance to you. In the case of the draughts game there, for example, there are three or four moves that white has, and if you know you are looking for a certain kind of pre-pattern, you will not even see them. This is literally the case - you play a game and there are literally dozens of moves you never even think about. Why not, because on the face of it they are all legal continuations of the game? Of course, what this means is that we have a set of heuristics to knock them down; looking at it from the point of view of 'what move can I make?' is a powerful heuristic. The real question is, however, 'can you characterize the heuristic formally?', not 'is it heuristic or isn't it?'.

Langridge: Near the end of a complicated game you adopt some heuristic for guiding the direction of play.

Clowes: I am sure that in a complicated game, what I said would mean an enormous weight of things. The fact that you are beginning to get the feeling perhaps that a certain pattern of moves is a better, in some vaguely numerical or ordered sense, way to go than other ways, because it may well be that's the only way you can handle all the possibilities. But again I would say that playing it that way does give a particular sense to the idea of creative play. Its creative play that I am interested in talking about - I get this out of Chomsky.

Abbreviations Used:

A.N.U.	Australian National University
C.S.I.R.O.	Commonwealth Scientific and Industrial Research Organization
A.C.T.	Australian Capital Territory
N.S.W.	New South Wales
Vic.	Victoria
S.A.	South Australia
Q'ld.	Queensland
W.A.	Western Australia

Mr. D.E. Abel
Division of Mathematical Statistics, C.S.I.R.O., Glen Osmond, S.A.

Professor M.W. Allen
Dept. of Electronic Computation, University of N.S.W., Sydney, N.S.W.

Mr. J.S. Armstrong
Forestry Department, A.N.U., Canberra, A.C.T.

Mr. A.J. Barlow
Department of National Development, Canberra, A.C.T.

Mr. K.J. Barnes
Division of Computing Research, C.S.I.R.O., Canberra, A.C.T.

Mr. C.J. Barter
Dept. of Electronic Computation, University of N.S.W., Sydney, N.S.W.

Mr. G.A.V. Bary
Department of National Development, Canberra, A.C.T.

Mr. N.W. Bennett
Australian Atomic Energy Commission, Lucas Heights, N.S.W.

Mr. P.R. Benyon
Weapons Research Establishment, Salisbury, S.A.

Mr. T.L. Blum
Division of Computing Research, C.S.I.R.O., Canberra, A.C.T.

Dr. D.G. Bobrow
Vice President, Information Science Division, Bolt, Beranek and Newman Inc., Cambridge, Massachusetts, U.S.A.

Dr. D. Beswick
Department of Psychology, A.N.U., Canberra, A.C.T.

Dr. I.N. Capon
University of Adelaide, Adelaide, S.A.

Dr. M.B. Clowes
Division of Computing Research, C.S.I.R.O., Canberra, A.C.T.

Mr. J.H. Coghlan
Basser Computing Department, University of Sydney, Sydney, N.S.W.

Mr. B.G. Cook
Division of Land Research, C.S.I.R.O., Canberra, A.C.T.

Dr. M.L. Cook
Department of Psychology, A.N.U., Canberra, A.C.T.

Dr. M.B. Dale
Division of Plant Industry, C.S.I.R.O., Canberra, A.C.T.

Mrs. G. Dey
Bureau of Census & Statistics, Canberra, A.C.T.

Mr. J.S. Drabble
Division of Computing Research, C.S.I.R.O., Canberra, A.C.T.

Dr. L.O. Freeman
Defence Standards Laboratories, Maribyrnong, Vic.

Mr. I. Firth
Tertiary Education Research, University of N.S.W., Sydney, N.S.W.

Mr. C.T. Gates
Division of Tropical Pastures, C.S.I.R.O., St. Lucia, Brisbane, Q'ld.

Dr. J.B. Hext
Basser Computing Department, University of Sydney, Sydney, N.S.W.

Dr. J. Hiller
Dept. of Control Engineering, University of N.S.W., Sydney, N.S.W.

Mr. T.S. Holden
Division of Computing Research, C.S.I.R.O., Canberra, A.C.T.

Mr. K.H. Holywell
Plessey Central Research Laboratories, Melbourne, Vic.

Mr. P.V. Horne
Psychology Department, A.N.U., Canberra, A.C.T.

Mr. R.H. Hudson
Division of Computing Research, C.S.I.R.O., Canberra, A.C.T.

Mr. E.L. Jacks
Research Laboratories, Computer Technology Department, General Motors Corporation, Warren, Michigan, U.S.A.

Mr. C. L. Jarvis
Department of Pathology, University of W. A., Perth, W.A.

Dr. S. Kaneff
Department of Engineering Physics, A. N. U., Canberra, A.C.T.

Mr. T. Kinsella
Aeronautical Research Laboratories, Melbourne, Vic.

Mr. G.R. Knowles
Division of Computing Research, C.S.I.R.O., Canberra, A.C.T.

Mr. M. Kovarik
Division of Mechanical Engineering, C.S.I.R.O., Melbourne, Vic.

Dr. G.N. Lance
Division of Computing Research, C.S.I.R.O., Canberra, A.C.T.

Mr. D.J. Langridge
Division of Computing Research, C.S.I.R.O., Canberra, A.C.T.

Mr. T.J. Lawrence
Department of Engineering Physics, A. N. U., Canberra, A.C.T.

Mr. T. Liptak
Division of Building Research, C.S.I.R.O., Melbourne, Vic.

Mr. M. Macauley
Information Electronics Limited, Canberra, A.C.T.

Mr. I.D.G. Macleod
Department of Engineering Physics, A. N. U., Canberra, A.C.T.

Mr. H. Mackenzie
Division of Computing Research, C.S.I.R.O., Canberra, A.C.T.

Mr. G. Masters
Division of Computing Research, C.S.I.R.O., Sydney, N.S.W.

Mr. P.C. Maxwell
Department of Engineering Physics, A. N. U., Canberra, A.C.T.

Dr. D.J. Moore
Dept. of Electrical Engineering, University of Newcastle, Newcastle, N.S.W.

Professor R. Narasimhan
Computer Group, Tata Institute of Fundamental Research, Bombay, India.

Mr. M.J. Nicholls
Department of Works, Sydney, N.S.W.

Mr. J.F. O'Callaghan
Division of Computing Research, C.S.I.R.O., Canberra, A.C.T.

Mr. D.L. Overheu
Department of Defence, Canberra, A.C.T.

Mr. J. Paine
Division of Computing Research, C.S.I.R.O., Canberra, A.C.T.

Mr. G. Petru
Forest Research Institute, Canberra, A.C.T.

Mr. W.A. Phillips
Department of National Development, Canberra, A.C.T.

Mr. F.B. Power
Department of Defence, Canberra, A.C.T.

Mr. V. Pratt
Basser Computing Centre, University of Sydney, Sydney, N.S.W.

Mr. T. Quinlan
Department of National Development, Canberra, A.C.T.

Mr. J.H. Raeburn
Department of Defence, Canberra, A.C.T.

Mr. B. Rope
Commonwealth Public Service Board, Canberra, A.C.T.

Mr. P.R. Rundle
Weapons Research Establishment, Salisbury, S.A.

Mr. P.R. Rutter
Division of Computing Research, C.S.I.R.O., Adelaide, S.A.

Mr. R.M. Scott
Division of Land Research, C.S.I.R.O., Canberra, A.C.T.

Professor G.N. Seagrim
Department of Psychology, A.N.U., Canberra, A.C.T.

Mrs. K. Serkowska
Division of Computing Research, C.S.I.R.O., Canberra, A.C.T.

Mr. R.A. Simmons
Bureau of Census and Statistics, Canberra, A.C.T.

Mr. R. Skeivys
Basser Computing Department, University of Sydney, Sydney, N.S.W.

Mr. G.R. Small
Department of National Development, Canberra, A.C.T.

Mr. B.W. Smith
Computer Centre, A.N.U., Canberra, A.C.T.

Dr. N. Solntseff
Dept. of Electronic Computation, University of N.S.W., Sydney, N.S.W.

Mr. J.G. Speight
Division of Land Research, C.S.I.R.O., Canberra, A.C.T.

Miss L.E. Staines
English Department, University of Newcastle, Newcastle, N.S.W.

Mr. P.J. Staines
Department of Philosophy, University of N.S.W., Sydney, N.S.W.

Mr. R. Stanton
Division of Computing Research, C.S.I.R.O., Canberra, A.C.T.

Mrs. H. Steiger
Department of Psychology, A.N.U., Canberra, A.C.T.

Mr. R.J. Thomas
Department of Defence, Canberra, A.C.T.

Mr. A.A. Thompson
Department of Electronic Computation, University of N.S.W., Sydney, N.S.W.

Dr. P.G. Thorne
Computation Department, University of Melbourne, Melbourne, Vic.

Mr. B.A. Vazey
Department of Communication, School of Electrical Engineering,
University of N.S.W., Sydney, N.S.W.

Mr. A. Yezerski
Dept. of Electronic Computation, University of N.S.W., Sydney, N.S.W.

Mr. R.J. Zatorski
Languages Section, Faculty of Science, University of Melbourne,
Melbourne, Victoria.